New
KEY GEOGRAPHY
for GCSE

DAVID WAUGH & TONY BUSHELL

...es

...iness

9/2062263

D0537958

Key Geography for GCSE Book 1 first published in 1994.
Key Geography for GCSE Book 2 first published in 1994.

Key Geography for GCSE Book 1 New Edition published in 1998 and
Key Geography for GCSE Book 2 New Edition published in 1998 by:
Stanley Thornes (Publishers) Ltd

New Key Geography for GCSE published in 2002 by:
Nelson Thornes Ltd
Delta Place
27 Bath Road
CHELTENHAM
GL53 7TH
United Kingdom

06 / 10 9 8 7 6

A catalogue record for this book is available from the British Library

ISBN 0 7487 6581 6

Original design by Hilary Norman, revised and typeset for this edition by Clare Park
Illustrations by Kathy Baxendale, Jane Cope, Jo Dennis, Hardlines, Nick Hawken, Angela Lumley, Richard Morris, Tim Smith, and TTP International
Photo research by Julia Hanson and Penni Bickle
Edited by Katherine James

Printed and bound in China by Midas Printing International Ltd.

Acknowledgements

The author and publishers are grateful to the following for permission to reproduce photographs and other copyright material in this book.

Adams Picture Library: p 258A (ii); Aerofilms Ltd: p 31C, 39C, 157C, 289E; Andes Press Agency: p 121D; Art Directors/Trip Photography: p 24 right, 38B, 48D (Rogers), 98A , 167B (Rogers), 170A (iv) (Samuels), 175C (Rogers), 178B, 221C (Maddison), 222 B,C, 250A(i) (Rogers); Associated Press: p 58A, 103E, 264C; Chris Bentley: p 196B; Penni Bickle: p 28B, 29C, 49E, 56C, 160A, 161E, 178A, 178C; Tony Bushell: p 72B, 227C, 228C, 228D; J Allan Cash: p 53F; Centre for Church and Industry, Manchester: p 170/171B (i,ii,iii); Cephas Picture Library: p72A, 198B (Blythe); John Cleare Mountain Photography: p 238A; Commonwealth Institute: p 125D; Val Corbett: p 233D; Corbis: p 72D,102B (Arruza), 194E (Chromo Sohn INC), 206C (Benvie), 207D (Ecoscene), 270B(i) (Slone), 273E (Ecoscene); James Davis Travel Photography: p 140B, 248B; Dundee University Satellite Station: p 84A, 85C; Environmental Images: p 82B, 179D, 221D, 230C; Eye Ubiquitous: p 32D, 41C (Wren), 48C (Wren), 146A (Page), 161E (Pickering), 162/163 (iii), 181D (Cumming),183F, (Fairclough), 194C (Pickering); Frank Lane Picture Agency: p 113C, 197D; Geoscience Features Picture Library: p 7E, 21C,42C, 46B; Getty Images: p 7F , 35C; GettyOne: p 4C, 149C, 172A, 264B; Robert Harding Photo Library: p 149D, 159D; R. Humphries: p 32C; The Hutchison Library: p 70A (Francis), 79F (Francis), 135C (Errington) 270C (ii) (Horner); ICI: p 220B; The

Image Bank: p 10B; Images Nature Photography: p 82C (Smith; Intermediate Technology: p 124C, 183E, 271C; Intermediate Technology: p 229E (i,ii); Katherine James: p 219C; Landform Slides: p 43C, 44B, 45F, 52B; London Aerial Photo Library: p 219D; Nelson Copyright free: 48A, 48B, 49F, 52C, 57E, 86C, 86D, 87E, 115C,129E, 153B, 192A, 194A, 194B, 198C, 214A; Yiorgos Nikiteas: p160B; Nuclear Electric: p 247B; Panos Pictures: p 22A (Bolstad), 32B (Ewans), 132B (Stowers), 181C (Hartley), 181E, 187D (Cooper), 201D (Pugast),272A (Hartley), 272B (McKenna), 275B (i) (Taylor) (ii) (Page); Planet Earth Pictures: p 59C; Popperfoto: p 70B; Realistic Photos: p163C, 172B; Rex Features: p 64C (SIPA), 135D (Dunn), 249C, 69D; Dick Roberts: 194D; RSPB Picture Library: p 208A and B; Science Photo Library: p 1 (M-SAT Ltd), 8A (Spicer), 18B (ESC), 35E (Bartel), 59E ,67D (Parker, 72C (Mead); Scottish Hydro Electric: p 247C; Sealand Aerial Photography:p153C, 158A ,158B, 169B; Sheffield City Council: p 167C; Peter Smith Photography: p 182C, 183D,225C; Spectrum Colour Library: p 40B, 166A, 206A; Still Pictures: p 23C (Nooran), 25 top (Moti), 25 bottom, 57F (Garcia), 86A (Garcia), 87F (Edwards), 87G (Cancalosi), 87H (B

(Schytte), 204B (Edwards), 205C (Schytte), 242B (Nicollotti), 250A (iii) (Boulton), 253D (Grillo), 256B (Grillo), 257C (Glendell), 265E (Arbib), 265F (Schytte); Trafford Park Development corporation: p 170A; Stockwave: p158C; Syndication International: p 35D; Telegraph Colour Library: p 19D; Topham Picturepoint: p 68A; Tony Waltham Geophotos: p 16A, 32A, 45E, 54D, 56B; Simon Warner: p 46A, 54A,B, C,173D, 197C; David Waugh p 4A, 4B, 6D, 26B, 138A,B,C, 174A, 184B,C,D,E,F, 236C,D, 237E, 240C, 241D,E; Eric Whitehead: 234B; Woodfall Wild Images: p 5E, 17D, 177C; World Pictures: p 27E, 44C, 78E, 86B, 1299D, 230A,B, 230D, 240A, 242A.

The map extracts on pages 43, 47, 82–83 and 115 are reproduced from the 1995 Ordnance Survey map of Snowdonia (Landranger 115), the 1994 Ordnance Survey map of Wensleydale and Upper Wharfedale (Landranger 98), the 1995 Ordnance Survey map of Carlisle and Solway Firth (Landranger 85) and the 1991 Ordnance Survey map of Haltwhistle, Bewcastle and Alston (Landranger 86). Maps reproduced from the Ordnance Survey Landranger mapping with permission of The Controller of Her Majesty's Stationery Office © Crown copyright; Licence number 07000U.

The authors also wish to thank Gilbert Hitchen for his contribution to pages 122–124.

Every effort has been made to contact copyright holders and we apologise if any have been overlooked.

Contents

What are the main features of a river basin?

A **river** (or **drainage**) **basin** is an area of land drained by a river and its tributaries. The higher land which forms the boundary of the river basin, and which separates two river basins, is called the **watershed**. Most rain falls in mountainous areas. Rain falling on higher land near the watershed will flow slowly downhill either over the surface (photo **A**) or through any topsoil. In time the water will collect in a channel to form a small stream which, as it continues downhill, will increase in size to become a river.

A

The watershed of a river basin above its source in the Pennines. Rainfall is flowing over the land surface, but the water has yet to make a channel for itself.

Source of main river

Photo A

Watershed

Photo B

Steep valley sides

No valley floor

B The River Glasyn flowing through a steep-sided valley in Snowdonia

Watershed

Tributary

Confluence
Photo C

The confluence of the Amazon (reddish water) and the Rio Negro (black water). The waters are different densities and do not mix for about 50 km. C

The point at which a river begins is called its **source**. At first the channel is small but it increases rapidly as the river is joined by many **tributaries**. A tributary is a small stream or river flowing into the main river. The place where a tributary joins the main river is its **confluence**. The river valley, in highland areas, is usually steep-sided (photo **B**). As the river approaches lower land, its valley sides become less steep and its channel widens (photo **C**). Most rivers eventually flow into the sea, although a few end in lakes. The end of a river is known as its **mouth** (photo **E**).

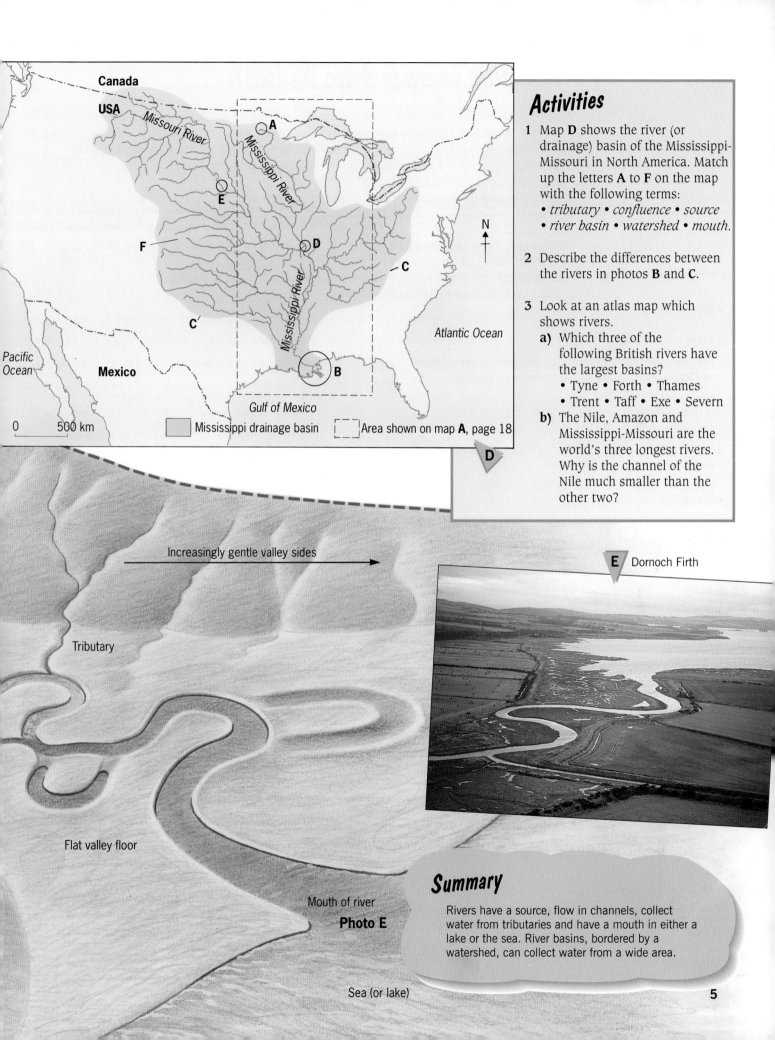

Canada

USA

Missouri River

Mississippi River

A

E

F

D

C

C´

Mississippi River

B

Pacific Ocean

Mexico

Gulf of Mexico

Atlantic Ocean

N

0   500 km

Mississippi drainage basin

Area shown on map **A**, page 18

D

Activities

1 Map **D** shows the river (or drainage) basin of the Mississippi-Missouri in North America. Match up the letters **A** to **F** on the map with the following terms:
   • *tributary* • *confluence* • *source*
   • *river basin* • *watershed* • *mouth.*

2 Describe the differences between the rivers in photos **B** and **C**.

3 Look at an atlas map which shows rivers.
   a) Which three of the following British rivers have the largest basins?
      • Tyne • Forth • Thames
      • Trent • Taff • Exe • Severn
   b) The Nile, Amazon and Mississippi-Missouri are the world's three longest rivers. Why is the channel of the Nile much smaller than the other two?

Increasingly gentle valley sides

Tributary

Flat valley floor

Mouth of river

Photo E

Sea (or lake)

E Dornoch Firth

Summary

Rivers have a source, flow in channels, collect water from tributaries and have a mouth in either a lake or the sea. River basins, bordered by a watershed, can collect water from a wide area.

5

How do rivers shape the land?

If water flows over the ground surface (photo **A**, page 4) it can pick up fine material. Where valleys have very steep sides (photo **B**, page 4), large rocks can break off and fall downhill under the force of gravity. In both cases the material can end up in the channel of a river. Once in its channel, the river can **transport** this material downstream. As the material is transported, it can cause **erosion**. Erosion is the wearing away of the land. As the rate of erosion increases then more material becomes available for the river to transport. A cycle is created in which erosion depends upon the river transporting material, and transportation depends upon the river producing more material by erosion.

There are four main processes by which a river can cause erosion (figure **A**), and four processes by which a river can transport material (figure **B**). Diagram **C** shows the relationship between the various processes of erosion and transportation.

A

Processes of erosion

Attrition – material is moved along the bed of a river, collides with other material, and breaks up into smaller pieces.

Corrasion – fine material rubs against the river bank. The bank is worn away, by a sand-papering action called abrasion, and collapses (photo **D**).

Corrosion – some rocks forming the banks and bed of a river are dissolved by acids in the water.

Hydraulic action – the sheer force of water hitting the banks of the river.

B

Processes of transportation

Traction – large rocks and boulders are rolled along the bed of the river.

Saltation – smaller stones are bounced along the bed of a river in a leap-frogging motion.

Suspension – fine material, light enough in weight to be carried by the river. It is this material that discolours the water.

Solution – dissolved material is transported by the river.

C

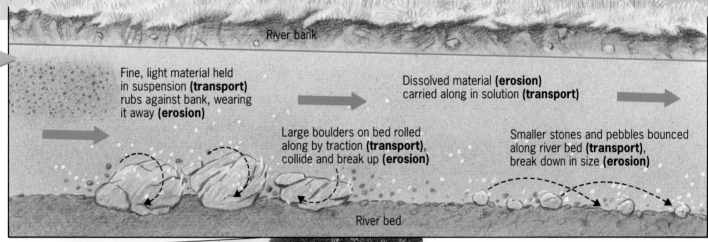

River bank

Fine, light material held in suspension **(transport)** rubs against bank, wearing it away **(erosion)**

Large boulders on bed rolled along by traction **(transport)**, collide and break up **(erosion)**

Dissolved material **(erosion)** carried along in solution **(transport)**

Smaller stones and pebbles bounced along river bed **(transport)**, break down in size **(erosion)**

River bed

D

Afon Glaslyn, Snowdonia

Most erosion occurs when a river is in flood. It can then carry huge amounts of material in suspension as well as being able to move the largest of boulders lying on its bed. Erosion can both deepen and widen a river valley (photo **D**). The valley deepens as a result of vertical erosion. This is more usual in mountainous areas nearer to the source of the river. Here the river forms a series of characteristic landforms which include a **V-shaped valley** with **interlocking spurs** (diagram **E**) as well as **waterfalls** and rapids (diagram **F**).

Snake Pass, Yorkshire

Iguaçu River and Falls, Brazil

E **F**

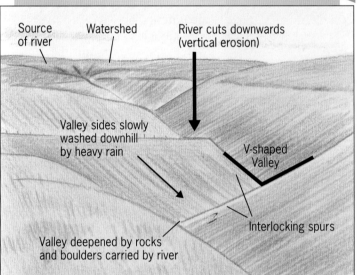

Source of river — Watershed — River cuts downwards (vertical erosion)

Valley sides slowly washed downhill by heavy rain

V-shaped Valley

Interlocking spurs

Valley deepened by rocks and boulders carried by river

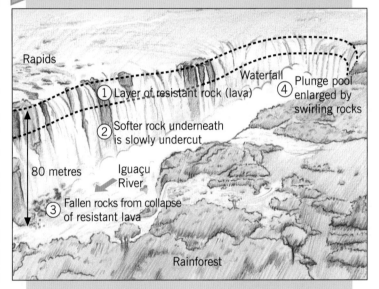

Rapids

① Layer of resistant rock (lava)

Waterfall

④ Plunge pool enlarged by swirling rocks

② Softer rock underneath is slowly undercut

80 metres

Iguaçu River

③ Fallen rocks from collapse of resistant lava

Rainforest

A river, especially when in flood, transports material along its bed. The material cuts downwards (vertical erosion) relatively quickly, deepening the bed of the river. After periods of heavy rain, soil on the valley sides slowly moves downhill under gravity. The valley forms a V-shape as it is deepened faster than it is widened.

The hard resistant surface rock is left unsupported as the underlying softer rock is eroded more quickly by the river. In time the resistant rock will collapse. This material will be swirled around by the river, widening and deepening the plunge pool at the foot of the waterfall. Over a period of time, as more rock collapses, the waterfall will slowly retreat leaving a steep-sided gorge.

Activities

1 Describe four processes by which a river might:
   a) erode its banks and bed
   b) transport material downstream.

2 a) Describe, with the help of neat and carefully labelled diagrams, how a river might form:
      i) a waterfall   ii) a V-shaped valley.
   b) For each answer, explain which processes of erosion and which processes of transportation affect its formation.

Summary

There are several processes by which rivers can erode the land and transport material. Together, these processes can produce a group of distinctive landforms which include V-shaped valleys and waterfalls.

How do meanders and oxbow lakes form?

Meander on North Slope River, Alaska

As rivers get nearer to their mouths they flow in increasingly wide, gentle-sided valleys. The channel increases in size to hold the extra water which the river has to receive from its tributaries. As the river gets bigger it can carry larger amounts of material. This material will be small in size, as larger rocks will have broken up on their way from the mountains. Much of the material will be carried in suspension and will erode the river banks by corrasion.

When rivers flow over flatter land, they develop large bends called **meanders** (photo **A**). As a river goes around a bend most of the water is pushed towards the outside causing increased erosion (diagram **C**). The river is now eroding sideways into its banks rather than downwards into its bed, a process called lateral erosion. On the inside of the bend, in contrast, there is much less water. The river will therefore be shallow and slow-flowing. It cannot carry as much material and so sand and shingle will be deposited. Diagram **B** is a cross-section showing the typical shape across a meander bend.

Due to erosion on the outside of a bend and deposition on the inside, the shape of a meander will change over a period of time (diagram **D**). Notice how erosion narrows the neck of the land within the meander. In time, and usually during a flood, the river will cut right through the neck. The river will take the new, shorter route (diagram **E**). The fastest current will now tend to be in the centre of the river, and so deposition is likely to occur in gentler water next to the banks. Eventually deposition will block off the old meander to leave an **oxbow lake** (photo **F**). The oxbow lake will slowly dry up, only refilling after heavy rain or during a flood.

Large rivers like the Mississippi and the Amazon have many oxbow lakes. Several new oxbow lakes were created following the Mississippi floods of mid-1993 (pages 18 and 19).

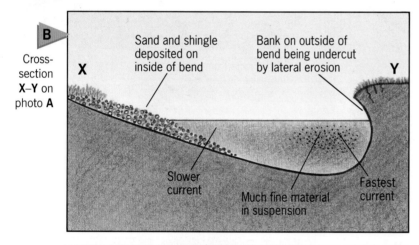

Cross-section **X–Y** on photo **A**

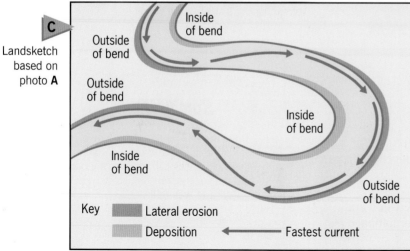

Landsketch based on photo **A**

D Changing shape of a meander

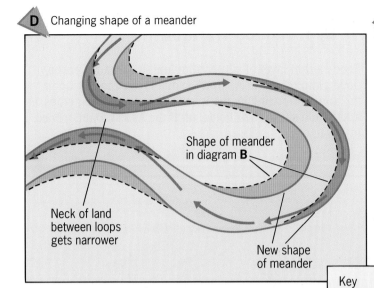

Neck of land between loops gets narrower

Shape of meander in diagram **B**

New shape of meander

E Formation of an oxbow lake

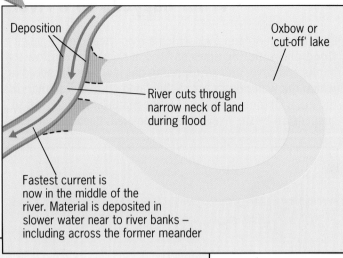

Deposition

Oxbow or 'cut-off' lake

River cuts through narrow neck of land during flood

Fastest current is now in the middle of the river. Material is deposited in slower water near to river banks – including across the former meander

Key

▨	Land lost to the river (eroded)
☐	Land gained from the river (deposited)

⟵ Fastest current

- - - - Earlier course of river

Activities

1 Using photo **A**, draw a cross-section from **P** to **R**. On your cross-section, label:
   a) the areas with the fastest current and the slowest current
   b) the places where erosion is taking place
   c) the places where deposition is taking place.

2 Describe, with the help of a diagram, what is likely to happen in the future at point **S** on photo **A**.

3 Diagram **G** is an incomplete cross-section of a meander. Copy and complete the diagram by using the following information:

Distance from left bank in metres	Depth of river in metres
0.5	1.0
0.75	2.0
1.0	3.0
2.0	3.25
3.0	3.0
4.0	2.5
5.0	2.0
6.0	1.0
7.0	0.5
8.0	0.0

4 If there is a small river or stream near to your school, take your own class measurements to produce a cross-section similar to the one in Activity **3**. But remember **to take care**. Serious accidents can occur in even small rivers and streams.

F Meanders and an oxbow lake, near Dalton in Yorkshire

G River meander

Surface of river

Depth of river (metres)

Distance from left bank (metres)

Summary As most rivers approach the sea they begin to meander and, in some cases, to form oxbow lakes.

What happens to a river as it approaches its mouth?

The flat area of land over which a river meanders is called a **flood plain**. During times of flood, a river will overflow its banks and cover any surrounding flat land. As the speed at which the water flows across the flood plain is less than in the main channel, some of the fine material transported in suspension by the river will be deposited. Each time a river floods a thin layer of silt, or alluvium, is spread over the flood plain (diagram **A**). The Egyptians used to rely upon the annual flooding by the River Nile to water their crops and to add silt to their fields, until the Aswan Dam opened in 1970.

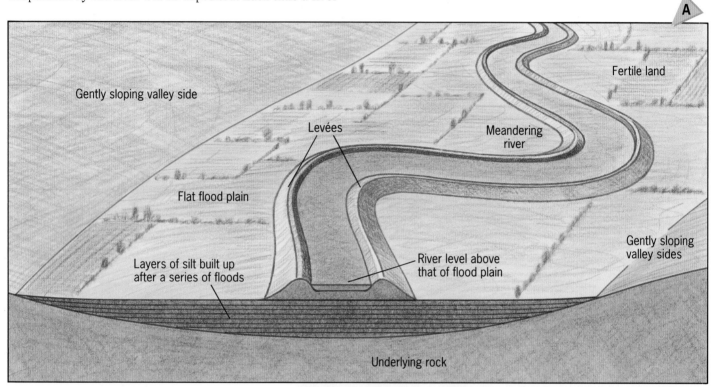

A

Fertile land

Gently sloping valley side

Levées

Meandering river

Flat flood plain

Layers of silt built up after a series of floods

River level above that of flood plain

Gently sloping valley sides

Underlying rock

When a river floods, it is the coarsest material that is deposited first. This coarse material can form small embankments alongside a river which the Americans call **levées**. Large rivers, like the Mississippi, carry tremendous quantities of material in suspension, especially in times of flood. However, during times when the river level falls and its speed is reduced, large amounts of silt fall out of suspension onto the bed of the river. In time the bed of the river builds up so that, at times when river levels are high again, the river is more likely to overflow its banks.

To try to prevent this from happening, large artificial levées are built. The Mississippi now flows at a much higher level than the surrounding flood plain, and cities like New Orleans and St Louis are protected by levées that are up to 16 metres high. The problem is, what happens if these levées break (pages 18 and 19)?

Mississippi delta **B**

Large rivers transport great amounts of fine material down to their mouths. If a river flows into a relatively calm sea, or lake, then its speed will reduce and the fine material will be deposited. The deposited material will slowly build upwards and outwards to form a **delta** (photo **B**). River deltas provide some of the best soils in the world for farming (e.g. River Nile) but they are also prone to serious flooding because the land is so flat (e.g. Bangladesh). The Mississippi delta is extending rapidly into the Gulf of Mexico. As in all deltas, deposition blocks the main channel of the river, so the Mississippi has to divide into a series of smaller channels called **distributaries**. These channels need constant dredging if they are to be used by ships.

C Part of OS map sheet 98

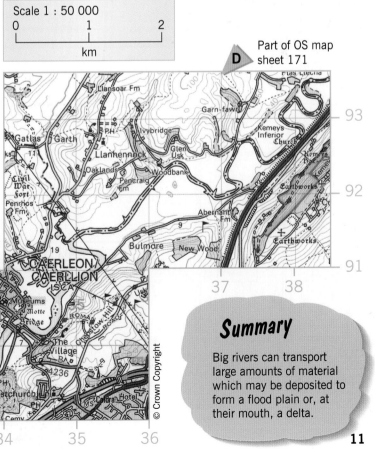

© Crown Copyright

Scale 1 : 50 000

0   1   2

km

D Part of OS map sheet 171

© Crown Copyright

Activities

1 Map **C** shows two tributaries of the River Ure near Hawes in the Yorkshire Dales National Park. Map **D** shows the River Usk near Caerleon in South Wales.

 a) Draw cross-sections (see pages 282–283) to show the differences between the two river valleys:
 i) on map **C** from the spot height at 855953 (Fossdale Moss) to the spot height at 879951 (Lovely Seat).
 ii) on map **D** from the public house at 353928 (91 m) to the hotel at 362911 (105 m).

 b) Describe the differences in the shape of the valley sides and valley floor of the two rivers.

2 a) Make a simple copy or tracing of maps **C** and **D**. Add, on the appropriate map and in the appropriate place, the following labels:
 • *gentle valley sides* • *steep valley sides* • *meanders* • *no flood plain* • *wide flood plain* • *V-shaped valley* • *waterfalls* • *deposition in main channel*.

 b) Give a grid square where i) an oxbow lake is most likely to form and ii) rapids are likely to be seen.

Summary

Big rivers can transport large amounts of material which may be deposited to form a flood plain or, at their mouth, a delta.

What is the hydrological cycle?

Hydrology is the study of water. The **hydrological cycle**, more commonly referred to as the water cycle, is the continuous transfer of water from the oceans into the atmosphere, then onto the land and finally back into the oceans. The cycle is complicated since it involves several processes which include evaporation, transpiration, condensation, precipitation and surface run-off. These processes, together with the various links within the cycle, are shown in diagram **A**.

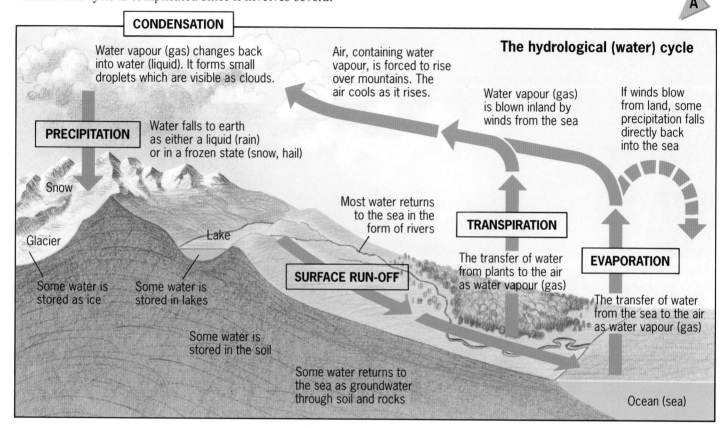

A

The hydrological (water) cycle

CONDENSATION
Water vapour (gas) changes back into water (liquid). It forms small droplets which are visible as clouds.

Air, containing water vapour, is forced to rise over mountains. The air cools as it rises.

Water vapour (gas) is blown inland by winds from the sea

If winds blow from land, some precipitation falls directly back into the sea

PRECIPITATION
Water falls to earth as either a liquid (rain) or in a frozen state (snow, hail)

Snow

Glacier

Lake

Most water returns to the sea in the form of rivers

TRANSPIRATION
The transfer of water from plants to the air as water vapour (gas)

EVAPORATION
The transfer of water from the sea to the air as water vapour (gas)

SURFACE RUN-OFF

Some water is stored as ice

Some water is stored in lakes

Some water is stored in the soil

Some water returns to the sea as groundwater through soil and rocks

Ocean (sea)

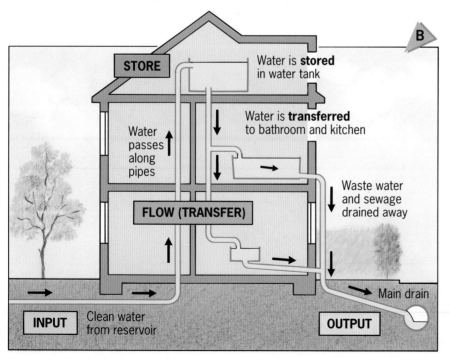

B

STORE
Water is **stored** in water tank

Water is **transferred** to bathroom and kitchen

Water passes along pipes

FLOW (TRANSFER)

Waste water and sewage drained away

Main drain

INPUT Clean water from reservoir

OUTPUT

The recycling of water in the hydrological cycle should mean that water is a sustainable resource. However, at times there are natural interruptions within the cycle. These can either create an extreme surplus of water on the land, resulting in flooding, or an extreme shortage of water, causing drought. Human interference with the natural cycle can also increase the risk of environmental disasters.

How does the river (drainage) basin system work?
The river basin system is that part of the hydrological cycle which operates on the land. Diagram **B** illustrates how a water **system** works in your own home. A system consists of **inputs** (entering the system), **flows** or **transfers** (movement through the system), **stores** (held within the system) and **outputs** (leaving the system). The river basin system is also complicated and is best illustrated as a diagram (diagram **C**).

C

The river basin system

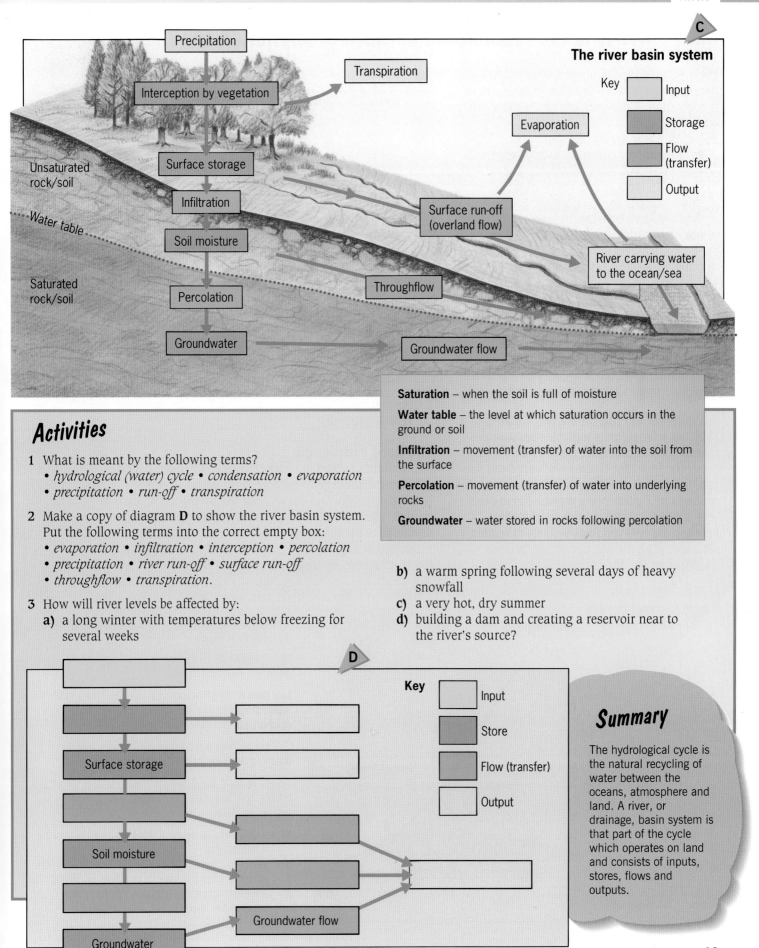

Precipitation

Transpiration

Interception by vegetation

Evaporation

Unsaturated rock/soil

Surface storage

Infiltration

Water table

Soil moisture

Surface run-off (overland flow)

River carrying water to the ocean/sea

Saturated rock/soil

Percolation

Throughflow

Groundwater

Groundwater flow

Key
- Input
- Storage
- Flow (transfer)
- Output

Saturation – when the soil is full of moisture

Water table – the level at which saturation occurs in the ground or soil

Infiltration – movement (transfer) of water into the soil from the surface

Percolation – movement (transfer) of water into underlying rocks

Groundwater – water stored in rocks following percolation

Activities

1   What is meant by the following terms?
    - *hydrological (water) cycle* • *condensation* • *evaporation*
    - *precipitation* • *run-off* • *transpiration*

2   Make a copy of diagram **D** to show the river basin system.
    Put the following terms into the correct empty box:
    - *evaporation* • *infiltration* • *interception* • *percolation*
    - *precipitation* • *river run-off* • *surface run-off*
    - *throughflow* • *transpiration*.

3   How will river levels be affected by:
    a)   a long winter with temperatures below freezing for several weeks

    b)   a warm spring following several days of heavy snowfall
    c)   a very hot, dry summer
    d)   building a dam and creating a reservoir near to the river's source?

D

Key
- Input
- Store
- Flow (transfer)
- Output

Surface storage

Soil moisture

Groundwater flow

Groundwater

Summary

The hydrological cycle is the natural recycling of water between the oceans, atmosphere and land. A river, or drainage, basin system is that part of the cycle which operates on land and consists of inputs, stores, flows and outputs.

What is the relationship between precipitation and run-off?

The systems diagram (page 13) shows what happens once water has fallen to the Earth as precipitation. Rainwater, or melted snow, will either:

- be lost to the system through **evapotranspiration** (i.e. evaporation and transpiration)
- be held in storage in lakes, the soil or underground, or
- flow into a river to return, eventually, to the sea as run-off.

In other words, the amount of rainwater which will become the run-off of a river will be

> Precipitation – (evapotranspiration + storage)

Under normal conditions, therefore, the run-off of a river will be less than precipitation. Precipitation and run-off figures for a year can be plotted graphically as in diagram **A**.

Precipitation and run-off are two variables. They are referred to as variables because the figures used to construct graph **A** are for one particular year. The figures will vary from one year to another. However, although the chances of the same figures being repeated in another year are highly unlikely, the relationship between the two variables is likely to remain similar, e.g. as the amount of precipitation increases so too does run-off.

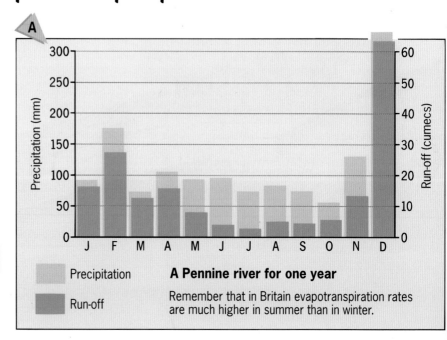

A Pennine river for one year

Precipitation
Run-off

Remember that in Britain evapotranspiration rates are much higher in summer than in winter.

Excess precipitation gives excess run-off which increases the risk of flooding.

The flood hydrograph

The amount of water in a river channel at a given time is called the **discharge**. Discharge is measured in cumecs (cubic metres of water per second). Following an increase in rainfall, there will also be an increase in the level and the discharge of the river. The relationship between precipitation and the level of a river is illustrated by the **flood** (or storm) **hydrograph** (diagram **B**).

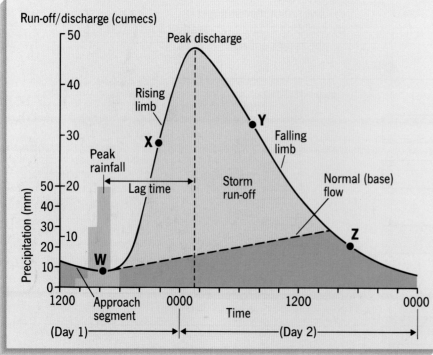

The **approach segment** shows the discharge of the river before it rains.

The **rising limb** results from a rapid increase in rainwater reaching the river.

Lag time is the difference between the time of the heaviest rainfall and the maximum level and/or discharge of the river.

The **falling limb** is when some rainwater is still reaching the river, but in decreasing amounts.

Points **W**, **X**, **Y** and **Z** are not usually shown on a flood hydrograph, but have been added here to help to explain its shape.

W – a very small amount of rain falls straight into the river channel.

X – water reaches the river rapidly by surface run-off.

Y – water reaches the river more slowly by throughflow.

Z – a limited amount of groundwater eventually reaches the river.

By showing the relationship between precipitation and run-off, the flood hydrograph indicates whether a particular river has a high or low flood risk. It therefore provides essential information for any river management scheme. The shorter the lag time and the steeper the rising limb, the greater is the flood risk. This is because much of the precipitation reaches the channel so quickly, mainly due to surface run-off, that the river has insufficient time to transport the excess water. In contrast, a river with a long lag time and a very gentle rising limb will have a very low flood risk. This is because rainwater reaches the channel slowly and over a longer period of time, allowing the river time to transport the excess water.

Activities

1 a) Using the information in table **C**, draw a graph to show the precipitation and run-off totals for a one-year period at a river recording site.

b) In which season are the figures at their highest for:
i) precipitation
ii) run-off?

c) Why is there a bigger difference between the precipitation and run-off totals in summer than in winter?

d) Suggest reasons why run-off was slightly higher than precipitation in March.

C

	J	F	M	A	M	J	J	A	S	O	N	D
Precipitation (mm)	116	164	71	103	83	75	74	79	86	81	130	148
Run-off (mm)	103	143	79	84	47	17	22	26	41	64	106	125

2 Diagram **D** shows the hydrograph for a British river for three days.

a) i) How many hours did storm **1** last?
ii) What was the time of peak rainfall in storm **1**?
iii) How many hours was the lag time?

b) i) Why is there lag time between peak rainfall and peak discharge?

ii) Why is the rising limb much steeper than the falling limb?

iii) Give two reasons why discharge was higher after storm **2** than after storm **1**.

c) If the level of the river reached the top of its banks with a discharge of 70 cumecs, what must have happened after storm **2**?

D

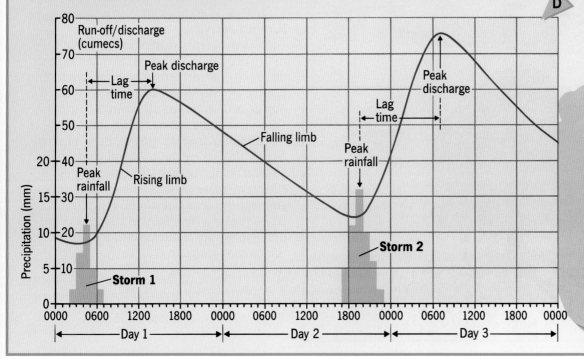

Summary

Precipitation and run-off are two variables. It is possible to identify a relationship between them showing that river run-off (discharge) depends upon the amount of precipitation. The flood hydrograph illustrates discharge and indicates the level of flood risk.

What are the main causes of river flooding?

Not every river has a high flood risk. However, those that do may flood for a combination of reasons. Often the four most important reasons are:

physical
- type and amount of precipitation
- type of soil and underlying rock

human
- land use of the river basin
- human activity.

Type and amount of precipitation

The most frequent cause of flooding is heavy rainfall which lasts over a period of several days. The ground will become saturated and infiltration will be replaced by surface run-off. The most severe cause of flooding usually follows short, but very intense, thunderstorms. In Britain these storms are more likely to occur after a hot, dry spell in summer. The ground becomes too hard for the rain to infiltrate, and the surface run-off causes river levels to rise rapidly causing a **flash flood**.

Heavy snowfalls over several days mean that water is held in storage (photo **A**). When temperatures rise, there will be a release of water. The flood risk is greater if there is a large rise in temperature, if the rise is accompanied by a period of heavy rain, or if the ground remains frozen preventing infiltration.

Type of soil and underlying rock

Rocks that allow water to pass through them, like chalk, limestone and sandstone, are said to be **permeable**. Rocks that do not let water pass through them, such as granite, are **impermeable**. It is the same with soil. Sandy soils are permeable and allow water to infiltrate, whereas clay soils are impermeable. Surface run-off, and the flood risk, are much greater in river basins where the soil and underlying rock are impermeable.

Land use

River basins that have little vegetation cover have a much higher flood risk than forested river basins. This is because trees intercept rainfall, delaying the time and reducing the amount of water reaching the river (diagram **B**).

Human activity

There is sufficient evidence to prove that the risk of flooding is increasing in many parts of the world. The increase in both the frequency and the severity of flooding is usually linked to human activity, especially when this activity changes the land use of a basin through either deforestation or urban growth. Bangladesh is one country where the already high flood risk has increased due to deforestation in the Himalayas. Elsewhere, as urban areas grow in size, impermeable tarmac and concrete surfaces replace fields and woodland. Infiltration and throughflow are reduced, while surface run-off is increased (diagram **C**).

A Water held in storage: winter in the Yorkshire Dales

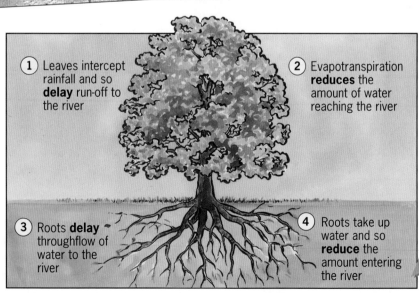

① Leaves intercept rainfall and so **delay** run-off to the river

② Evapotranspiration **reduces** the amount of water reaching the river

③ Roots **delay** throughflow of water to the river

④ Roots take up water and so **reduce** the amount entering the river

B How trees help to reduce the flood risk

Drains and gutters are constructed to remove surface water. This might decrease the time taken by rainwater to reach the river, but it increases the risk of a flash flood. Small streams are forced to travel along culverts (photo **D**) or underground pipes. Drains and underground pipes may not be large enough to cope with rainwater falling during thunderstorms.

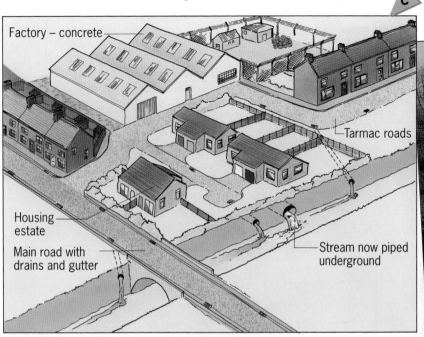

C

Factory – concrete
Tarmac roads
Housing estate
Main road with drains and gutter
Stream now piped underground

D

Activities

1 Explain why there is a high flood risk:
   a) after a long period of heavy rainfall
   b) after a summer thunderstorm
   c) when a heavy snowfall follows a few days when temperatures remained below freezing
   d) in an area with an impermeable underlying rock
   e) in a river basin that has just been deforested.

2 Copy and complete diagram **E** to show how the river basin system (diagram **C**, page 13) is changed by urban growth.

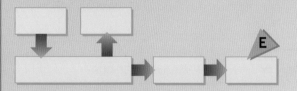

E

3 Diagram **F** shows flood hydrographs following a rainstorm for a stream in a wooded rural area and a stream in a nearby urban area.

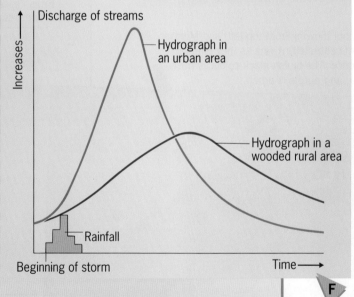

Discharge of streams

Increases

Hydrograph in an urban area

Hydrograph in a wooded rural area

Rainfall

Beginning of storm

Time

F

   a) Give three differences between the shape of the two hydrographs.
   b) Give reasons for the differences between the two hydrographs.

Summary

The risk of a river flooding depends upon several factors including the type and amount of precipitation, soil, underlying rock, and land use. Recently the risk has increased due to human activities such as deforestation and urban growth.

17

What were the causes and effects of the Mississippi flood?

The Mississippi River is 3800 km in length and flows through ten states. It receives over 100 major tributaries, including the Missouri which joins it at St Louis (map **A**). Its drainage basin covers one-third of the USA and a small part of Canada (map **D**, page 5).

Frequent flooding by the Mississippi has created a wide flood plain. The flood plain is 200 km at its widest point, and consists of fertile silt deposited by the river at times of flood. Even before the area was settled by Europeans, the river flowed above the level of its flood plain and between natural levées (page 10). Nineteenth-century Americans considered the Mississippi to be 'untameable', and a major flood in 1927 caused 217 deaths. Since then over 300 dams and storage reservoirs have been built, and natural levées have been heightened and strengthened to protect major urban areas. The levée at St Louis is 18 km long and 16 metres high. Flooding continued throughout the 1950s and 1960s, but the last big flood was in 1973. The Americans believed that, due to large investments of money and modern technology, they had at least 'controlled' the river. Certainly, the danger to human life and damage to property had been considerably reduced. . . but that was before the events of summer 1993.

The Mississippi (flowing from top left) and Missouri (flowing from centre left) Rivers as they approach their confluence. The bluish-black colour is the flooded area, and purple is urban.

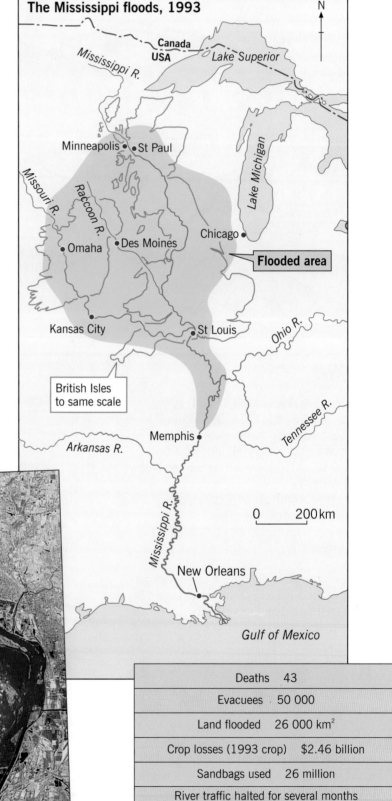

A

The Mississippi floods, 1993

N

Canada
USA
Lake Superior

Mississippi R.

Minneapolis • • St Paul

Lake Michigan

Missouri R.

Raccoon R.

• Omaha • Des Moines

Chicago •

Flooded area

Kansas City

St Louis •

Ohio R.

British Isles to same scale

Memphis •

Tennessee R.

Arkansas R.

Mississippi R.

0   200 km

New Orleans

Gulf of Mexico

Deaths	43
Evacuees	50 000
Land flooded	26 000 km²
Crop losses (1993 crop)	$2.46 billion
Sandbags used	26 million
River traffic halted for several months (oil, cereals, coal, etc.)	
Overall estimated damage	**$12 billion (£8 bn)**

B

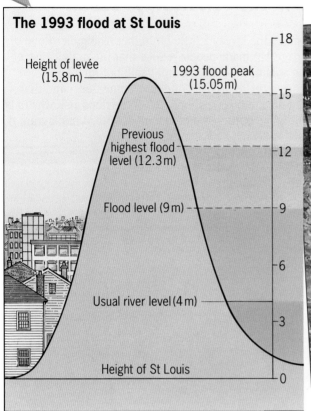

The 1993 flood at St Louis

Height of levée (15.8 m)

1993 flood peak (15.05 m)

Previous highest flood level (12.3 m)

Flood level (9 m)

Usual river level (4 m)

Height of St Louis

D Mississippi river floods, 1993

Heavy rain in April 1993 saturated the upper Mississippi basin. Thunderstorms throughout June caused rapid surface run-off and flash floods (page 16). During July the thunderstorms increased in severity with one giving 180 mm of rain in a few hours. By mid-July the level of the Mississippi had reached an all-time high (diagram **C**). Levées surrounding towns were put under tremendous pressure from the weight of water in the river, and in many places they collapsed (photo **B**). Away from towns the river spread across its flood plain up to a width of 25 km (photo **D**). An area larger than the British Isles was affected by flooding (map **A**). Only one road bridge, and no rail bridge, remained open for 400 km north of St Louis. River traffic on one of the world's busiest highways had long since been brought to a stop. The Mississippi proved it had not been tamed, as it claimed lives and destroyed property (map **A**). Many Americans felt that the effects of the flood had been made worse unintentionally, because people had interfered with nature in trying to manage and control the river.

The effects of the flood did not end when the river levels began to fall. It took several months for the water to drain off the land. Although the land was covered in fertile silt, the ground was too wet for planting crops. The contents of houses and factories, even if not the buildings themselves, were ruined. Cleaning-up operations took months. Where sewage had been washed into waterways, there was a threat of disease. Stagnant water attracted mosquitoes and rats. Insurance claims were high and numerous.

Activities

1 a) Why is the Mississippi a high flood risk river?
   b) What caused the flood of 1993?
   c) What were the immediate effects of the flood?
   d) What might be some of the long-term effects of the flood?

2 How had human activity unintentionally increased the flood risk?

Summary

When rivers flood they can put lives in danger, damage property and disrupt people's normal way of life. Sometimes attempts to reduce the flood risk can unintentionally make the effects of flooding worse.

How might the flood risk be reduced in the Mississippi basin?

In 1718 the site for a town was chosen near to the mouth of the Mississippi River. To protect this town, the present-day New Orleans, from the risk of flooding, a levée one metre high was built. For the next 200 years the Mississippi flooded parts of its basin between St Louis and its mouth almost annually. The response was always the same: make the levées a little bit higher. The major flood of 1927 (page 18) made people realise that the main cause of flooding was not the Mississippi itself, but its main tributaries – the Missouri, Ohio and Tennessee – and that more drastic methods were needed if the flood risk was to be reduced. Some of these methods are described in diagram **A**.

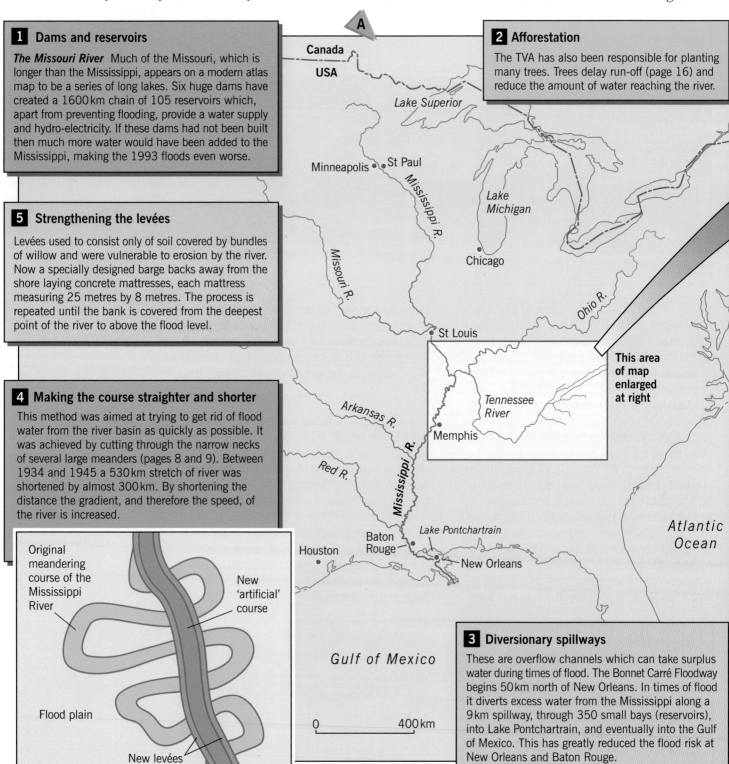

A

1 Dams and reservoirs

The Missouri River Much of the Missouri, which is longer than the Mississippi, appears on a modern atlas map to be a series of long lakes. Six huge dams have created a 1600 km chain of 105 reservoirs which, apart from preventing flooding, provide a water supply and hydro-electricity. If these dams had not been built then much more water would have been added to the Mississippi, making the 1993 floods even worse.

5 Strengthening the levées

Levées used to consist only of soil covered by bundles of willow and were vulnerable to erosion by the river. Now a specially designed barge backs away from the shore laying concrete mattresses, each mattress measuring 25 metres by 8 metres. The process is repeated until the bank is covered from the deepest point of the river to above the flood level.

4 Making the course straighter and shorter

This method was aimed at trying to get rid of flood water from the river basin as quickly as possible. It was achieved by cutting through the narrow necks of several large meanders (pages 8 and 9). Between 1934 and 1945 a 530 km stretch of river was shortened by almost 300 km. By shortening the distance the gradient, and therefore the speed, of the river is increased.

2 Afforestation

The TVA has also been responsible for planting many trees. Trees delay run-off (page 16) and reduce the amount of water reaching the river.

3 Diversionary spillways

These are overflow channels which can take surplus water during times of flood. The Bonnet Carré Floodway begins 50 km north of New Orleans. In times of flood it diverts excess water from the Mississippi along a 9 km spillway, through 350 small bays (reservoirs), into Lake Pontchartrain, and eventually into the Gulf of Mexico. This has greatly reduced the flood risk at New Orleans and Baton Rouge.

Original meandering course of the Mississippi River

New 'artificial' course

Flood plain

New levées

This area of map enlarged at right

Canada
USA
Lake Superior
Minneapolis
St Paul
Mississippi R.
Lake Michigan
Chicago
Ohio R.
Missouri R.
St Louis
Tennessee River
Arkansas R.
Memphis
Red R.
Mississippi R.
Baton Rouge
Lake Pontchartrain
Houston
New Orleans
Atlantic Ocean
Gulf of Mexico

0   400 km

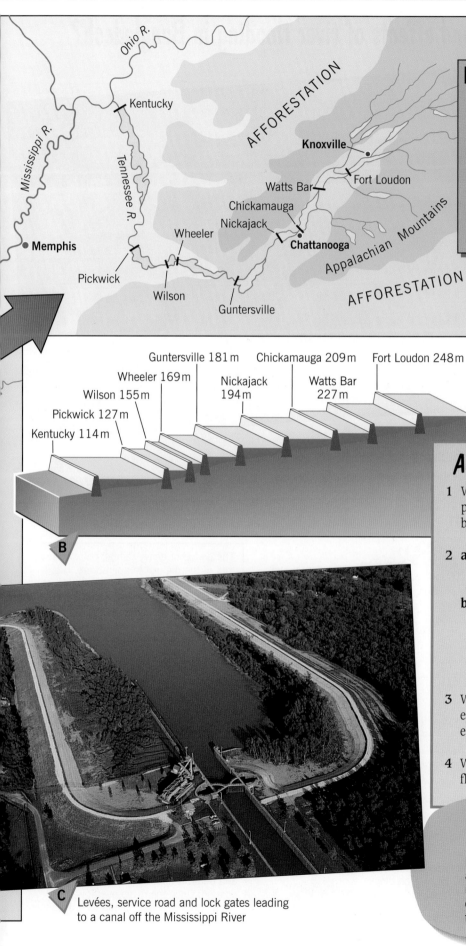

1 **Dams and reservoirs**

The Tennessee River The Tennessee Valley Authority (TVA) was set up in the 1930s. It had many functions (diagram **B**), one of which was to control the flooding of the river. Nine reservoirs were created on the main river and 10 on its tributaries. Dams hold back water during times of flood and release it when river levels are lower. One measure of success came in 1957. Instead of the river rising to a dangerous peak of 16.5 metres, the dams and reservoirs limited the level to a harmless 9.8 metres.

The TVA is a multipurpose scheme which:
• controls flooding
• provides a water supply
• produces hydro-electricity
• improves navigation
• increases afforestation
• reduces soil erosion
• encourages industry
• encourages tourism.

Ohio R.

Kentucky

Mississippi R.

Tennessee R.

AFFORESTATION

Knoxville

Watts Bar — Fort Loudon

Chickamauga

Nickajack

Wheeler

Chattanooga

Appalachian Mountains

Memphis

Pickwick

Wilson

Guntersville

AFFORESTATION

Guntersville 181 m   Chickamauga 209 m   Fort Loudon 248 m

Wheeler 169 m   Nickajack 194 m   Watts Bar 227 m

Wilson 155 m

Pickwick 127 m

Kentucky 114 m

B

Activities

1 What was the only method used to try to prevent the Mississippi River from flooding before 1927?

2 **a)** Describe four of the methods used since 1927 to try to reduce the flood risk on the Mississippi River.
   b) Which of the methods do you think will be:
   i) the cheapest to use
   ii) the most expensive to use
   iii) the most successful in reducing the flood risk?

3 Why is it harder to reduce the flood risk in economically less developed countries than in economically more developed countries?

4 What attempts have been made to reduce the flood risk on a river near to where you live?

C Levées, service road and lock gates leading to a canal off the Mississippi River

Summary

It often needs considerable amounts of money and high levels of technology to reduce the flood risk. It is therefore economically more developed countries that can make a positive response to the river flood hazard.

What are the causes and effects of river flooding in Bangladesh?

Flooding is an annual event in Bangladesh. The monsoon rains cause rivers such as the Brahmaputra and Ganges to overflow their banks between July and mid-August. Most of the country's 125 million inhabitants live on the flood plains and the delta created by these rivers. For most of them, the seasonal flood is essential for their survival because it provides water in which to grow the main crops of rice (page 200) and jute, as well as depositing silt to fertilise their fields. Flooding of 20 per cent of the country is considered beneficial for crops and the ecological balance. However, a figure much less than that can result in food shortages, while an inundation much in excess can cause considerable loss of life, ruin crops and seriously damage property and communications. In 1998, nearly 70 per cent of the country was flooded, in some places for over 70 days. This flood was unprecedented in terms of size and duration.

A

What were the causes of the 1998 flood?

B

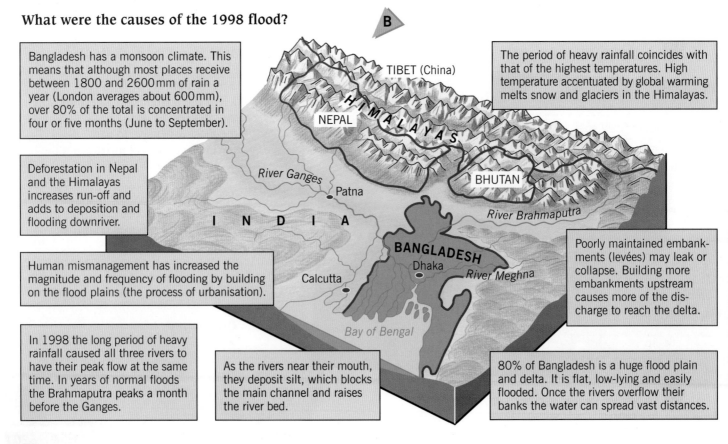

Bangladesh has a monsoon climate. This means that although most places receive between 1800 and 2600 mm of rain a year (London averages about 600 mm), over 80% of the total is concentrated in four or five months (June to September).

The period of heavy rainfall coincides with that of the highest temperatures. High temperature accentuated by global warming melts snow and glaciers in the Himalayas.

Deforestation in Nepal and the Himalayas increases run-off and adds to deposition and flooding downriver.

Human mismanagement has increased the magnitude and frequency of flooding by building on the flood plains (the process of urbanisation).

Poorly maintained embankments (levées) may leak or collapse. Building more embankments upstream causes more of the discharge to reach the delta.

In 1998 the long period of heavy rainfall caused all three rivers to have their peak flow at the same time. In years of normal floods the Brahmaputra peaks a month before the Ganges.

As the rivers near their mouth, they deposit silt, which blocks the main channel and raises the river bed.

80% of Bangladesh is a huge flood plain and delta. It is flat, low-lying and easily flooded. Once the rivers overflow their banks the water can spread vast distances.

What were the effects of the flood?

The floods were devastating to Bangladesh. At their peak, they covered almost 70 per cent of the country and affected two-thirds of the population. In many places only the tops of trees and buildings could be seen. The water in Dhaka, the country's capital, was 2 metres deep and covered three-quarters of the city. The electricity supply was cut off for several weeks and, because wells were flooded and the water in them polluted, there was no safe drinking water.

Estimates suggested that, across the country, 7 million homes were destroyed and over 25 million people were made homeless. The official death toll was put at over 1300, but many more were simply reported missing and unaccounted for. Most deaths were due to drowning, but others died from diseases such as cholera and dysentery.

Both during and after the floods there were shortages of food and medicines. Two million tonnes of rice, a quarter of a normal year's crop yield, were destroyed. Most of the jute, sugar cane and vegetable crops were also lost, as were up to half a million cattle and poultry. Thousands of kilometres of roads, a third of the railways, and Dhaka's international airport, were all flooded. Numerous bridges were also destroyed. The disruption of the transport network meant that it was impossible to deliver emergency food and medical help to those in greatest need. Damage was estimated at US$1.5 billion.

C

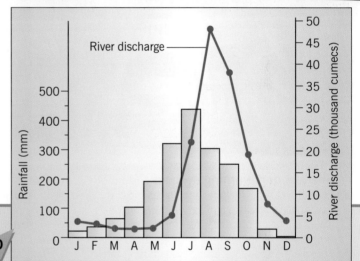

D

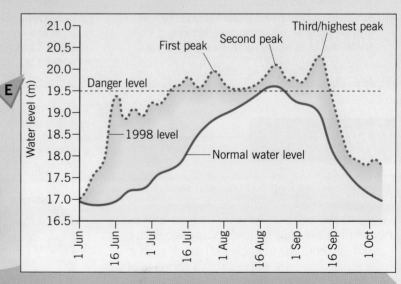

E

Activities

1 Graph **D** is the flood hydrograph for the Brahmaputra River at Dhaka.
   a) What is the rainfall for July and the discharge for August?
   b) Which three months have most rainfall and which three have the highest discharge?
   c) Describe the link between rainfall and discharge.

2 **a)** List the causes of the 1998 flood using the headings below.

Possible causes of flooding	
Natural	Human

   b) Briefly describe the problems caused by the 1998 floods.
   c) With the help of graph **E**, explain why the 1998 floods were the worst ever experienced in Bangladesh.

Summary

Flooding in Bangladesh is a combination of natural factors and human mismanagement. Although flooding normally brings benefits, extreme floods, especially in a developing country, can be disastrous.

Can Bangladesh be protected from flooding?

The Plan

In 1989 the government of Bangladesh began working with several international agencies to produce a Flood Action Plan (FAP). This huge scheme contained 26 action points which together, it was hoped, would provide a long-term solution to the country's serious flooding problem.

A
Managing flooding in Bangladesh

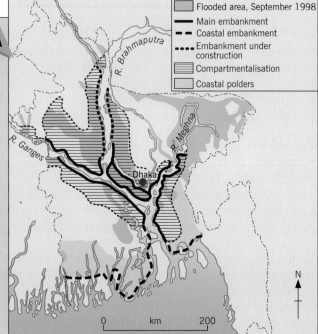

Flooded area, September 1998
Main embankment
Coastal embankment
Embankment under construction
Compartmentalisation
Coastal polders

R. Brahmaputra
R. Ganges
R. Meghna
Dhaka

N

0 km 200

Shelters and warning systems
- Build 5000 flood shelters in the areas most at risk. These are cheap and easy to construct and would provide a place of safety for almost everyone.
- Improve the flood forecasting system using satellite and computer technology.
- Prepare flood disaster management plans which provide early warning and clear, effective instructions as to what people should do before, during and after a flood.

Dams
- Build dams to control river flow and hold back the monsoon rainwater in reservoirs. These would be concentrated in Bangladesh but the plan could be extended to India and Nepal. The water would be used for irrigation and generating electricity. The cost would be more than £500 million.

Chauba Dam, India

The Problems

B

The Flood Action Plan has not been welcomed by everyone and indeed has attracted considerable criticism internationally as well as from within Bangladesh itself. Many people are concerned about the unknown effects of such a large scheme and are worried that a shortage of money will result in only the urban areas being protected, leaving the very poor rural inhabitants still at risk.

The embankments will trap rainwater and make the flooding worse.

How can we provide solutions if we don't really know the causes?

Dam construction could increase the build-up of silt and make flooding worse.

Flood control

Divide the land into compartments and control water flow through a system of channels by sluice gates and water pumps. In the dry season water can be moved to farming areas requiring irrigation. Before the monsoon, water would be drained away to leave room for the floodwaters.

Embankments

Complete and strengthen the embankments along all main river channels to a height of up to 7 metres. More than 7500 km of embankment is already in place but repairs, heightening and new building would cost over $6 billion. This scheme should prevent serious flooding from river overflow.

Activities

1 What are the main aims of the Flood Action Plan?

2 Draw up a table like the one below to describe the main points of the Flood Action Plan.

Proposal	Description	Good points	Bad points

3 Which of the proposals do you think will be:
a) the cheapest
b) the most expensive
c) the most successful in reducing the flood risk?
Give reasons for your answers.

4 Why is it harder to reduce the flood risk in economically less developed countries like Bangladesh than in economically more developed countries like the USA? (*Clues:* funding, materials, workforce, organisation, equipment, expertise.)

Summary

There is no easy solution to Bangladesh's flooding problem. The enormous size of the problem, the extreme poverty of the country, and the difficulty of identifying the exact causes of flooding make the task almost impossible. The Flood Action Plan tries to give protection from disastrous floods whilst still retaining the benefits of normal flooding. Not everyone agrees with the plan.

The embankments will restrict river access for fishing people.

Flood shelters save lives but don't help protect our property and livelihood.

These plans are far too expensive.

Flood control systems may damage the environment.

Up to half a million people will lose their land to reservoirs and embankments

2 Coasts

How do waves wear away the land?

Although waves may sometimes result from submarine earth movements, they are usually formed by the wind blowing over the sea. The size of a wave depends upon the:

- strength of the wind
- length of time that the wind blows
- distance of sea that the wind has to cross.

As a wave approaches shallow water near to the coast, its base is slowed down by friction against the sea-bed. The top of the wave will therefore move faster, increase in height and will eventually break ('tumble over') onto the beach.

Coastal erosion

There are four main processes by which the sea can erode the land. These are similar to those of a river (page 6).

- **Hydraulic pressure** is the sheer force of the waves, especially when they trap and compress air in cracks and holes in a cliff.
- **Corrasion** results from large waves hurling beach material against the cliff.
- **Attrition** is when waves cause rocks and pebbles on the beach to bump into each other and to break down in size.
- **Corrosion** is when certain types of cliff are slowly dissolved by acids in the sea-water.

There are three main groups of landforms which result from erosion by the sea.

Headlands and bays These form along coasts that have alternating resistant (harder) and less resistant (softer) rock. Where there is resistant rock the coast will be worn away less quickly leaving a **headland** which sticks out into the sea. Where there is softer rock, erosion will be more rapid and a **bay** will form (diagram **A** and page 50). As the headland becomes more exposed to the full force of the wind and waves, it will become more vulnerable to erosion than the sheltered bay.

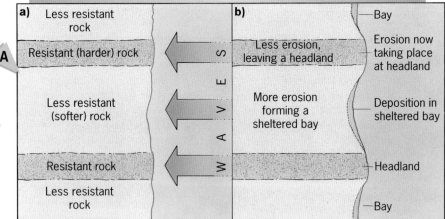

Cliffs and wave-cut platforms Erosion is greatest when large waves actually break against the foot of a cliff. The foot of the cliff is undercut to form a **wave-cut notch** (diagram **B**). As the notch gets larger, the cliff above it will become increasingly unsupported and in time will collapse. As this process is repeated the cliff will slowly retreat and, usually, increase in height. The gently sloping land left at the foot of the retreating cliff is called a **wave-cut platform** (diagram **B**).

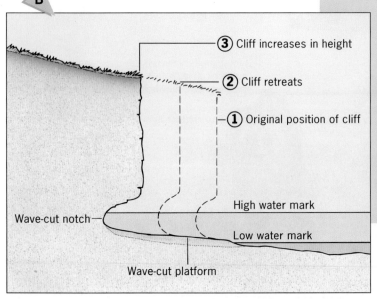

Flamborough Head, Yorkshire

Caves, arches and stacks Although cliffs, especially where they form headlands, consist of resistant rock they are still likely to contain areas of weakness. Areas of weakness will be the first to be worn away by the sea. Diagram **C** shows a typical sequence in which a weakness is enlarged to form a **cave** and, later, an **arch** where the sea cuts right through the headland. The arch is widened by the sea undercutting at its base. As the rock above the arch becomes unsupported it collapses to form a **stack**. Further undercutting causes the stack to collapse leaving only a stump.

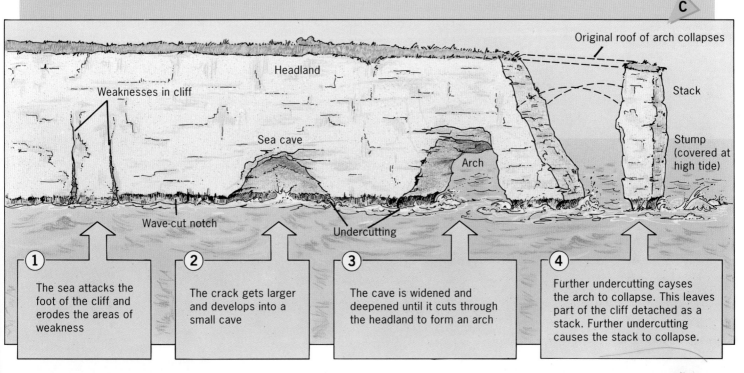

C

Headland
Weaknesses in cliff
Sea cave
Wave-cut notch
Undercutting
Arch
Original roof of arch collapses
Stack
Stump (covered at high tide)

1. The sea attacks the foot of the cliff and erodes the areas of weakness

2. The crack gets larger and develops into a small cave

3. The cave is widened and deepened until it cuts through the headland to form an arch

4. Further undercutting cayses the arch to collapse. This leaves part of the cliff detached as a stack. Further undercutting causes the stack to collapse.

Activities

1 Give three reasons why, on diagram **D**, the waves at **X** are likely to be higher and more powerful than the waves at **Y**.

2 Describe briefly four processes by which the sea can erode the land.

3 Photo **E** shows several coastal features.
   a) Make a landsketch of the photo and add the following labels:
      • *corrasion by waves*
      • *wave-cut notch*
      • *wave-cut platform*
      • *cave* • *arch* • *stacks.*
   b) Use broken lines to show the position of two collapsed arches.

D

Strong winds from south-west have been blowing for three days, after crossing 1000 km of ocean

Gentle winds from south-west have blown for one day after crossing 10 km of sea

X

Y

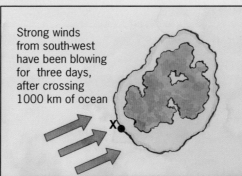

E

Old Harry Rocks, Dorset

Summary

Waves are caused when the wind blows over the surface of the sea. There are four processes by which the sea can erode the land to produce such landforms as headlands, bays, cliffs, wave-cut platforms and stacks.

How does the sea transport material?

The sea can be very powerful and is able to transport large amounts of material either along a beach (diagram **A** and photo **B**) or up and down a beach (photo **C**).

Transportation along a beach Waves rarely approach a beach at right-angles. They usually approach at an angle that depends upon the direction of the wind (diagram **A**). The water that rushes up a beach after a wave breaks is called the **swash**. The swash, which picks up sand and shingle, travels up the beach in the same direction as the breaking wave. When this water returns down the beach to the sea it is called the backwash. Due to gravity the **backwash**, and any material it is carrying, tends to be straight down the beach (diagram **A**). The result is that material is transported along the beach in a zig-zag movement. This movement of beach material is called **longshore drift**. Longshore drift is usually in one direction only, that of the prevailing wind. For example, the prevailing wind in Britain is from the south-west and so material is moved from west to east along the south coast of England.

A

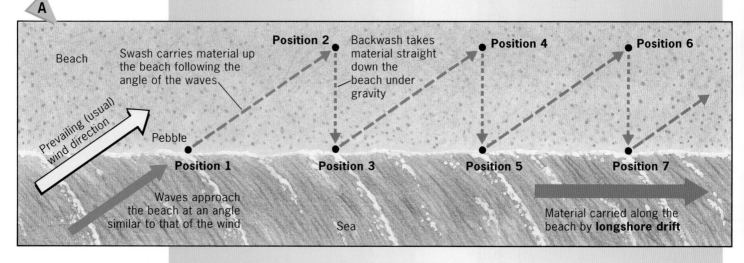

Beach

Swash carries material up the beach following the angle of the waves

Position 2 Backwash takes material straight down the beach under gravity

Position 4

Position 6

Prevailing (usual) wind direction

Pebble

Position 1

Position 3

Position 5

Position 7

Waves approach the beach at an angle similar to that of the wind

Sea

Material carried along the beach by **longshore drift**

Longshore drift can sometimes affect human activities. In response, people may erect wooden breakwater fences down the beach (photo **B**). The fences, called **groynes**, reduce the force of the waves and cause sand or shingle to pile up on their windward side (the side facing the prevailing wind). This is an advantage to people living in a seaside resort who do not wish to lose their sand, and to sailors in a small port who do not want their harbour to become blocked.

B Worthing, West Sussex

Transportation up and down a beach

Under normal conditions waves tend to move material up a beach. Photo **C** shows how shingle has been piled up at Hove in Sussex. Shingle, when piled at the foot of a cliff protects it from erosion. However, under storm conditions, larger waves often move material back down the beach.

What happens when people remove beach material?

We have already seen that the building of groynes can slow down the transport of material along the beach. Sea-walls have been constructed along many parts of our coastline to try to reduce the force of the waves and to protect cliffs from erosion (photo **C** page 32). Sometimes, however, human activity can unintentionally speed up the rate at which cliffs are eroded. During one year at the end of the last century, 660 000 tonnes of shingle were removed from the beach at Hallsands in Devon.

C

It was used for construction of the naval dockyard at Plymouth. The speed at which the shingle was removed was far greater than the rate at which nature could replace it. The cliff was exposed to erosion and within a century it had retreated by almost 10 metres. Buildings in Hallsands were threatened as the cliff retreated, and the village has now been left virtually abandoned and in ruins.

Activity

- What name is given to the process by which material is moved along a stretch of coastline?
- How can waves transport sand from point **A** to point **X** on diagram **D**?
- Why might people in each of **a)** place **R** and **b)** place **S** want to reduce the movement of material along the beach?

- How might they stop material from point **A** moving to point **X**?
- How might waves move material from point **A** to point **Y**?
- A building firm applied to the local council for permission to remove shingle from point **Y**. Why do you think that their application was turned down?

D

Summary   The movement of material along a beach is called longshore drift. Human activities can affect the rate at which coastal landforms develop.

How do landforms result from deposition by the sea?

Deposition occurs in sheltered areas where the build-up of sand and shingle is greater than its removal. The most widespread coastal deposition feature is the beach. Although rocky beaches are formed by erosion (wave-cut platforms, page 26), sand and shingle beaches result from deposition. Diagram **A** shows the differences in steepness between sand and shingle beaches, and how material of different sizes is distributed on those beaches.

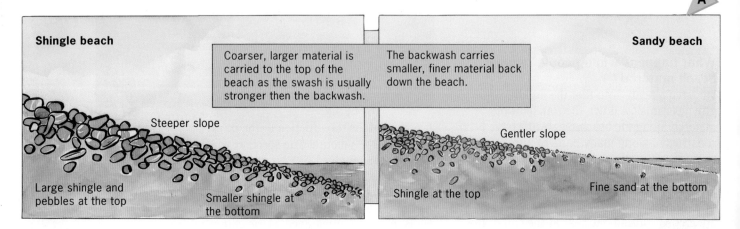

A

Shingle beach

Coarser, larger material is carried to the top of the beach as the swash is usually stronger then the backwash.

The backwash carries smaller, finer material back down the beach.

Sandy beach

Steeper slope

Gentler slope

Large shingle and pebbles at the top

Smaller shingle at the bottom

Shingle at the top

Fine sand at the bottom

Spits

A **spit** is an area of sand or shingle which either extends at a gentle angle out to sea or which grows across a river estuary (diagrams **B** and **C**). Many spits are characterised by a hooked, or curved, end. Spits only develop in places where:

- longshore drift moves large amounts of material along the beach
- there is a sudden change in the direction of the coastline
- the sea is relatively shallow and becomes progressively more sheltered.

Diagram **B** shows how a typical spit forms. The line **X–Y** marks the position of the original coastline. The prevailing wind, in this example, is from the south-west and so material is carried eastwards by longshore drift (**A**). Where the coastline changes direction (**B**), sand and shingle are deposited in water which is sheltered by the headland. This material builds upwards and outwards (**C**) forming a spit. Occasionally strong winds blow from a different direction, in this case the south-east. As waves will now also approach the land from the south-east, then some material will be pushed inland causing the end of the spit to curve (**D**). When the wind returns to its usual direction the spit will continue to grow eastwards (**E**), developing further hooked ends during times of changed wind direction (**F**). The spit cannot grow across the estuary (**G**) due to the speed of the river carrying material out to sea. Spits become permanent when the prevailing wind picks up sand from the beach and blows it inland to form sand dunes. Salt marsh (**H**) develops in the sheltered water behind the spit.

B

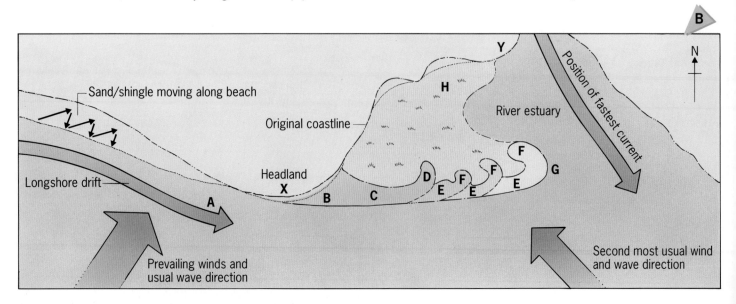

N

Sand/shingle moving along beach

Y

H

River estuary

Position of fastest current

Original coastline

Headland

Longshore drift

X

A

B

C

D

E

F

E

F

E

G

Prevailing winds and usual wave direction

Second most usual wind and wave direction

C Dawlish Warren, Devon

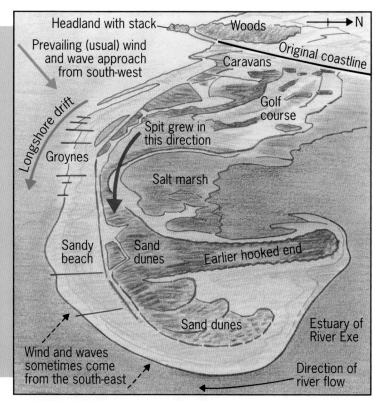

Bars

A **bar** is a barrier of sand stretching across a sheltered bay. It is only able to extend across the bay due to the absence of any large river (diagram **D**). Bars may form in several ways. One way is when a spit is able to grow right across a bay. A second is when a sand bank develops some distance off the shore, but parallel to it. Waves slowly move the sand bank towards the coast until it joins with the mainland. In both cases a lagoon is usually found to the landward side of the bar.

D

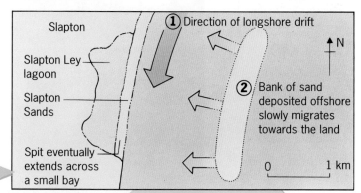

Activities

1 How does the appearance of a sandy beach differ from that of a shingle beach?

2 The following sketches show several stages in the formation of a spit. Unfortunately they are not in the correct order.
   a) Redraw the sketches putting the stages into the correct order.
   b) Describe how each stage developed.

E

Summary

Sand and shingle are deposited where the sea is calm and gentle. Beaches, spits and bars are examples of landforms which result from deposition.

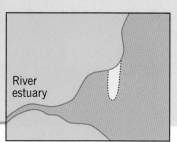

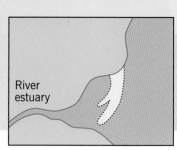

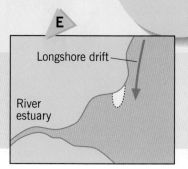

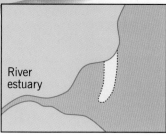

What are the causes, effects and human responses to cliff erosion?

Causes

- Where cliffs consist of resistant rock, as at Flamborough Head, waves erode at their base causing them to become unstable and to collapse (page 26).
- Where cliffs consist of less resistant rock, as in Holderness in Yorkshire, rain can wash loose material down to the cliff base. Together with the softer material there, it can be rapidly removed by waves (photo **A**).
- Sand and shingle provide a natural protection at the foot of cliffs, but if this is removed by human activity then erosion and cliff retreat are accelerated (page 29).
- Buildings built on cliff tops add weight which can contribute to cliff collapse (e.g. a hotel near Scarborough was destroyed in this way in 1993).

Effects

The Holderness coastline is retreating by an average of 2 metres a year. It is now some 3 km further west than it was in Roman times, and some 50 villages mentioned in the Domesday Book of 1086 have been lost to the sea. Villages, farms and campsites situated in places that a few years ago were considered safe, have been abandoned and lost (photo **B**).

Human responses

The Holderness towns of Hornsea and Withernsea have been protected by building concrete sea-walls. Sea-walls often curve towards the top to divert the force of the waves back to sea, and are an example of hard defences (photo **C**). Elsewhere concrete rip-rap (photo **D**) and submerged offshore breakwaters both help to reduce the power of waves. Unfortunately the protection of one part of the coastline, especially by hard defences, often only increases the rate of erosion elsewhere. People argue over the high cost of constructing and maintaining sea-walls. Those at risk from cliff collapse want their property protected, while others feel that nature should be allowed to run its natural course.

Activities

1 a) What are the main causes of cliff erosion?

b) What are some of the effects of cliff erosion?

c) What are some of the reasons for and against constructing sea defences?

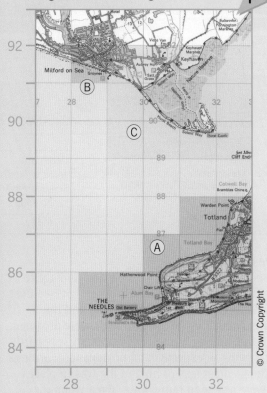

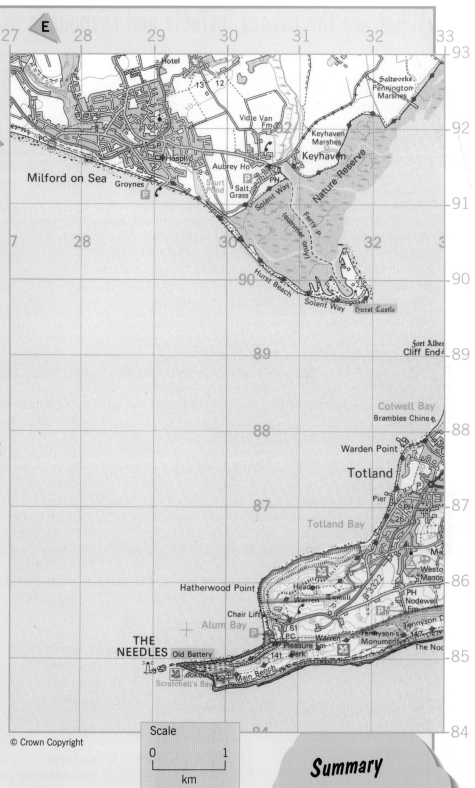

© Crown Copyright

2 Map **E** shows part of the coastline of Hampshire and the Isle of Wight in southern England. Map **F**, which is map **E** reduced in size, highlights three areas.

a) i) Draw an enlarged sketch map (see page 288) of area Ⓐ and, using information on map **E**, locate and label:
4 bays, 2 headlands, 2 areas of cliffs, stacks, and *an area of sand.*

ii) Name four features of coastal erosion *not* shown on map **E** that you might expect to find on a field visit to an area of coastal erosion.

b) What evidence is there in area Ⓑ of the transportation of beach material by the process of longshore drift?

c) Draw an enlarged sketch map of area Ⓒ and, again using information from map **E**, locate and label:
• *a sandy beach* • *sand dunes* • *a salt marsh* • *a spit with a hooked end.*

Summary

The natural rate of cliff erosion can be accelerated by human activity. There are arguments for and against trying to protect cliffs from erosion.

What are the causes, effects and responses to coastal flooding in Britain?

The worst coastal flood in recent years in Britain occurred during the night of 31 January/1 February 1953. The worst affected area was between the estuaries of the Humber and the Thames.

Causes

Most of the area between the Humber and the Thames is low-lying. Indeed, some parts surrounding The Wash are actually below sea-level. These areas were protected by small sea walls and embankments, many of which were in a state of disrepair due to a lack of attention during and after the Second World War. Although people realised that there was a high flood risk, they were totally unprepared for that night in 1953. Four main factors combined to cause a **storm** (or **tidal**) **surge**. A storm surge is when the level of the sea rises rapidly to a height well above that which was predicted.

1 An area of low atmospheric pressure, called a depression, moved southwards into the North Sea. Air rises in a depression. As the air rose it exerted less pressure, or weight, upon the sea. The reduction in pressure was enough to allow the surface of the sea to rise by half a metre.

2 The severe northerly gale created huge 6-metre high waves. These waves 'pushed' sea-water southwards down the North Sea to where it gets shallower and narrower, as shown on map **A**. As the extra water 'surged' southwards, it was unable to escape fast enough through the Straits of Dover. The result was a further 2-metre rise in sea-level, especially in river estuaries.

3 It was a time of spring tides. These occur every month and are when the tides reach their highest level.

4 Rivers flowing into the North Sea were in flood but could not discharge their water due to the high sea-levels.

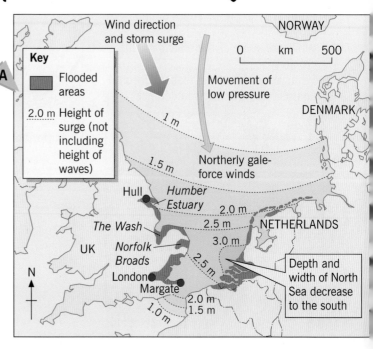

A

Key

▨ Flooded areas

2.0 m Height of surge (not including height of waves)

Wind direction and storm surge

Movement of low pressure

NORWAY

DENMARK

Northerly gale-force winds

NETHERLANDS

Depth and width of North Sea decrease to the south

Hull · Humber Estuary

The Wash

UK · Norfolk Broads

London · Margate

1 m · 1.5 m · 2.0 m · 2.5 m · 3.0 m · 2.5 m · 2.0 m · 1.5 m · 1.0 m

N

0 — km — 500

Consequences

The flood caused the deaths of 264 people and damage to 25 000 homes (map **B** and photo **E**). Sea water covered over 1000 km² of land in Lincolnshire, East Anglia and the Thames Estuary. Thousands of farm animals were drowned. The high death rate among humans and animals was partly due to the flood being unexpected and no advance warnings being given. The greatest loss of human life was in places where people were asleep in bungalows. Even when the flood subsided, sea water was still able to penetrate gaps in sea defences, and farmland remained contaminated by salt water. The disaster was even worse across the North Sea where, in the Netherlands, more than 1800 people died.

B

Low pressure

Northern gales and surge tides

Aberdeen

River Tees

Mablethorpe
Skegness
Hunstanton
Southend
London
Canning Town

Tilbury

Harwich

Canvey Island

Isle of Sheppey

F r a n c e

TIMETABLE		
Saturday 31 January 1953:		
a.m.	**Aberdeen**	Gales with gusts of 130 km/h (80 mph)
15.30	**River Tees**	Water overflows banks
17.00	**Lincolnshire**	Flooding along coast, 16 die and 1600 evacuated at Mablethorpe. 20 die as Skegness is flooded.
19.15	**Norfolk**	Train forced back to Hunstanton as waves break through sand dunes, engulfing village and killing 65
22.18	**Southend**	Tanker *Kosmos V* runs aground
Midnight	**Isle of Sheppey**	Much of island, including naval dockyard, flooded
Sunday 1 February:		
00.30	**Harwich**	1200 homes flooded, eight die
00.30	**Southend**	600 homes engulfed, two killed
00.40	**Tilbury**	Thames overflows, killing one and making 6102 homeless
01.10	**Canvey Island**	11 000 homeless, 58 dead
01.55	**Canning Town**	Sea breaks in, making 150 homeless, killing one

C The 1953 floods

The Thames Barrier
at Woolwich D

Causes of coastal flooding

Coastal flooding elsewhere can result from a combination of factors.

- Land behind the coastline is flat and low-lying (Towyn, North Wales 1990) or may even lie below sea-level (Netherlands 1953).
- Severe storms can cause exceptionally high waves and create storm surges (south-east England 1953, and Bangladesh, see pages 36–37).
- Very high tides can flood areas that are not protected either naturally (by sand dunes) or artificially (by sea-walls).
- Global warming (page 104) is causing sea-level to rise, and so increasing the flood risk in many places around the world (Bangladesh, Nile delta and Florida).

Precautions

- The most obvious of these would appear to be the building of higher and stronger sea-walls. However, many people feel that these are expensive, unsightly and, because they take the full force of the waves, need fairly constant maintenance.
- Build tidal barriers across river estuaries (the Thames, photo **D**, and the Netherlands).
- Stop building on coastal flood plains.
- Improve weather forecasting and early flood warning systems.
- Try to reduce the increase in global warming.

Activities

1 Parts of south-east England and the Netherlands have a high risk of coastal flooding resulting from storm surges in the North Sea (map **A** and photo **E**).
   a) What is a storm surge?
   b) Give four reasons why there was a storm surge in the North Sea in 1953.
   c) Which places were worst affected by the storm surge?
   d) What were the consequences of the storm surge?

2 a) Name three places in the world, other than south-east England and the Netherlands, that are at risk from coastal flooding.
   b) Give five reasons why these places are at risk.
   c) Describe five precautions that can be attempted to try to reduce the risk.

E

Summary

Storm surges are a major cause of coastal flooding. They can cause serious disruption to human activity, and prevention schemes are expensive to implement.

35

What are the causes, effects and responses to coastal flooding in Bangladesh?

Bangladesh is one of the poorest and most densely populated countries in the world. Most of the land is either a river delta or a coastal plain – both of which are flat and low-lying. The rich soils and hot, wet climate support millions of rice farmers and their families (page 200). Unfortunately, many of them live where flooding by rivers (pages 22–23) or the sea is an annual risk. The causes of coastal flooding in Bangladesh are summarised in map **A**.

Bangladesh lies in an area that, each year, experiences several tropical storms (page 100). These violent storms can cause the deaths of thousands of people as they sweep up the Bay of Bengal flooding villages and farmland, and destroying everything before them. Diagram **B** describes the effects of one such storm.

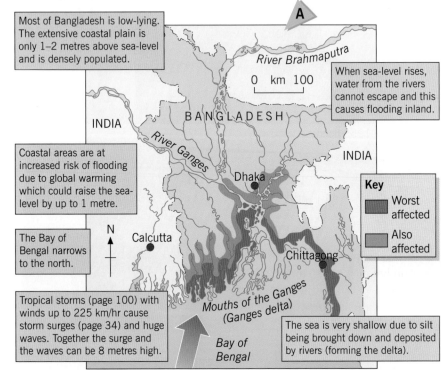

A

Most of Bangladesh is low-lying. The extensive coastal plain is only 1–2 metres above sea-level and is densely populated.

River Brahmaputra

0 km 100

When sea-level rises, water from the rivers cannot escape and this causes flooding inland.

INDIA

BANGLADESH

River Ganges

INDIA

Dhaka

Coastal areas are at increased risk of flooding due to global warming which could raise the sea-level by up to 1 metre.

Key

Worst affected

Also affected

N

Calcutta

Chittagong

The Bay of Bengal narrows to the north.

Tropical storms (page 100) with winds up to 225 km/hr cause storm surges (page 34) and huge waves. Together the surge and the waves can be 8 metres high.

Mouths of the Ganges (Ganges delta)

Bay of Bengal

The sea is very shallow due to silt being brought down and deposited by rivers (forming the delta).

B

Crops were destroyed, food supplies lost and drinking water contaminated (with raw sewage and dead bodies). Half a million head of cattle drowned. Fishing boats were destroyed. Over 4 million people faced starvation and disease.

Thousands of poorly constructed houses were either blown or washed away. Hundreds of villages were wiped out. Over 130 000 people were killed – many bodies were washed out to sea and were never recovered.

Many roads and bridges were destroyed and electricity supplies were cut off for several months.

Emergency electricity supplies and telephone links failed to work. The early warning system proved ineffective. The shelters provided by the government were insufficient. Relief could not reach us due to poor transport.

Our country is very poor and we cannot afford to prepare properly for a flood disaster. The effects of flooding are therefore a lot worse than they would be for a rich country.

What precautions could be taken to reduce the flood risk?

- Build more shelters (photo, page 24). These would:
  - give protection against the wind
  - be above the level of the sea floodwater (they are built on stilts)
  - contain emergency food and medical supplies.
- Improve the early warning system by:
  - using satellites to track storms and to predict their path and impact
  - using mosques' loudspeaker systems to advise people when to evacuate their homes
  - increase the number of cyclist volunteers who warn the many families who have no access to warnings given on radio or TV.
- Prepare flood disaster plans that provide clear instructions as to what people should do before, during and after flood (the instructions need to be visual because many farmers are unable to read).
- Improve transport so that:
  - warning messages can reach villages
  - evacuation is made easier
  - emergency services and relief aid can arrive more quickly.
- Construct a higher and stronger embankment along the whole of the coast and around the many islands. Earlier embankments have proved to be too small in height and too weak in strength against the waves. In many places the embankment was either badly maintained or was non-existent.

Places like Bangladesh can never be safe from the threat of coastal flooding, and people living near to the sea are likely to remain at risk. If a rich country like the United States finds it difficult to provide full protection for its people and their property against the ferocity of a tropical storm in low-lying coastal areas, then what chance have poor countries like Bangladesh? It is Bangladesh's poverty that puts local people at risk and makes protection by its government so difficult (diagram **C**).

C

The government finds if very difficult to find and provide:
- adequate early warning systems before the event
- protection during the event
- emergency services and aid (food, water and medicines) after the event.

Poor families simply cannot afford to leave the danger areas. Even if we could, where would we go?

D

What are the effects and responses to coastal flooding in Bangladesh?
- What are the causes of coastal flooding?
- Which places are most at risk?
- What are the likely effects?
- Why are precautions often insufficient: *before*, *during*, and *after* the event?
- What precautions could be taken to reduce the risk of future flooding?

Activities

These activities are in the form of a mini-enquiry. The questions shown on the clipboard should help you structure your work. Use maps, labelled diagrams and facts and figures where appropriate.

1 Many precautions have been suggested to reduce the impact of the flood risk, but Bangladesh is a poor country.
  a) Describe four precautions that the country might be able to afford.
  b) Describe three that would be too expensive.

2 Give reasons for your answers.

Summary

There is no easy solution to Bangladesh's flooding problem. Extreme poverty, the high population density and the size of the problem make it difficult to plan for future disasters.

Why does the Wessex coast need to be protected?

The coastline between the towering cliffs at Durlston Head in Dorset and the low-lying spit at Hurst Castle in Hampshire (see OS map on page 33) is shown on map **A**. Like most other coastlines in Britain it is dynamic and always changing. These changes are caused by:

- natural (physical) processes of weather, waves and tides
- human intervention and activity.

The problems created by these changes vary from one part of the coastline to another (places A to F on map **A**).

A Swanage Bay

This stretch of coastline is relatively stable apart from some erosion of the softer clay cliffs south of Ballard Point (diagram **A** page 50). The adjacent limestone and chalk cliffs form a habitat for birds and lime-loving plants. Most of the problems in this area result from the settlement at Swanage and the influx of tourists in summer. Coastal footpaths become eroded, litter is left, and there is some sewage outflow into the sea.

B South Haven peninsula and Poole Harbour

The peninsula is a major sand dune ecosystem (habitat) which is slowly growing seawards (photo **B**). Behind it are salt and freshwater marsh (wetlands), and heath and woodland ecosystems noted for their seabirds, wildfowl, lizards and butterflies. The main problem is tourists, with over one million people a year trampling over the fragile South Haven dunes. Poole Harbour is also threatened by pollution from a freight and cross-Channel terminal as well as from several marinas.

C Bournemouth

Bournemouth is a major tourist resort built on top of cliffs. There is little natural coastline left and wildlife habitats are limited. The sea is affected by some sewage outflow, the hotels add pressure to the cliffs, and tourists and residents create noise and litter.

D Hengistbury Head and Christchurch Harbour

The headland and harbour consist of several wildlife habitats – salt marsh, freshwater marsh, heath, sand dunes and rocky foreshore – all with an abundance of wildlife (photo **C**). The harbour suffers from silting and water sports.

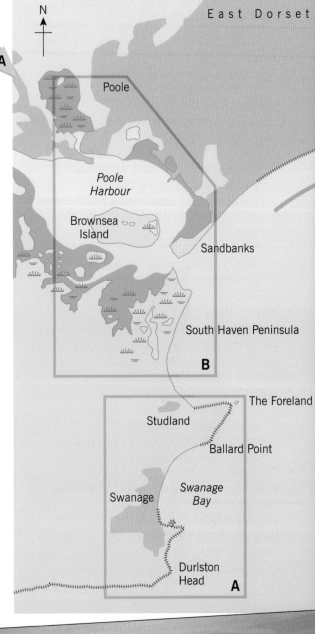

Sand dunes at South Haven

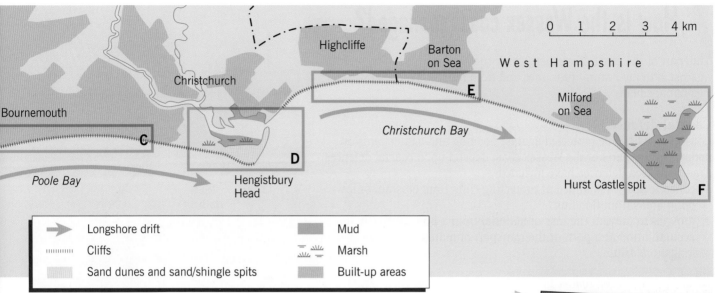

Longshore drift

Cliffs

Sand dunes and sand/shingle spits

Mud

Marsh

Built-up areas

E Christchurch Bay

The 30-metre high cliffs between Highcliffe and Barton on Sea have retreated by over 60 metres since 1971. The sand and clay cliffs readily become waterlogged after heavy rain, causing mudflows and landslips (photo **D**). This material is then removed by waves, especially during storms and at high tide.

F Hurst Castle spit

The shingle spit was formed by eroded cliff material from Christchurch Bay being carried eastwards by longshore drift (diagram **A** page 28, pages 30–31 and OS map **E** on page 33). Before 1954 the spit was never breached by the sea. Since then, however, the supply of material has been reduced by the introduction of groynes to the west (photo **B** page 28). This, together with the increased frequency and intensity of storms and higher tides (attributed to global warming), has meant that the spit is now breached by the sea several times a year. The spit, with its sand dunes and salt marsh behind, is an important wildlife habitat – but a habitat with a threatened existence.

D Landslip near Barton on Sea

C Hengistbury Head and coast protection

Activities

1 Using pages 26–27, 30–31, 38–39 and 50, describe how the coastal landforms in Swanage Bay (A on map **A**) differ from those at Hurst Castle (F on map **A**).

2 a) Which parts of the coast shown on map **A** experience the greatest:
   i) natural (physical) problems
   ii) human problems?
   b) Give a brief description of these problems.

Summary

Coastlines are dynamic and constantly changing. These changes, which may be due to natural processes or to human activity, can create problems.

How is the Wessex coast managed?

This part of the coast, like other British coastlines, needs to be protected against extreme natural processes and from human misuse and mismanagement.

Coastal defence

Coastal areas can be protected against flooding and against erosion. Today, several parts of the Wessex coast depend upon coastal defences. Traditional sea defences have included:

- concrete sea-walls aimed at protecting cliffs from erosion, or low-lying areas from flooding
- groynes to prevent the loss of material from a holiday beach, its accumulation at a port, or the transport of material through longshore drift.

However, it has been realised that these methods have themselves created problems. For example:

- Concrete sea-walls absorb, rather than deflect, wave energy. Without fairly constant maintenance, these defences become breached and, as at Barton on Sea, cliff erosion is renewed.
- Groynes can have damaging effects on neighbouring coastlines. These places often experience increased erosion as there is no longer a supply of beach material to protect their cliffs or, in the case of Hurst Castle, spit.
- Older schemes were not in harmony with the environment – either visually or with local habitats (ecosystems).
- Traditional-type defences are expensive to build and to maintain and do not react to short-term or long-term changes in coastal processes.

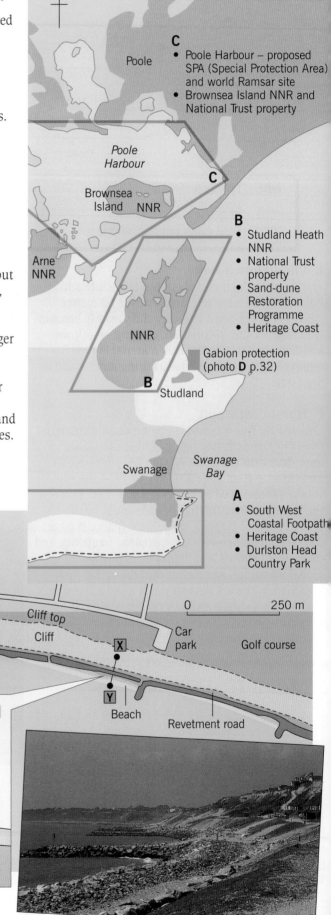

A

N

East Dorset

C
- Poole Harbour – proposed SPA (Special Protection Area) and world Ramsar site
- Brownsea Island NNR and National Trust property

Poole

Poole Harbour

Brownsea Island NNR

Arne NNR

NNR

C

B
- Studland Heath NNR
- National Trust property
- Sand-dune Restoration Programme
- Heritage Coast

Gabion protection (photo **D** p.32)

B Studland

Swanage

Swanage Bay

A
- South West Coastal Footpath
- Heritage Coast
- Durlston Head Country Park

0   250 m

B Recent sea defences at Barton on Sea

Marine Drive

1975 landslip

Cliff top

Cliff

Car park

Golf course

1993 slip area

Rock revetment

Rock strongholds

X

Y

Beach

Revetment road

X

Top level of revetment 4.3 m

Y

Cliff

Revetment roadway

Large rock boulders

Beach, to try to protect revetments

Existing timber pile (earlier sea defence)

Core stones

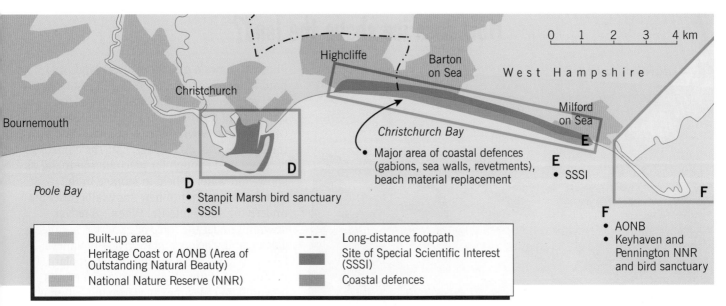

- Major area of coastal defences (gabions, sea walls, revetments), beach material replacement

D
- Stanpit Marsh bird sanctuary
- SSSI

E
- SSSI

F
- AONB
- Keyhaven and Pennington NNR and bird sanctuary

	Built-up area	- - - -	Long-distance footpath
	Heritage Coast or AONB (Area of Outstanding Natural Beauty)		Site of Special Scientific Interest (SSSI)
	National Nature Reserve (NNR)		Coastal defences

The modern approach (the last ten years):
- accepts the need to understand the natural processes at work
- realises that new schemes are likely to be more cost-effective if they work with, rather than against, nature
- appreciates that while rigid schemes may protect specific locations they can have disastrous effects elsewhere
- accepts that creating a natural scheme can retain wildlife and enhance the quality of the environment.

Diagram **B** describes recent schemes aimed at protecting the unstable cliffs at Barton on Sea. A similar scheme, to the west of Barton on Sea, was later rejected. This was due to its high cost, a realisation that eroded cliff material from here did partly protect the cliff-foot at Barton, and because by then the coast had become a Site of Special Scientific Interest (SSSI – map **A**) due to the presence of important fossils in the cliffs.

Coastal management
Almost one-quarter of this coastline is considered 'developed'. The developed areas include permanent settlements (Bournemouth–Poole), tourist amenities (caravan parks) and farmland. This development has increasingly led to conflict, especially between people who live and work here, tourists who visit the area, and conservationists (figure **C**).

Government policy states that:
- public access to the coast should be a basic right, unless it would either be damaging to the environment or impractical
- local authority management plans should protect the coast from unnecessary development
- coastal recreational activities should be in harmony with the environment. New or expanded activities should only be allowed if the coast has the capacity to absorb the increase in pressure put upon it.

Map **A** lists and locates some of the varied and more important management schemes along this part of the Wessex coast.

Activities

1 a) Describe the traditional methods of protecting coasts from erosion.
  b) Why are these methods no longer in favour?
  c) How do present-day protection schemes work with, rather than against, nature?

2 a) How do the needs of local people, tourists and conservationists differ when it comes to managing a stretch of coastline (figure **C**)?
  b) Using map **A**, describe some of the ways in which part of the Wessex coastline has been managed.

WE NEED MORE . . .
- caravan parks → roads → jobs
- water sports → sea defences
- information centres → bird sanctuaries
- parking areas → litter collection
- paths to the beaches → picnic areas
- sand dune restoration schemes → houses
- water quality monitoring → marinas
- protection of farmland → SSSIs

C

Summary

Although it is now debatable as to whether or not coastlines should be protected from the sea, it is agreed that their use needs careful management.

How does ice shape the land?

A · A glacier in Greenland

Photo **A** shows a **glacier** in Greenland. It is hard to imagine that only a few thousand years ago, very recently in the life history of the Earth, all of northern Britain looked like this. Glaciers form when there is an interruption in the hydrological cycle (page 12). The climate becomes cold enough for precipitation to fall as snow, and water is held in storage in the system. The weight of new snowfalls turns the underlying snow into ice. When ice moves downhill under the force of gravity it is called a glacier, and glaciers replace rivers in valleys. Like rivers, ice picks up and transports large amounts of material. As the material moves downhill it erodes the land, forming extremely scenic landforms in highland areas. Later, it will be deposited in valleys and across lowlands.

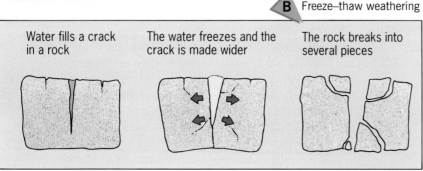

B · Freeze–thaw weathering

Water fills a crack in a rock	The water freezes and the crack is made wider	The rock breaks into several pieces

Much of the material carried by a glacier results from a process called **freeze–thaw** weathering (or **frost shattering**). Freeze–thaw weathering occurs when temperatures are often around freezing point. Water gets into cracks in rocks and freezes (diagram **B**). As the water turns to ice it expands, putting pressure on the surrounding rock. When the ice melts back into water the pressure is released. Repeated freezing and thawing widens the cracks and causes jagged pieces of rock to break off (page 49). Photo **C** shows how material resulting from freeze–thaw has fallen onto a glacier. This material, called **moraine**, is then transported by the glacier either on its surface, within it (if covered by later snowfalls or avalanches), or dragged along under it (diagram **D**). Moraine is able to erode the sides and floor of the valley just as a river erodes its banks and bed.

Glacial erosion

Glaciers erode much faster than rivers. The two main processes of glacial erosion are **abrasion** and **plucking**. Abrasion is when the material carried by a glacier acts like sandpaper on a giant scale, rubbing against and wearing away the sides and floor of a valley. It is a similar process, therefore, to corrasion by rivers and waves. Plucking is when ice freezes and sticks to rock. When the ice moves, large pieces of rock are pulled away with it. These two processes are perhaps better understood when explaining the formation of one typical landform of a glaciated highland area, the **corrie** (diagram **D**).

Glaciers in the Swiss Alps · C

Landforms of glacial erosion – corries

Corries are also known as cirques or cwms. They are deep, rounded hollows with a steep back wall and a rock basin (photo **F**). Some corries contain a deep, rounded lake, or tarn. Corries develop when snow collects in pre-glacial hollows. As more snow accumulates, it turns to ice, and begins to move downhill. Freeze–thaw and plucking cut into the back wall of the hollow making it steeper (diagram **D**).

Abrasion, underneath the glacier, deepens the floor to give a basin shape. A rock lip marks the place where there was less erosion and, sometimes, where moraine was deposited. After the ice melts, water may be trapped by the rock lip creating a small lake, or tarn (diagram **E**). An example of a corrie (cwm) in Snowdonia is seen in photo **F**, Llyn (lake) Glaslyn (grid square 6154, page 47).

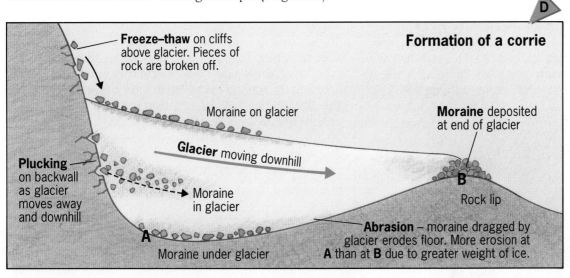

D Formation of a corrie

Freeze–thaw on cliffs above glacier. Pieces of rock are broken off.

Moraine on glacier

Moraine deposited at end of glacier

Glacier moving downhill

Plucking on backwall as glacier moves away and downhill

Moraine in glacier

B Rock lip

A Moraine under glacier

Abrasion – moraine dragged by glacier erodes floor. More erosion at **A** than at **B** due to greater weight of ice.

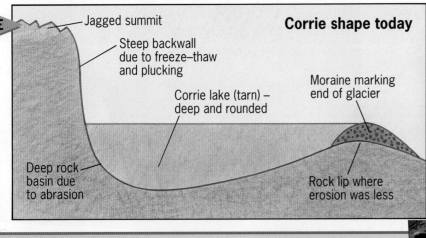

E Corrie shape today

Jagged summit

Steep backwall due to freeze–thaw and plucking

Corrie lake (tarn) – deep and rounded

Moraine marking end of glacier

Deep rock basin due to abrasion

Rock lip where erosion was less

F Llyn Glaslyn, Snowdonia

Activities

1 **a)** How does snow eventually become a glacier?
  b) What causes a glacier to move downhill?

2 **a)** What is the name given to material transported by a glacier?
  b) Why is this material sometimes transported
    i) on the surface of
    ii) inside
    iii) underneath a glacier?

3 **a)** Name the glacial process likely to be found at each of **A**, **B** and **C** on diagram **G**.

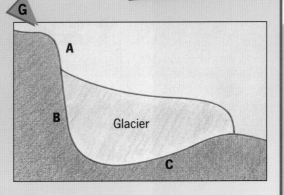

G

A

B Glacier

C

b) Describe each of the three processes.
c) Describe, using a labelled sketch only, the shape and appearance of a corrie.

Summary

Erosion, transport and deposition by glaciers produces a distinctive group of highland landforms. Freeze–thaw weathering and erosion by plucking and abrasion, are three of the most significant glacial processes.

What landforms result from glacial erosion?

Arêtes and pyramidal peaks

Corries are rarely found in isolation. Several usually form within the same mountainous area. Corries sometimes form in adjacent river valleys. In this case erosion of the sidewalls between them creates a narrow knife-edge ridge. At other times they may form back to back in valleys on either side of a watershed. On this occasion erosion of the backwalls narrows the distance between them until, again, a narrow knife-edged ridge is formed (diagram **A**). This narrow knife-edged ridge is called an **arête** (photo **B**). Crib-Goch is one of several arêtes on Snowdon (grid square 6255, page 47).

When three or more corries cut back into the same mountain, a **pyramidal peak** is formed. The most famous, and scenic, is the Matterhorn in Switzerland (photo **C**). Arêtes radiate from pyramidal peaks (diagram **D**).

Glacial troughs, truncated spurs and hanging valleys

Glaciers tend to follow the easiest route when moving downhill which, in most cases, is along an existing river valley. As glaciers move downwards, they erode both the floor and the sides of the valley to form a **glacial trough**. The characteristic V-shape of a river valley (page 7) is turned into the typical U-shape of a glaciated valley. A glacial trough is deep, and has a wide, flat valley floor with steep valley sides (photo **E**). The Glaslyn Valley (Afon Glaslyn) is an example of a glacial trough in Snowdonia (grid square 6553, page 47).

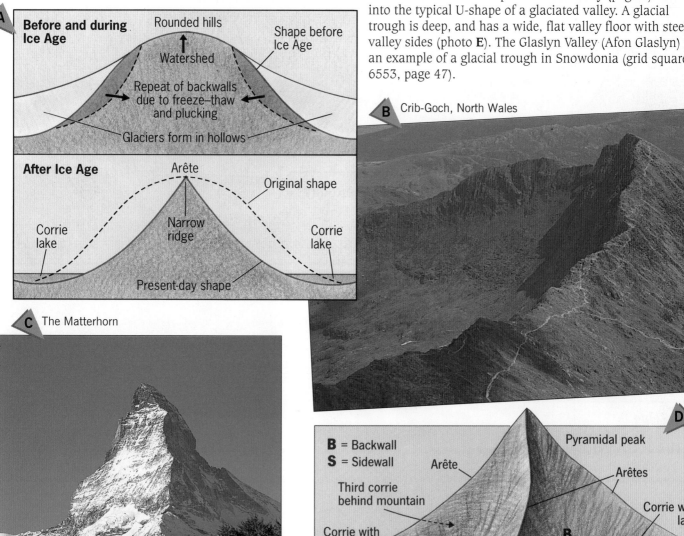

A

Before and during Ice Age

Rounded hills

Watershed

Shape before Ice Age

Repeat of backwalls due to freeze–thaw and plucking

Glaciers form in hollows

After Ice Age

Arête

Original shape

Corrie lake

Narrow ridge

Corrie lake

Present-day shape

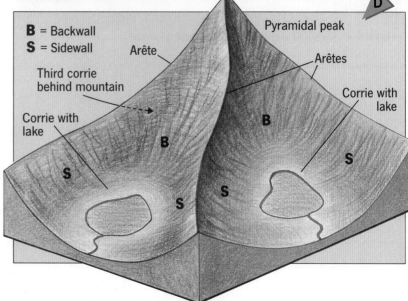

B Crib-Goch, North Wales

C The Matterhorn

D

B = Backwall
S = Sidewall

Arête

Pyramidal peak

Arêtes

Third corrie behind mountain

Corrie with lake

Corrie with lake

The valley is usually straightened by the glacier moving down it. The ends of interlocking spurs (page 7) are removed by abrasion to leave cliff-like features called **truncated spurs**. **Hanging valleys** result from differences in the rate of erosion between glaciers in the main and in a tributary valley (photo **F**). The floor of the smaller tributary is deepened more slowly than the floor of the main valley. When the ice melts, the tributary valley is left 'hanging' above the main valley, and its river has to descend by a single waterfall, or by a series of them (grid square 6251, page 47).

Because glaciation creates spectacular scenery, places such as Snowdonia (page 47) and the Lake District attract large numbers of tourists (pages 233–235). Such landscapes have, therefore, to be managed to preserve their natural characteristics while at the same time meeting the needs of tourists and local people.

E Nant Ffrancon, Snowdonia

F Jotunheim Mountains, Norway

Activities

1 Landsketch **G** shows a typical highland area which has been glaciated. The numbers refer to one glacial process (freeze–thaw) and six glacial landforms (corries, arêtes, pyramidal peaks, a glacial trough, truncated spurs and hanging valleys). Match up the numbers with the correct process or landform.

2 Explain how glaciation turns the:
   a) V-shape of a river valley into a U-shape
   b) interlocking spurs of a river valley into truncated spurs
   c) rounded watersheds of a river basin into jagged pyramidal peaks and arêtes.

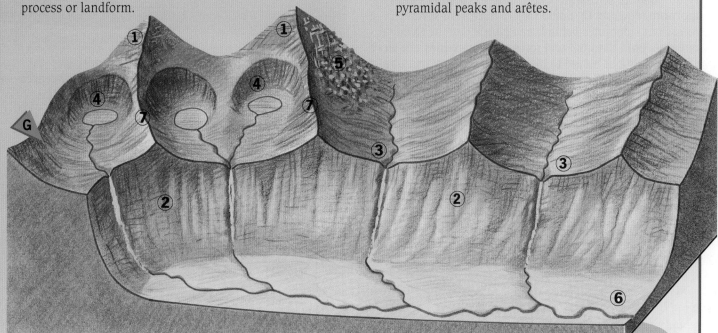

G

Summary Glacial erosion turns a gently rounded landscape formed by water into one that has steep and jagged landforms.

What landforms result from glacial deposition?

We have already seen that glaciers erode and transport large amounts of material. This material will be deposited either on the floor of glacial troughs in highland areas or across lowlands at the foot of highland areas. Deposition occurs when a rise in temperature causes ice to melt and the glacier is no longer able to carry as much material. A group of deposition landforms develop, mainly at the snout (end) of the melting glacier.

Erratics

Erratics are rocks and boulders picked up and transported many kilometres by the glacier, and deposited in an area of different rock (photo **A**). Rocks from Norway have been found in coastal cliffs in East Anglia, and granite from the Lake District on Anglesey in Wales.

Terminal and recessional moraines

Moraine is material that is transported and later deposited by a glacier. **Terminal moraine** marks the furthest, or maximum, point that a glacier reached (photo **B**). It is deposited at the end, or snout, of the glacier. If the glacier remains in the same position for a long time, terminal moraine can build up into a sizeable ridge which will extend across a valley (diagram **C**). The moraine is a mass of unsorted rocks, clays and sands. At the end of the Ice Age, glaciers began to melt and to retreat. This retreat was, however, rarely even but took place in several stages. At times the ice would stop melting and the glacier would remain in one position long enough for moraine to once again build up across the valley. This is a **recessional moraine** (diagram **C**). Some valleys may have several recessional moraines behind, and parallel to, the terminal moraine. Both terminal and recessional moraines can block valleys, acting as a dam to meltwater and to present-day rivers. Long, narrow **ribbon lakes** have formed on the floors of many glacial troughs in highland Britain. One example of a ribbon lake in Snowdonia is Llyn Gwynant (grid square 6451, page 47).

Following the retreat of the ice, various post-glacial processes continue to shape and alter the landscape. The main process results from meltwater which is released in enormous quantities as ice higher up in the mountains continues to melt. Meltwater can pick up angular, unsorted material originally left by the ice. The water rounds and sorts the material, depositing the largest particles first and the smallest last.

A glacial erratic near Ingleborough, Yorkshire **A**

Terminal moraine, Greenland **B**

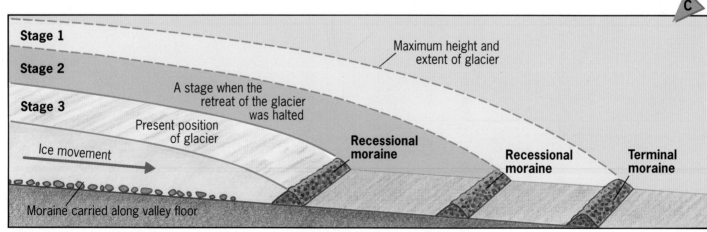
C

Stage 1

Stage 2

Stage 3

Maximum height and extent of glacier

A stage when the retreat of the glacier was halted

Present position of glacier

Ice movement

Recessional moraine

Recessional moraine

Terminal moraine

Moraine carried along valley floor

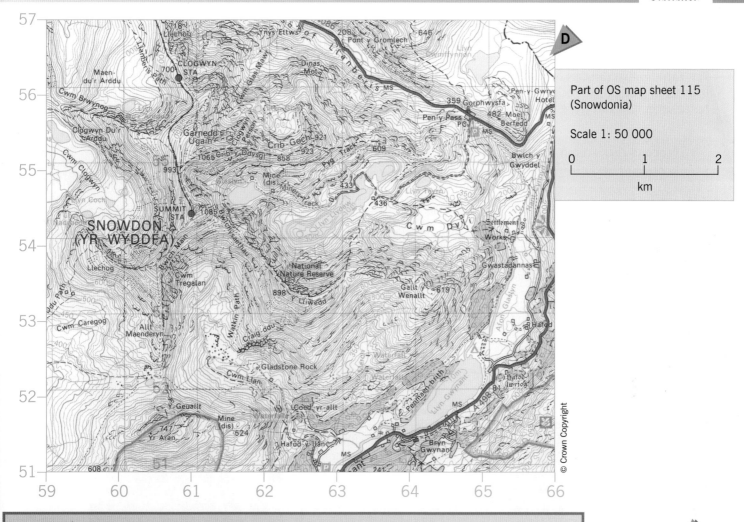

Part of OS map sheet 115 (Snowdonia)

Scale 1: 50 000

© Crown Copyright

Activities

1 a) Under what conditions will a glacier deposit the materials which it is carrying?

b) What are erratics?

c) What is the difference between a terminal moraine and a recessional moraine?

2 The OS map (map **D**) shows a glaciated highland area around Snowdon in North Wales.

a) Locate the following glacial landforms on the OS map, and then copy and complete table **E** by adding examples of the landforms to it.

- Afon Glaslyn (6553)
- Waterfalls (6251)
- Llyn Gwynant (6451)
- Waterfalls (6352)
- Cliffs along the side of Llanberis (6356)
- Bwlch Main (6053)
- Crib-Goch (6255)
- Llyn Glaslyn (6154)
- The summit of Snowdon (6054)
- Pass of Llanberis (6356)
- Llyn Du'r Arddu (6055)

b) What glacial feature might you find at the exit to Llyn Gwynant (637514)? Give a reason for your answer.

3 Which of the two cross-valley profiles in diagram **F** do you think represents Afon Glaslyn (6553)? Give two reasons for your answer.

Landform	Example
Corrie (cirque, cwm)	
Arête	
Pyramidal peak	
Glacial trough	
Truncated spur	
Hanging valley	
Ribbon lake	

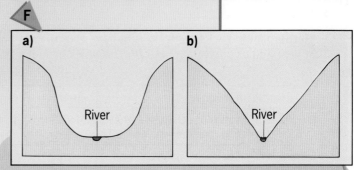

a) River

b) River

Summary Deposition by ice also creates landforms although these are far less spectacular than those formed by glacial erosion.

What are the main types of rock?

The Earth's crust consists of many different types of rock. It
is usual to group these rocks into three main categories
based upon how each type was formed.

1 **Igneous rocks** result from volcanic activity. They consist of tiny
crystals which formed as the volcanic rock cooled down (photo **A**).
Two examples are granite (page 57) and basalt.

A

	Formation	*Uses*
Granite (page 57)	Magma rises from the Earth's mantle and cools within the crust. The slow rate of cooling produces large crystals.	Building materials (houses and roads), sites for reservoirs, and can attract tourists.
Basalt	Magma reaches the Earth's surface as lava. As it cools rapidly, it creates small crystals.	Foundations for roads, weathers into a fertile soil, can attract tourists.

2 **Sedimentary rocks** are those that have been laid down in layers.
They usually consist either of small particles of other rocks that
have been eroded and transported (e.g. sandstone, photo **B**) or of
the remains of plants and animals (e.g. coal, chalk and limestone,
photo **C**).

B

	Formation	*Uses*
Limestone (pages 52–55)	Remains of shells and skeletons of small marine organisms (e.g. coral) that lived in warm, clear seas.	Quarried for lime and cement, used for stone walls, attracts tourists, and provides pastureland for sheep.
Chalk (page 56)	A form of limestone.	Thin soils used for cereals, spring-line provides water.
Coal	Fossilised remains of trees and plants that grew under hot, wet conditions.	Thermal energy – power stations, industry and domestic uses.
Sandstone	Grains of sand being compressed and cemented together.	Building materials, and can produce a fertile soil.

C

3 **Metamorphic rocks** are rocks that have been altered by extremes
of heat and/or pressure (photo **D**). Examples include marble and
slate.

D

	Formation	*Uses*
Marble	Limestone changed by heat and pressure.	Monuments.
Slate	Shales and clays changed by pressure.	Roofing materials.

What is weathering?

Rocks that are exposed to air, water, changes in temperature and vegetation, become vulnerable to **weathering**. Weathering includes the breaking up (disintegration) and decay (decomposition) of rocks in places where they formed. Unlike erosion, weathering need not involve the movement of material.

There are three main types of weathering: physical, chemical and biological.

1 **Physical weathering** is when rock is broken into smaller pieces by physical processes. It is most likely to occur in areas of bare rock where there is no vegetation to protect the rock from extremes of weather.

   • **Freeze–thaw** or **frost shattering** is common when the temperature is around freezing point and where exposed rocks contain many cracks. Water, entering cracks during the day, freezes during colder nights. As the water turns to ice it expands and, due to the increase in pressure, causes the cracks to widen (diagram **B** page 42). When the temperature rises, the ice melts and pressure is released. This repeated process weakens the rock until pieces break off. Where the broken-off rock collects at the foot of a cliff it is called **scree** (photo **E**).

   • **Exfoliation** or **onion weathering** occurs in very warm climates where exposed rock is repeatedly heated and cooled. During the day, the surface layers of rock are heated and expand. At night, they cool and contract. In time this causes the outer layers to peel off, like those of an onion, to leave steep-sided, rounded hills (photo **F**) and boulders.

2 **Chemical weathering** is when water and air activate chemical changes that cause rock to rot and decay. Chemical reactions are greatest where the climate is very warm and wet. **Limestone solution** (page 52) is an example of chemical weathering. It occurs when carbonic acid, which occurs naturally as a weak solution in rainwater, reacts with rocks such as limestone that contain calcium carbonate. As the limestone slowly dissolves, it is removed by running water to create distinctive landforms (pages 52–55).

3 **Biological weathering** occurs when either tree roots penetrate and widen cracks in a rock (physical) or acids, released by decaying vegetation, attack the rock (chemical).

E Lilloet, British Columbia, Canada

F Uluru (Ayers Rock), Australia

Activities

1 a) Give two examples for each of igneous, sedimentary and metamorphic rock.
   b) Describe the formation of each of your named examples.
   c) Give one use of each of the examples.

2 a) What is meant by the term *weathering*?
   b) Describe two types of physical weathering.
   c) Describe one type of chemical weathering.

Summary   Rocks can be grouped together, according to their formation, as being igneous, sedimentary or metamorphic, and can be broken down by physical or chemical weathering.

How do differences in rock type affect landforms?

The structure of a rock can affect its **resistance** to erosion and its **permeability** to water.

Resistance

Rocks have different strengths which enable them to produce different landforms. For example:

- The harder the rock, the more resistant it is likely to be to erosion. Harder rock is less likely to be broken up or worn away than is softer rock. This means that hills and mountains tend to form in areas of harder rock while valleys are found on softer rock. Valley sides are steeper where the rock is harder.
- On coasts, resistant rock forms steep cliffs and protrudes as headlands while softer rocks are eroded to create bays (diagram **A** page 26).

Permeability

An **impermeable** rock is one that does not allow water to pass through it. In contrast, a **permeable** rock does let water through it. Impermeable rock has numerous surface rivers and may be badly drained, in contrast to permeable rock which has few surface rivers. Permeable rocks may:

- consist of numerous tiny pores through which water can pass – such rocks, which include chalk (page 56), are said to be **porous**, or

- contain areas of weakness, such as bedding planes, along which water can flow – bedding planes, which separate different layers of rock occur in Carboniferous limestone (diagrams **D** and **E** on pages 52 and 53).

The **Isle of Purbeck** is not an island but a small peninsula on the south coast of England. It consists of four main types of rock – limestone, clays, chalk and sands (diagram **A**). All four are more or less parallel to one another and all reach the coast of the peninsula at right-angles. Compared with the clays and sands, the limestone and chalk are:

- much harder and more resistant to erosion
- permeable rocks.

Since the formation of the rocks, the area has been affected by different types of erosion. The sea and the rain have mainly been responsible for erosion on the coast, and rivers for the erosion of places inland. The result has been the formation of contrasting types of landform and scenery (table **B**).

Rock structure therefore affects the landforms of an area and plays a major role in producing distinctive types of scenery, e.g. Carboniferous limestone (pages 52–55), chalk page 56) and granite (page 57).

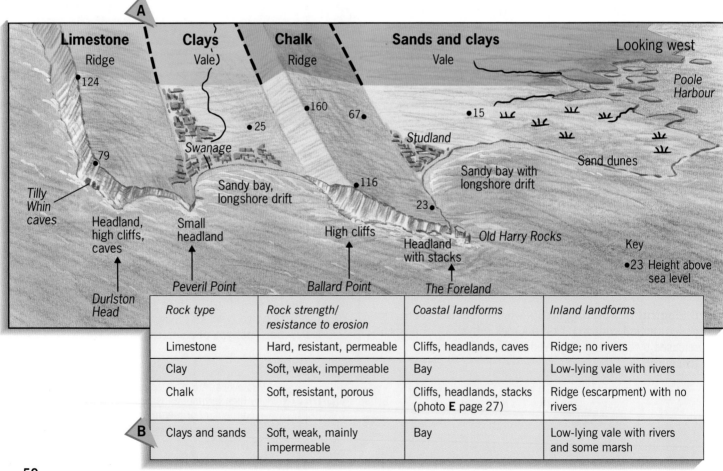

Rock type	Rock strength/ resistance to erosion	Coastal landforms	Inland landforms
Limestone	Hard, resistant, permeable	Cliffs, headlands, caves	Ridge; no rivers
Clay	Soft, weak, impermeable	Bay	Low-lying vale with rivers
Chalk	Soft, resistant, porous	Cliffs, headlands, stacks (photo **E** page 27)	Ridge (escarpment) with no rivers
Clays and sands	Soft, weak, mainly impermeable	Bay	Low-lying vale with rivers and some marsh

Quarrying in National Parks

People have quarried since the earliest of times – initially for flint for axes, later for stone for buildings (see page 150). Quarries are still worked in several National Parks, including slate in Snowdonia and the Lake District, and Carboniferous limestone in the Peak District and Yorkshire Dales (photo **C**). However, quarrying has become a major conflict in the National Parks, because while it can bring economic benefits to some local communities, it can also cause considerable environmental damage (diagram **D**).

National Park Authorities try to ensure that working quarries are landscaped and screened and that disused quarries are restored, whenever possible, to their pre-quarry appearance.

C

D

Benefits
- Quarrying provides jobs and increases local income, often in areas of limited employment.
- Local roads are improved for the increase in traffic.
- Local councils get money in rates from quarry firms.
- Quarries provide raw materials that help the national economy.

Disadvantages
- Spoil heaps, scarred hillsides and quarry buildings cause visual pollution.
- Heavy lorries cause noise, raise dust and block narrow country lanes.
- Blasting operations cause noise, dust and ground vibration.
- Disused quarries may flood and become dangerous.
- Wildlife is frightened away and habitats are lost.

Activities

1 **a)** Explain how chalk and limestones:
   i) form cliffs where they reach the coast, and ridges and hills inland
   ii) have very little surface drainage.
   b) Explain how clays:
   i) form bays where they reach the coast and valleys inland
   ii) have much surface drainage.

2 Quarrying in National Parks is said to have 'economic advantages but causes environmental loss'.
   a) Describe how quarrying can benefit the local community.
   b) Describe how it can harm the environment.

3 Copy and complete diagram **E** by adding labels to show what can be done to reduce environmental damage caused by quarries:
   a) while they are still being worked
   b) when they are no longer being used.

Summary

Rock resistance and rock permeability can affect landforms. The quarrying of rock causes conflict in National Parks and other areas of attractive scenery.

E

| Whilst the quarry is in operation | After quarrying has finished |

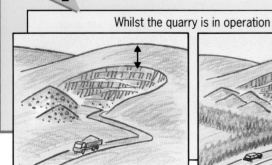

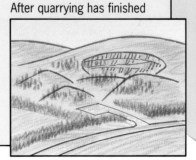

Why do Carboniferous limestone areas have a special type of scenery?

Limestone is a rock consisting mainly of calcium carbonate. Calcium carbonate comes from the remains of sea shells and coral. This means that limestone was formed on the sea-bed. There are several types of limestone (map **A**). Each type produces its own special type of scenery, with **karst** landforms developing in areas of Carboniferous limestone (photo **B**).

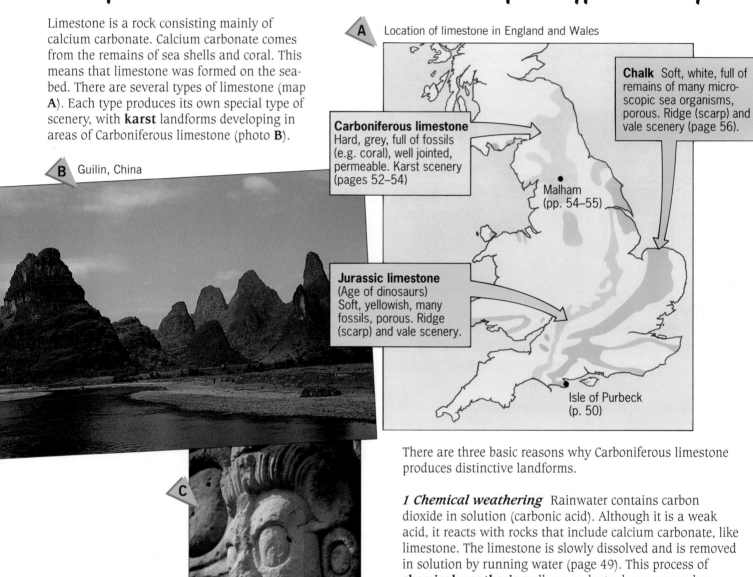

A Location of limestone in England and Wales

Carboniferous limestone Hard, grey, full of fossils (e.g. coral), well jointed, permeable. Karst scenery (pages 52–54)

Chalk Soft, white, full of remains of many microscopic sea organisms, porous. Ridge (scarp) and vale scenery (page 56).

Malham (pp. 54–55)

Jurassic limestone (Age of dinosaurs) Soft, yellowish, many fossils, porous. Ridge (scarp) and vale scenery.

Isle of Purbeck (p. 50)

B Guilin, China

C

D

Joints

Bedding planes

There are three basic reasons why Carboniferous limestone produces distinctive landforms.

1 Chemical weathering Rainwater contains carbon dioxide in solution (carbonic acid). Although it is a weak acid, it reacts with rocks that include calcium carbonate, like limestone. The limestone is slowly dissolved and is removed in solution by running water (page 49). This process of **chemical weathering** allows rocks to decompose where they are located. It is different to the actions of water, waves, ice and the sea as these include the transport of material. The effects of chemical weathering can be seen on tombstones, statues and buildings made from limestone (photo **C**).

2 Rock structure Limestone is a sedimentary rock, which means it was laid down in layers. Each layer is separated by a bedding plane (diagram **D**). At right-angles to bedding planes are joints. Bedding planes and joints are areas of weakness which are dissolved and widened by chemical weathering.

3 Permeability Carboniferous limestone is pervious, unlike chalk which is porous. A porous rock consists of many small pores which allow water to pass through it. A pervious rock restricts water to flowing along bedding planes and down joints.

Carboniferous limestone (karst) landforms

Rivers flow over the ground's surface across impermeable rocks until they reach an area of limestone. When a river reaches limestone, it begins to dissolve joints and bedding planes. In time it will 'disappear' down a **swallow hole**, or sink (diagram **E**, and photo **A** page 54). Sometimes solution is so active that **underground caves** form. Where a series of caves develop, they are linked by narrow passages – ideal for potholers. Water drips constantly from the roof of these caves. As the water drips, some of it will slowly evaporate and calcium carbonate is deposited. In time a **stalactite** will form. A stalactite is an icicle-shaped feature which hangs downwards from the roof. Stalactites, in Yorkshire caves (photo **F**), grow by only 7 mm a year. As the water drips to the floor, further deposition of calcium carbonate forms **stalagmites**, features that grow up from the cave floor.

Meanwhile, the underground river will be constantly looking for new routes through the limestone. It will seek lower levels until eventually it meets an underlying layer of impermeable rock. It will then flow over the impermeable rock until it reaches the surface. The place where it reappears is called a **resurgence** (photo **D** page 54).

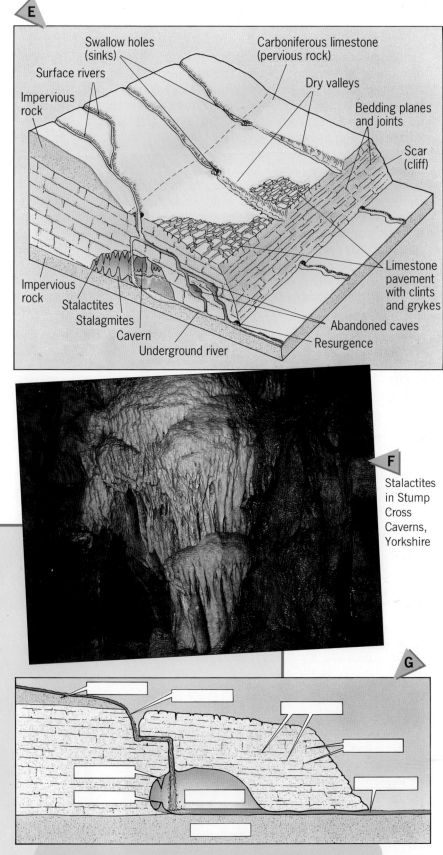

E

Swallow holes (sinks)
Surface rivers
Impervious rock
Carboniferous limestone (pervious rock)
Dry valleys
Bedding planes and joints
Scar (cliff)
Impervious rock
Stalactites
Stalagmites
Cavern
Underground river
Limestone pavement with clints and grykes
Abandoned caves
Resurgence

F Stalactites in Stump Cross Caverns, Yorkshire

G

Activities

1 **a)** What evidence is there that limestone was formed on the sea-bed?
 b) Name three different types of limestone found in Britain.
 c) What other name is given to landforms developed on Carboniferous limestone?
 d) How do the following affect the development of landforms on Carboniferous limestone?
  i) Chemical weathering
  ii) Rock structure
  iii) Pervious rock

2 **a)** Rivers in Carboniferous limestone areas usually flow underground. Copy sketch **G**, and complete it by adding appropriate labels to the empty boxes.
 b) What is the difference between:
  i) a bedding plane and a joint
  ii) a stalactite and a stalagmite
  iii) a swallow hole and a resurgence?

Summary

A very distinctive group of landforms develops on Carboniferous limestone. These landforms occur because limestone is vulnerable to chemical weathering, has a structure composed of bedding planes and joints, and is pervious.

What does a Carboniferous limestone area look like?

Landforms of Carboniferous limestone areas are often very scenic, and attract many visitors. The area around Malham Cove (map **A** page 52), in the Yorkshire Dales National Park, is a honeypot (page 234). A walk, or a transect across the OS map (map **E**), between Malham Tarn (8966) and Malham Village (9063) illustrates most surface limestone features.

Malham Tarn lies on impermeable rock. A small stream flows southwards from the tarn. After several hundred metres it comes to limestone and 'sinks' underground (photo **A**, 894657). Another stream, with its source at 871665, also sinks underground on reaching the limestone (882659). The confluence of the streams used to be at 893650. South of here is the steep-sided Watlowes **dry valley** (photo **B**). This valley indicates that a river once flowed on the surface, perhaps during the Ice Age when meltwater could not infiltrate into the frozen ground.

The Watlowes valley widens onto a **limestone pavement**, a flat area of exposed rock (photo **C**, 896642). A limestone pavement appears flat as it corresponds to the surface of a bedding plane. In reality it is very uneven. Rainwater has dissolved the joints in the rock to form **grykes**, leaving detached blocks of limestone called **clints**. The limestone pavement ends abruptly at the top of Malham Cove where, in the distant past, the surface river used to drop as a 90-metre waterfall. The foot of the cove marks the junction of limestone and impermeable rock, and the resurgence of the river (photo **D**, 897641) which disappeared at 882659.

The stream from Malham Tarn which sinks at 894657 resurfaces further south at Aire Head 909622. At nearby Gordale (9164) the roof of an underground cavern is believed to have collapsed leaving a steep-sided gorge.

A Water sinks

B Watlowes dry valley

C Limestone pavement with clints and grykes above Malham Cove

D Resurgence at the foot of Malham Cove

Land use

Limestone is quarried for use in blast furnaces in the steel industry and to produce lime either to spread on fields or for making cement. It can also be used as a building material, including for drystone walls. Due to a lack of surface drainage, limestone areas have limited settlement, apart from villages such as Malham that have grown up near to resurgences (page 53), or along river valleys. Farming is restricted mainly to sheep due to the heavy rainfall, thin soils and poor grass. Tourists are attracted by the scenery and the villages, for walking and to visit underground caves.

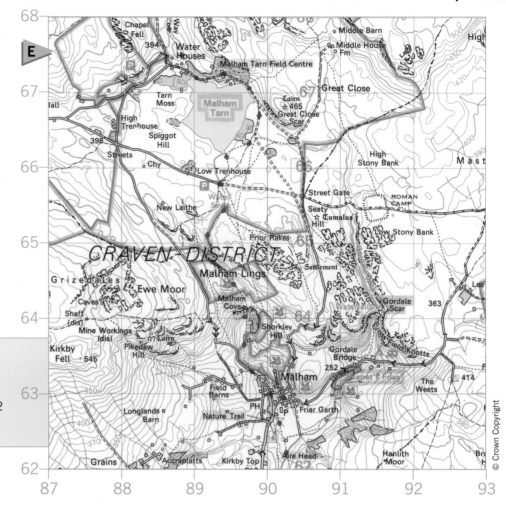

Part of OS map sheet 98

Scale 1: 50 000

0 1 2
km

© Crown Copyright

Activities

1 Make an enlarged copy of grid **F**, which covers the same area as the OS map (figure **E**).
   a) Mark on in blue and name:
      • *Malham Tarn* • *Gordale Beck*
      • *two streams before they disappear underground*
      • *the river which flows through Malham village.*
   b) Add the following labels in their correct places:
      • *source* • *two swallow holes* • *dry valley*
      • *limestone pavement with clints and grykes*
      • *Malham Cove* • *resurgence*
      • *collapsed underground cavern.*
   c) Colour, in yellow, the part of the map where there is Carboniferous limestone.

2 Refer to the OS map above.
   a) Give names and map references of other pieces of evidence (to those asked for in Activity **1**) to suggest that much of the area is limestone.
   b) List some limestone landforms which are not found on this particular OS map.

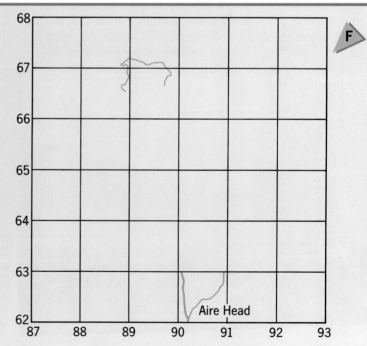

Summary Carboniferous limestone areas have distinctive landforms which make them attractive to visitors.

What do chalk and granite areas look like?

Chalk landscapes

Chalk, which is a soft limestone, occurs in south-eastern England where it forms its own characteristic landforms (diagram **A** page 52). As chalk is permeable and allows water to pass through it (page 52) it is relatively resistant to erosion and can form low-lying, gently rounded hills, e.g. near Stonehenge on the Salisbury Plain. Elsewhere, and especially where the rock has been tilted, the porous and more resistant chalk alternates with impermeable and less resistant clays. Here the chalk is likely to form **ridges** while the clays, which are more easily worn away by rivers, become **vales**, e.g. the so-called 'ridge and vale' scenery of the North and South Downs (diagram **A**).

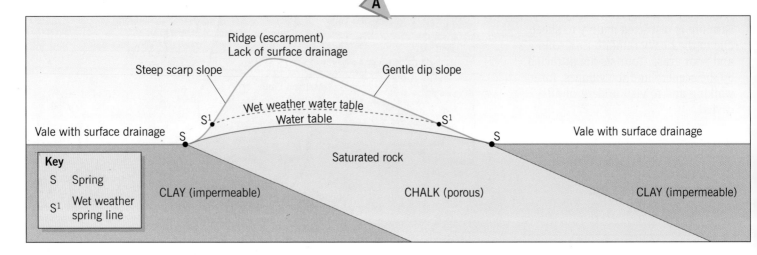

Ridge (escarpment)
Lack of surface drainage

Steep scarp slope   Gentle dip slope

Wet weather water table
Water table

S^1 S^1

Vale with surface drainage   Vale with surface drainage

S   S

Saturated rock

Key
S   Spring
S^1 Wet weather spring line

CLAY (impermeable)   CHALK (porous)   CLAY (impermeable)

Chalk ridges are also referred to as **escarpments**. Escarpments have a steep scarp slope and a more gentle dip slope (photo **B**). Rivers and streams are usually absent on the chalk itself. In contrast, surface drainage, with rivers emerging as springs at the junction of the permeable and impermeable rocks, is often abundant in the vales. Springs occur where the water table reaches the surface (diagram **C** page 13). The water table, below which the ground is saturated, is liable to rise after wet weather and to fall during times of drought.

Dry valleys (photo **C**) are another characteristic landform of chalk areas. These valleys are believed to have formed during the Ice Age when the ground was frozen and so acted as an impermeable rock. Although dry valleys in chalk have steep sides, they are not as steep as those found on Carboniferous limestone (photo **B** page 54).

Soils on chalk are thin. They used to provide grass for sheep but they are now also used for cereals. There is little settlement on the chalk itself but the springs, with their reliable water supply, provided an ideal site (page 150) for villages, i.e. **spring-line settlements**. Chalk is quarried for lime and cement, and flints within the chalk were used as early tools and, later, as a building material.

Granite landscapes

Granite forms when magma from inside the Earth rises towards the surface and slowly cools within the crust (diagram **D**). It is the slow rate of cooling that allows:

- large crystals to form (diagram **A** page 48) and
- large cracks, or joints, to develop (photo **E**).

At a later date the less resistant rocks that overlie the granite are removed by erosion. Where this removal is by running water, a moorland develops, e.g. Dartmoor. Where it includes ice, jagged mountains are formed, e.g. the islands of Skye and Arran (Scotland).

The most distinctive granite landform is the **tor** (photo **E**). Tors are thought to result from the joints within the granite being widened by chemical weathering (page 49) when the rock is near to, but still underneath, the surface. The weathered rock is only later exposed.

In south-west England, places like Dartmoor have acidic soils that are badly drained. Dartmoor is used as grazing land for sheep and ponies, by the army as a training ground, by tourists (walking and visiting tors), and is the location for several reservoirs. Granite is also quarried, either for roadstone or as china clay (photo **F**) which is used in the making of pottery. Granite areas are generally inhospitable for settlement.

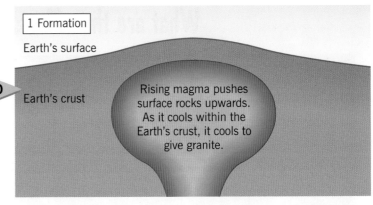

D

1 Formation

Earth's surface

Earth's crust

Rising magma pushes surface rocks upwards. As it cools within the Earth's crust, it cools to give granite.

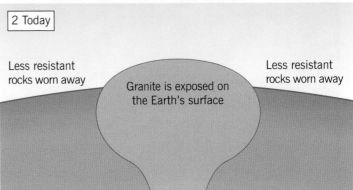

2 Today

Less resistant rocks worn away

Less resistant rocks worn away

Granite is exposed on the Earth's surface

E

① Joints formed as:
  i magma cooled near to surface
  ii pressure released when overlying rocks were later removed

② Joints widened by physical (freeze–thaw) and chemical weathering

③ Resultant blocks of granite

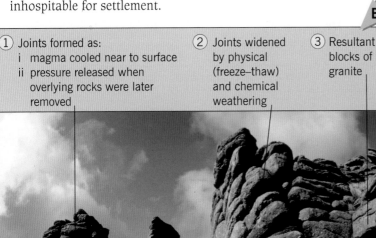

F

Activities

1 a) Draw a simple sketch of photo **B**. Add these labels:
      • *chalk ridge* • *scarp slope* • *dip slope* • *clay vale*.
   b) How did these landforms develop?
   c) What differences do photos **B** and **C** show in land use between the chalk ridge and the clay vale? Suggest reasons for these differences.

2 a) How did granite moorlands such as Dartmoor form?
   b) What are the major types of land use on Dartmoor? Give reasons for your answer.

Summary

The landforms of an area are often the result of the varying resistance and permeability of rocks together with processes that operate over long periods of time.

What are the effects of earth movements?

A

Gujarat in India, January 2001. Several towns were destroyed and over 30 000 people killed by an earthquake that lasted just 45 seconds

Earth movements usually result from a release of pressure. Over a period of time pressure can build up within the Earth. Often it is released so slowly that it is hardly noticed by people living on the surface. Sometimes the release of pressure can be very sudden. This can cause a violent movement on the Earth's surface, and can result in considerable damage, and even loss of life. Pressure can be released through **earthquakes** and **volcanoes**.

Earthquakes

It is estimated that there are over 150 000 earth movements every year. However, most of these are so gentle that they can only be detected by highly sensitive instruments called **seismometers**. The 10 000 or so earth movements per year which are strong enough to be felt are called tremors. (Of these 1000 occur in Japan – an average of three per day!) Earthquakes are violent earth movements where the ground actually shakes. Each year there are something like 20 to 50 earthquakes which are violent enough to cause serious damage and loss of life (photo **A**). The strength of an earthquake is measured on the **Richter scale** (diagram **B**). At first glance the scale is misleading as each point is actually ten times greater than the one below it. That means an earthquake which registers 6 on the scale is ten times greater than one measuring 5, and one hundred times greater than one measuring 4.

B

Richter scale	Effects
Down to 0	
3	Only detected by instruments (seismometers)
3.4	Noticed only by very sensitive people
3.5	
	Felt by most people sitting down, like the rumble of traffic
4	
4.2	Felt by walkers; windows and doors rattle
4.4	
	Suspended objects swing; sleeping people are wakened
4.8	
5	Furniture moves, objects fall, plaster begins to crack
5.4	
	Chimneys fall, walls crack, difficult to stand up
6 6.0	
	Severe structural damage to houses
6.7	
6.9	Some houses collapse; underground pipes crack
7	Many houses collapse; landslides; cracks in the ground
7.3	
	Most buildings collapse; bridges destroyed; large gaps appear in ground
8 8.1	
	Total destruction; ground actually rises and falls
9 8.9	Lisbon earthquake (strongest recorded)
Up to 12	

Volcanoes

Pressure under the Earth's crust keeps the rock there in a semi-solid state. If pressure is released, the rock turns to liquid and rises. Volcanoes form where the liquid rock, or lava, escapes onto the Earth's surface. Lava can escape by either a gentle or a violent movement. If the lava is very 'runny' it will find its way through cracks in the crust and spread out slowly across the surrounding countryside (photo **C**). The result is either a **lava flow** or a low, gentle-sided volcano (diagram **D**). Where there are no natural cracks, lava can only escape if there is a violent explosion, or eruption (photo **E**). Materials such as lava, ash and rock shoot out into the air from a single opening called a **crater**. Most of the material falls back around the crater, building up into a high, steep-sided volcanic **cone** (diagram **F**).

C Volcanic lava flowing from Kilauea in Hawaii

Mount St Helens, USA **E**

Lava flow

Crack

D

Gentle-sided

Ash and rock

Steep-sided cone shape

Layers of lava and ash from previous eruptions

Lava flow

F

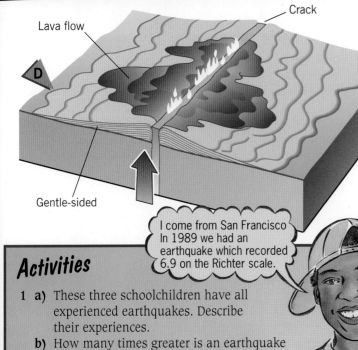

I come from San Francisco In 1989 we had an earthquake which recorded 6.9 on the Richter scale.

In Tokyo we get many earthquakes. The one we had this morning was 4.9.

I live in Mexico City. We had an earthquake in 1985 which measured 7.8.

G

Activities

1 a) These three schoolchildren have all experienced earthquakes. Describe their experiences.
   b) How many times greater is an earthquake registering 6.9 than one measuring 4.9?
   c) What is the highest ever recorded earthquake reading?

2 a) Why do gentle eruptions of lava produce gentle-sided volcanoes?
   b) Why do violent eruptions form steep-sided volcanoes?

Summary

Earth movements occur frequently. The most violent movements involve earthquakes and volcanic eruptions, both of which can cause severe damage and loss of life.

Where do earthquakes and volcanoes occur?

One task undertaken by geographers is to plot distributions on maps, and then to see if the maps show any recognisable patterns. Map **A** shows the distribution of world earthquakes and map **B** the distribution of volcanoes. In both cases there are some very obvious patterns.

Earthquakes around the world

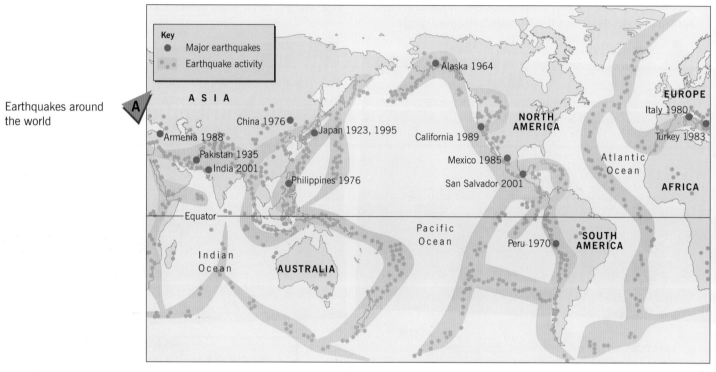

Earthquakes

Earthquakes occur in long narrow belts. The largest belt is the one that goes around the entire Pacific Ocean. The second most obvious is the one that runs through the middle of the Atlantic Ocean for its entire length. A third belt stretches across the continents of Europe and Asia from the Atlantic to the Pacific. There are several other shorter belts, including one going westwards from the west coast of South America.

Volcanoes

Volcanoes also appear in long narrow belts. The largest belt is the one that goes around the entire Pacific Ocean, the so-called 'Pacific Ring of Fire'. The second most obvious is the one that runs through the middle of the Atlantic Ocean for its entire length. Three other notable locations are in southern Europe, the centre of the Pacific Ocean, and eastern Africa.

Volcanoes around the world

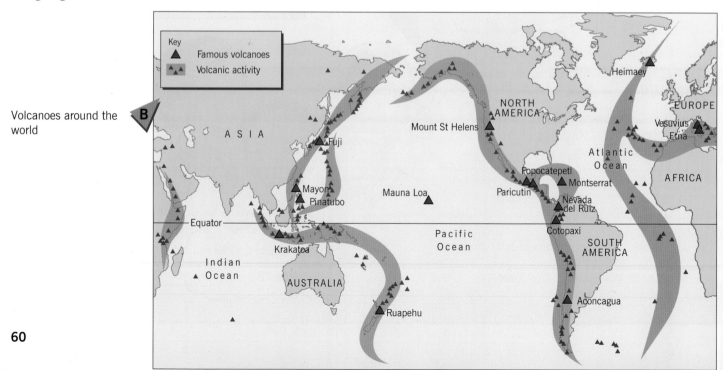

Geographers noticed that when the two maps were laid side by side or, better still, when one was laid on top of the other, earthquakes and volcanoes both seemed to occur in narrow belts and in the same places. These narrow belts are referred to as zones of activity. Having discovered both a pattern and a relationship between the two maps, the geographer then has to ask the question, 'Why?' In this case, why are earthquakes and volcanoes found within the same narrow belts?

Diagram **C** is a cross-section through the Earth. It shows the crust to be the very thin surface layer of cooled rock. It also shows that the crust is not one single piece but is broken into several slabs of varying sizes, called **plates**. Plates float, like rafts, on the molten (semi-solid) mantle. There are two types of crust, and it is important to accept three main differences between them:

- Continental crust is lighter, it cannot sink, and it is permanent (i.e. it is neither renewed nor destroyed).
- Oceanic crust is heavier (denser), it can sink, and it is continually being renewed and destroyed.

Heat from the centre of the Earth sets up convection currents in the mantle (diagram **D**). Where these currents reach the surface they cause the plates above them to move. Most plates only move a few millimetres a year. In some places two plates move towards each other. In others places they may either move apart or pass sideways to each other. A **plate boundary** is where two plates meet. It is at plate boundaries that most of the world's

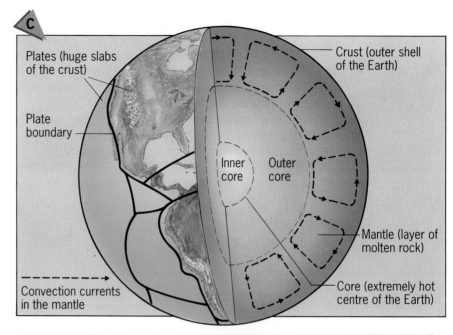

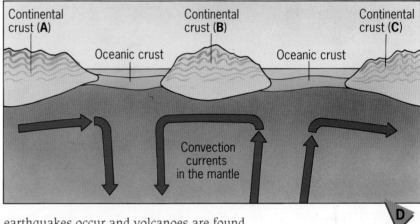

earthquakes occur and volcanoes are found (maps **A** and **B**). Pages 62 and 63 explain why there are so many earthquakes and volcanic eruptions, and why some are gentle and others violent, at plate boundaries.

Activities

1 Earthquakes and volcanoes seem to occur in long narrow belts.
   a) Name two belts (areas) where both earthquakes and volcanic eruptions occur.
   b) Name an area in Europe where both earthquakes and volcanic eruptions occur.

2 **a)** Look on the Internet for up-to-date information about earthquakes and volcanic eruptions. Try Volcano World at **http://volcano.und.nodak.edu/** and the National Earthquake Information Center (USA) at **http://earthquake.usgs.gov/**
   b) List the ten most recent earthquakes and the ten most recent eruptions.

   c) Find out where they happened and check on maps **A** and **B** to see if they are close to other earthquakes and volcanoes.

3 **a)** Give three differences between continental crust and oceanic crust.
   b) i)   What are plates?
          ii) Why do they move?
          iii) What happens at plate boundaries?
   c) On diagram **D**, what do you think will happen between:
          i)   continents **A** and **B**
          ii) continents **B** and **C**?

Summary

The Earth's crust is broken into several plates. Convection currents cause these plates to move about slowly. Earthquakes and volcanic eruptions occur at plate boundaries.

What happens at plate boundaries?

Map **A** shows the major plates and their boundaries (margins). The map key indicates that there are four types of plate boundary – destructive, collision, constructive and conservative. Earthquakes occur at all four types of boundary, but are more violent at some than others. Volcanic eruptions tend to occur at only two types of plate boundary, being violent at one and more gentle at the other.

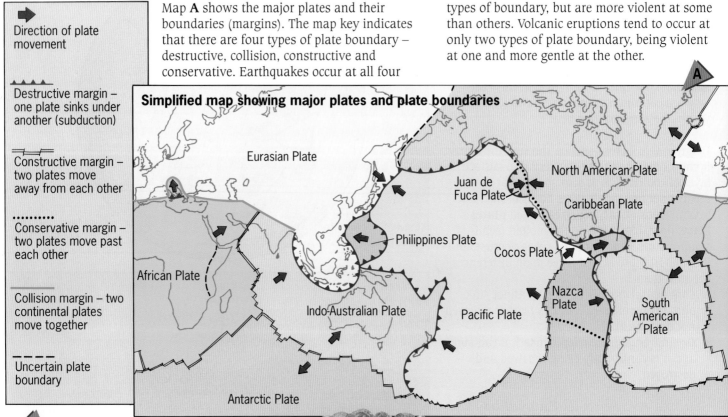

Key

➡️ Direction of plate movement

〰️ Destructive margin – one plate sinks under another (subduction)

Constructive margin – two plates move away from each other

⋯⋯ Conservative margin – two plates move past each other

Collision margin – two continental plates move together

– – – Uncertain plate boundary

A

Simplified map showing major plates and plate boundaries

Eurasian Plate

North American Plate

Juan de Fuca Plate

Caribbean Plate

Philippines Plate

Cocos Plate

African Plate

Nazca Plate

South American Plate

Indo-Australian Plate

Pacific Plate

Antarctic Plate

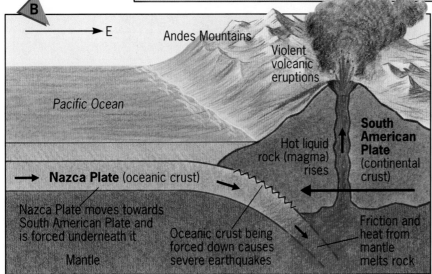

B

E →

Andes Mountains

Violent volcanic eruptions

Pacific Ocean

South American Plate (continental crust)

Hot liquid rock (magma) rises

Nazca Plate (oceanic crust)

Nazca Plate moves towards South American Plate and is forced underneath it

Oceanic crust being forced down causes severe earthquakes

Friction and heat from mantle melts rock

Mantle

Destructive margins

A destructive margin is when oceanic crust moves towards continental crust, for example the Nazca Plate moving towards the South American Plate (diagram **B**). As the oceanic crust is heavier it is forced downwards. As it is forced downwards pressure increases which can trigger extremely violent earthquakes. At the same time the heat produced by friction turns the descending crust back into liquid rock called magma. The hot magma tries to rise to the surface. Where it succeeds there will be violent volcanic eruptions. Notice that most of the Pacific Ocean is bounded by destructive margins where oceanic crust is being destroyed.

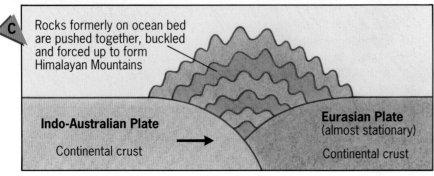

C

Rocks formerly on ocean bed are pushed together, buckled and forced up to form Himalayan Mountains

Indo-Australian Plate

Continental crust

Eurasian Plate (almost stationary)

Continental crust

Collision margins

Collision margins occur when the two plates moving together are both continental crust (diagram **C**). As continental crust cannot sink or be destroyed, then the land between them is buckled and pushed upwards to form high **fold mountains**, such as the Himalayas. Although pressure created by the plates moving together can cause severe earthquakes, there are no volcanic eruptions at collision margins.

Constructive margins

A constructive margin is where two plates move apart, for example the North American Plate moving away from the Eurasian Plate (diagram **D**). As a 'gap' appears between the two plates, then lava can easily escape either in the form of a relatively gentle eruption or as a lava flow. The lava creates new oceanic crust and forms a mid-ocean ridge. While not all earthquakes and volcanic eruptions on Iceland are 'gentle', they are gentle when compared with those at other plate margins.

Conservative margins

At conservative margins two plates try to slide slowly past each other, as in the case of the North American and Pacific Plates (diagram **E**). When the two plates stick, as they often do along the notorious San Andreas Fault in California, pressure builds up. When it is finally released, it creates a severe earthquake. As crust is neither created nor destroyed at conservative margins, there are no volcanic eruptions.

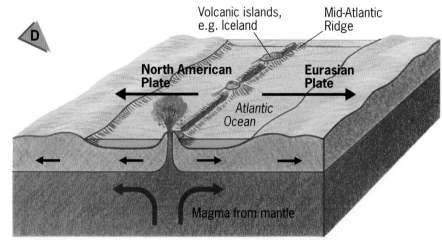

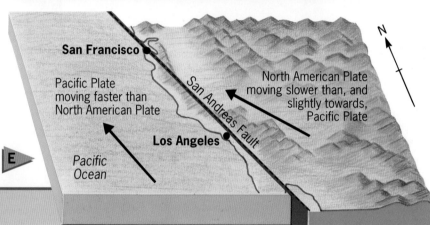

Activities

1 **a)** Describe the differences between *destructive*, *collision*, *constructive* and *conservative* plate margins.

   b) What type of plate margin can be found between the plates listed below?

   - Nazca and South American
   - North American and Eurasian
   - Nazca and Pacific
   - Indo-Australian and Eurasian
   - North American and Pacific
   - African and Eurasian
   - Pacific and Eurasian

   C – continental crust
   O – oceanic crust
   X – where new crust is being formed
   Y – where crust is being destroyed
   V – violent volcanic eruptions
   G – less violent volcanic eruptions
   E – severe earthquakes
   F – fold mountains
   M – a mid-ocean ridge

2 Make a copy of diagram **F**. Add the appropriate letters to the empty boxes.

3 Copy and complete table **G** by putting one tick in the earthquake column and one tick in the volcanic eruption column for each of the four types of plate boundary.

Plate margin	Earthquakes			Volcanic eruptions		
	Violent	Less violent	Rare	Violent	Fairly gentle	None
Destructive						
Collision						
Constructive						
Passive						

Summary
Plates can either move towards, away from or sideways past other plates. The resultant earth movements can cause earthquakes and volcanic eruptions of differing severity.

What are the effects of an earthquake?

Japan is located on a destructive plate margin. It has many active volcanoes and is constantly affected by earthquakes. At 5.46 a.m. on 17 January 1995, an earthquake measuring 7.2 on the Richter scale struck the city of Kobe on Japan's east coast (map **A** and photo **C**).

By Japanese standards, the earthquake was not large, yet the loss of life and damage to the area was enormous. Two factors help to explain the devastation the earthquake caused. First, the 'quake was shallow with its **focus** near to the Earth's surface and second, its **epicentre** was very close to Kobe (diagram **B**).

The earthquake itself was caused by the Philippines Plate being forced downwards and below the Eurasian Plate. Plates are not pushed down easily or smoothly. The pushing down often needs considerable force which comes from a build-up of pressure. If this pressure is released suddenly, then the plate may jerk forwards in a violent movement. This is what happened at Kobe.

C Earthquake damage: collapse of a motorway

A Location of Kobe

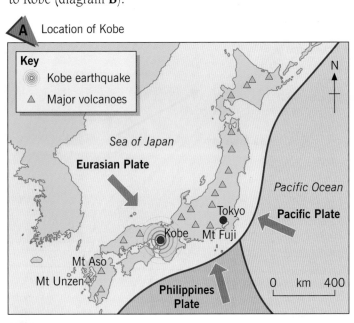

B Causes of the Kobe earthquake

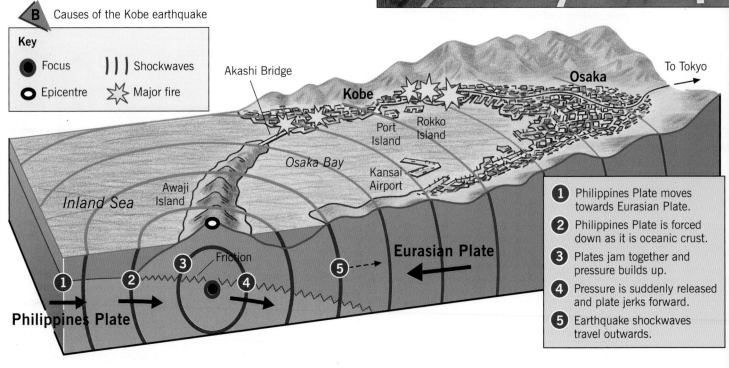

Key
- ● Focus
- ◎ Epicentre
-))) Shockwaves
- ✦ Major fire

1. Philippines Plate moves towards Eurasian Plate.
2. Philippines Plate is forced down as it is oceanic crust.
3. Plates jam together and pressure builds up.
4. Pressure is suddenly released and plate jerks forward.
5. Earthquake shockwaves travel outwards.

The damage caused by earthquakes is normally divided into two types (diagram **D**). **Primary effects** happen immediately and are due to the shaking of the ground. **Secondary effects** happen afterwards and are a result of the damage done by the initial tremors.

D

Primary effects

- Nearly 200 000 buildings were destroyed.
- A 1 km stretch of the elevated Hanshin Expressway collapsed (photo **C**).
- Numerous bridges along the bullet train route were severely damaged.
- Several trains on minor lines were derailed.
- 120 of the 150 quays in the port of Kobe were destroyed.

Secondary effects

- Electricity, gas and water supplies were disrupted.
- Fires caused by broken pipes and ruptured electricity lines, swept the city.
- Massive traffic jams blocked roads, delaying ambulances and fire engines.
- An estimated 230 000 people were made homeless.
- Industries including Mitsubishi and Panasonic were forced to close.
- The number of deaths was put officially at 5500.
- At least 40 000 people suffered serious injury.

In the months afterwards

- Water, electricity, gas and telephone services were fully operational by July.
- The fire-affected area had been cleared and commercial buildings repaired by August.
- All rail services were back to normal by August.
- 80% of port facilities were functional within a year.
- The Hanshin Expressway remained closed.
- The total cost of damage was estimated at £56 billion.
- New regulations were introduced to make buildings more resistant to earthquakes (page 67).

Activities

1 **a)** Make a larger copy of cross-section **E** and add the following labels:

> - *Philippines Plate* • *Eurasian Plate*
> - *Inland Sea* • *Osaka Bay* • *Awaji Island*
> - *Kobe* • *Osaka*.

b) Draw arrows to show the directions of plate movement and add symbols to show the epicentre and focus of the earthquake.

c) Give your diagram a title and explain how the earthquake happened.

E

2 Look at photo **C**. Imagine that you were standing nearby when this happened. Write a report to read out on your local radio station. Describe what you saw, felt and heard.

3 Describe the effects of the earthquake using the headings in table **F**.

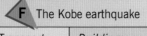 **F** The Kobe earthquake

Transport	Buildings	Industry	People

4 What is the difference between:
   a) the *focus* and *epicentre* of an earthquake
   b) *primary* and *secondary* effects?

5 In what ways do you think the Kobe earthquake would have been more or less damaging if it had occurred three hours later?

Summary

Earthquakes are a major natural hazard as they are difficult to predict and usually occur without any advance warning. This can cause considerable damage and loss of life.

How can the effects of an earthquake be reduced?

It is impossible to prevent earthquakes but there is much that can be done to reduce the damage caused by them. In Kobe, for example, whilst the earthquake caused considerable damage and loss of life, its effects were limited by measures taken by the government and local authorities well beforehand.

Measures to reduce the effects of earthquakes are usually in two parts. The first is to **predict** where and when the event will happen and the second is to **prepare** local people and emergency services for the disaster.

1 Predicting future earthquakes

Scientists still cannot say exactly when or where an earthquake will strike. As we have seen on pages 60–63, most earthquakes occur close to plate boundaries so it is in these areas that most research and monitoring is centred. These include:

- Setting up sensitive instruments to measure known signs of an on-coming earthquake. Just before a 'quake, small cracks develop in rocks and radon gas seeps out. The ground begins to bulge, and local water levels change.

- Mapping centres of previous earthquakes and identifying 'gaps' where the next earthquakes are most likely to happen. Map **A** shows the Tokai Gap. An earthquake here would probably destroy Tokyo and kill several million people.

- Plotting earthquake regularity. Diagram **B** shows that earthquakes in the Tokai area have occurred on average every 72.5 years. This would suggest that there should have been one in 1995/1996.

- Observing unusual animal and fish behaviour. This may include mice fleeing houses, dogs howling and fish jumping just before an earthquake strikes.

2 Preparing for earthquakes

Good preparation and planning should involve local authorities and emergency services as well as people living in the area. These include:

- Preparing disaster plans and carrying out regular practices. In Japan, people in schools, industry and public buildings have to practise earthquake drills at least once a month.

- Training emergency services such as police, fire and ambulance crews.

- Organising emergency supplies of water, food and power in advance.

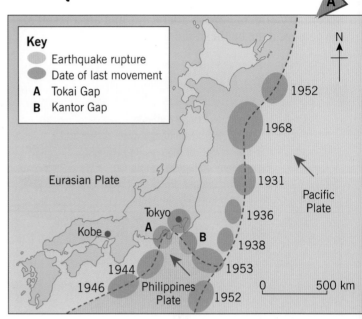

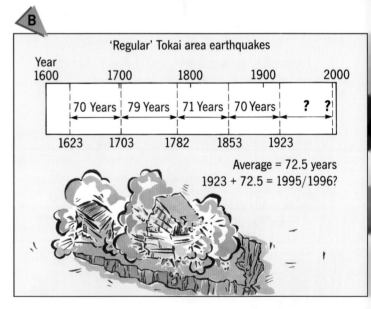

- Setting up an efficient earthquake warning and information system using television and radio. Earthquake preparation is taught in Japanese schools as part of the curriculum.

- Drawing-up and enforcing strict building regulations. These should ensure that buildings are earthquake-resistant and provide protection rather than cause danger in an earthquake. They should also restrict building on unstable surfaces like clay and reclaimed land, where earth movement is greatest and building collapse is most likely.

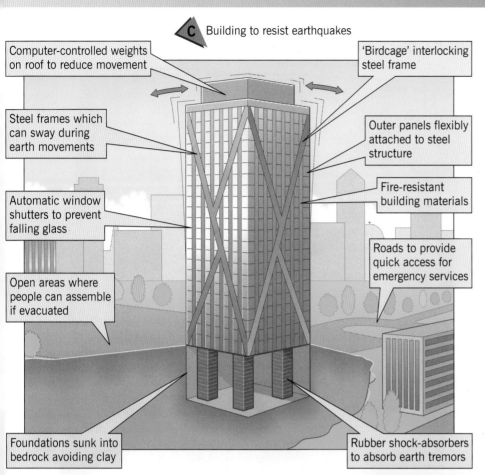

C Building to resist earthquakes

Computer-controlled weights on roof to reduce movement

'Birdcage' interlocking steel frame

Steel frames which can sway during earth movements

Outer panels flexibly attached to steel structure

Automatic window shutters to prevent falling glass

Fire-resistant building materials

Roads to provide quick access for emergency services

Open areas where people can assemble if evacuated

Foundations sunk into bedrock avoiding clay

Rubber shock-absorbers to absorb earth tremors

Why do people still get killed?

Since 1900 nearly 3 million people have lost their lives in earthquakes and the death toll continues to grow. This is mainly because it is impossible to predict exactly where or when an earthquake will strike, and how powerful it will be.

Places like Japan and California have, to some extent, reduced the earthquake danger but other places in the world have not been so fortunate. Many countries are simply too poor to protect themselves against natural disasters. Some have no disaster plans at all, whilst others are unable to enforce building safety codes or provide adequate resources and training for emergency services.

It is in these less developed countries where the damage to property and loss of life caused by earthquakes is greatest.

Activities

1 It is not easy to predict an earthquake.
   a) What does *predict* mean?
   b) Describe two ways that help scientists predict where and when an earthquake might happen.
   c) Draw a star diagram to show five signs that an earthquake may be about to happen.

2 Look at photo **D** showing schoolchildren practising an earthquake drill. Describe what they are doing and suggest reasons for their actions.

3 Imagine that your school is in an earthquake zone. Prepare an Earthquake Disaster Plan that will help reduce loss of life, injury and damage to the school. Divide your plan into **Before**, **During** and **After**. Include advice on the following.
   • building design and improvements
   • information and warning systems
   • classroom procedures
   • training and emergency needs.
   Use information from this page and try the Internet at **http://earthquake.usgs.gov/** for further advice.

D

4 Developed countries like Japan are better able to cope with earthquakes than developing countries like India. Suggest reasons for this.

Summary

Good preparation and planning can help reduce the worst effects of earthquakes. Richer countries can afford more effective disaster plans and are better able to cope with earthquakes than poorer countries.

What are the effects of a volcanic eruption?

In 1990 the US Geological Survey claimed that there were 540 active volcanoes in the world. Three-quarters of those were in the Pacific Ring of Fire. The list did not include a little-known volcano in the Philippines which had not erupted since 1380. On 9 June 1991, Mount Pinatubo hit the headlines. It became one of the largest eruptions in the world in the twentieth century (photo **A**).

The Philippines lie on a destructive plate margin. The Philippines Plate, composed of oceanic crust, moves north-westwards towards the Eurasian Plate, which is continental crust. Where they meet, the Philippines Plate is forced to dip steeply down under the Eurasian Plate (map **B**). The oceanic crust is turned into magma, rises, and erupts on the surface. The Philippines owe their existence to the almost constant ejection of lava over a period of several million years. Even before Pinatubo erupted, there were over 30 active volcanoes spread across the country's many islands.

A

Eruption of Mount Pinatubo, 1991

B

0 ___ 2000 km

N

China

Japan

Eurasian Plate
(continental crust)

Pacific Plate
(oceanic crust)

Philippines Plate
(oceanic crust)

Philippines

Indonesia

Indo-Australian Plate
(continental crust)

Australia

0 100
km

Pacific
Ocean

LUZON
ISLAND

Mt
Pinatubo▲ ● Angeles City

● Manila

South
China
Sea

What were the immediate effects?

Fortunately there were several advance warnings of a possible eruption. On 7 June the Americans evacuated all 15 000 personnel from their nearby airbase. From 9 June there were many eruptions, but none matched that of 12 June. An explosion sent a cloud of steam and ash 30 km into the sky. As the ash fell back to Earth, it turned day into night. Up to 50 cm of ash fell on nearby farmland, villages and towns. Over 10 cm fell within a 600 km radius, and some even reached as far away as Australia. The eruptions continued for several days. They were accompanied by earthquakes and torrential rain – except that the rain fell as thick mud. The weight of the ash caused buildings to collapse, including 200 000 homes, a local hospital, most of the schools and many factories. Power supplies were cut off for three weeks and water supplies were contaminated. Roads became unusable and bridges were destroyed, making relief operations even more difficult.

What were the longer-term effects?

The area surrounding Mount Pinatubo was excellent for rice growing. However, the thick fall of ash ruined the harvest in 1991, and made planting for 1992 impossible. Over one million farm animals died, either through starvation (no grass to eat) or from drinking contaminated water. Hundreds of farmers and their families were forced to move to cities to seek shelter and food. Huge shanty-type refugee camps were set up. Disease spread rapidly, especially malaria, chicken-pox and diarrhoea. Within a few days the monsoon rains started. Normally these rains are welcomed as they bring water for the rice crop. In 1991 they were so heavy that they caused flooding and lahars (mud flows). Lahars form when heavy rain flows over, and picks up, large amounts of volcanic ash. Lahars and landslides covered many low-lying areas in thick mud (photo **C**). Finally, ash ejected into the atmosphere encircled the Earth within a few days. It blocked out some of the sun's heat for several months, and lowered world temperatures. Scientists believe the eruption may have delayed global warming by several years.

The eruption and after-effects caused the deaths of about 700 people. Although only six died as a direct result of the initial eruption, over 600 died later through disease, and another 70 were suffocated by lahars.

C Angeles City covered in thick mud from the Mount Pinatubo eruption

D

Nearby populated areas		
Place	Distance	Population
Several small farming villages	5 km	30 000
Clark Air Base	20 km	15 000
Angeles City	25 km	250 000
Manila	100 km	10 million

Activities

1 Make a copy of diagram **E**. On it label:
   • *Philippines Plate* • *Eurasian Plate*
   • *Mount Pinatubo* • *continental crust*
   • *oceanic crust* • *magma* • *earthquakes*
   • *cloud of steam and ash.*

2 Write a front-page article for a newspaper about the Pinatubo eruption. Include the following and add an interesting headline.

• Cause of the eruption
• What happened
• Immediate effects
• Long-term effects

E

Summary

A volcanic eruption is a major natural hazard and can cause considerable damage and loss of life. The effects can be both immediate and long-lasting and are most serious in populated areas.

How can the effects of a volcanic eruption be reduced?

The eruption of Mount Pinatubo was the second biggest explosion ever recorded yet the damage and number of deaths that it caused were relatively small. The reason for this was mainly that scientists were able to forecast the eruption and put plans into operation which helped save lives and protect property. The two-part approach of **predict** and **prepare** used in helping to reduce the more damaging effects of earthquakes (page 66) is just as effective when used with volcanic eruptions.

A American scientists collecting gas samples

1 Predicting future volcanic eruptions

Although we cannot stop volcanoes from erupting, they are easier to predict than earthquakes. When a volcano is about to erupt it gives out several warning signs. These signs are monitored by scientists who can then alert emergency services to the danger.

- When magma is on the move it causes hundreds of small earthquakes which can be measured with **seismometers**. On Pinatubo, these tremors were recorded some two months before the volcano erupted.
- Days before an eruption, hot magma starts to move upwards causing ground temperatures to increase. This warming process can be detected by satellites using heat-seeking cameras.
- Rising magma causes the volcano to swell and bulge. The Americans at Clark Air Base used **tiltmeters** to measure slope changes. They also used a satellite **global positioning system (GPS)** to detect movement.
- Immediately before an eruption the moving magma gurgles and belches. The more gas and steam the volcano gives out, the nearer it is to erupting. It was at this stage that a 24-hour warning was given that Pinatubo was about to erupt.

2 Preparing for a volcanic eruption

The preparations needed for a volcanic eruption are very similar to those for an earthquake described on pages 66 and 67. With an eruption, however, there is almost always some warning of the approaching event. This gives time for emergency services to act and, if necessary, to **evacuate** the danger area.

- Set up a monitoring and warning system to forecast where and when an eruption will happen and how powerful it may be.
- Prepare an evacuation plan and organise transport, accommodation and food for those people who need to be moved.
- Train emergency services such as police, fire and ambulance crews.
- Organise emergency supplies of water, food and power in advance. This should be done both for the local area and in possible evacuation areas.

B Local people fleeing from Mount Pinatubo

- Provide an 'after-eruption plan' to return the area to normal and help people who are most badly affected by the eruption. This will include clearing up the mess, repairing damaged buildings and bridges and providing support for local people.

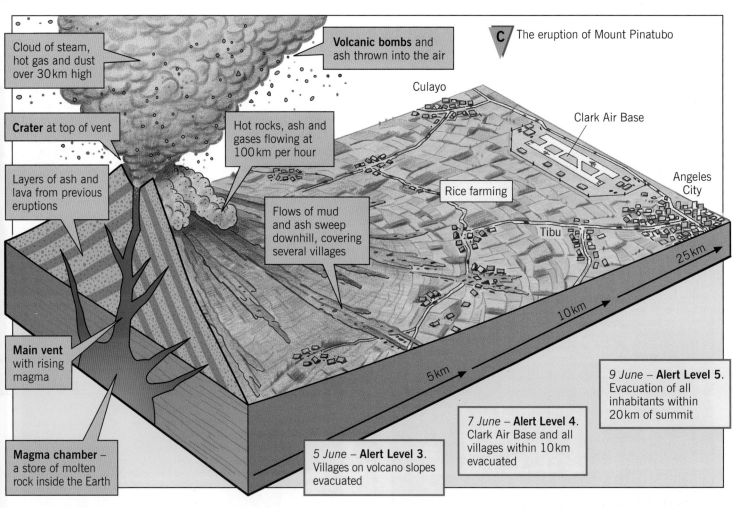

C The eruption of Mount Pinatubo

Cloud of steam, hot gas and dust over 30 km high

Volcanic bombs and ash thrown into the air

Crater at top of vent

Hot rocks, ash and gases flowing at 100 km per hour

Layers of ash and lava from previous eruptions

Flows of mud and ash sweep downhill, covering several villages

Culayo

Clark Air Base

Rice farming

Angeles City

Tibu

Main vent with rising magma

25 km

10 km

Magma chamber – a store of molten rock inside the Earth

5 km

9 *June* – **Alert Level 5**. Evacuation of all inhabitants within 20 km of summit

7 *June* – **Alert Level 4**. Clark Air Base and all villages within 10 km evacuated

5 *June* – **Alert Level 3**. Villages on volcano slopes evacuated

The monitoring of dangerous volcanoes requires complex technology and skilled scientists and is therefore expensive. It is for these reasons that only the richer, more economically developed countries can afford such monitoring. Japan has 22 fully manned volcanic observatories, whilst the volcanoes on Hawaii, USA have been continuously monitored for over 40 years.

The Philippines is one of the world's poorer countries but was fortunate to have American airforce and naval bases nearby. The Americans became heavily involved in monitoring the volcano and were able to provide accurate forecasts of when the eruption would happen and how powerful it would be. With this help, the worst effects of the eruption were reduced.

Other less developed countries are less fortunate and often have little warning of an eruption. They are rarely well prepared for such a disaster and have to rely on aid from the richer countries arriving some time after the disaster. For this reason, poorer countries are less able to cope with volcanic eruptions than richer countries.

Activities

1 a) Describe four warning signs that a volcano is about to erupt.
   b) In what ways can satellites be used to help predict eruptions?

2 Look at photo **B** showing people being evacuated from the Pinatubo area.
   a) What does *evacuate* mean?
   b) What problems does the photo show?
   c) Why are people often reluctant to be evacuated?

3 Why are poorer countries less good at predicting eruptions than richer countries?

4 'With good planning you can prevent loss of life and limit the damage caused by a volcanic eruption.' Do you agree with this? Explain your answer.

Summary

Predicting and preparing for volcanic eruptions can help reduce the damage that they cause. Poorer countries cannot afford to monitor volcanoes properly and may have less effective emergency plans.

Why live in a danger zone?

It has been estimated that 500 million people now live in areas that are likely to be affected by earthquakes and volcanic eruptions. In some places, like Mexico City, Tokyo and San Francisco, populations are actually increasing despite the knowledge that another earthquake will inevitably happen. The slopes of active volcanoes such as Mount Pinatubo and Mount Etna are well populated despite the obvious dangers of such locations.

So why do people live in such danger zones? The most obvious reason is that some of the areas, particularly those that are volcanic, can provide many benefits to people living there. Some of these are shown below. Another reason is simply that some people have always lived there and are happy to do so. Others may not be able to afford to move anywhere else. Drawing **B** shows the views of some people towards living in such places.

A

Good soil
Over a long period of time, volcanic rocks break down to form some of the most fertile soils on Earth. On Mount Etna, where grapes and other fruits are grown, yields are five times higher than the national average.

Tourism
Tourists are attracted to volcanic areas to watch eruptions, see geysers, relax in hot springs or have mud or sand baths (as in Japan, below). Tourism provides jobs and brings money into these areas.

Geothermal energy
Heat from the Earth can be used to generate electricity. This power station in Iceland also supplies hot spring water for the local spa resort. Geothermal heat warms more than 70 per cent of homes in Iceland.

Valuable raw materials
Gold, silver, copper, lead and many other useful minerals are found in the remains of extinct volcanoes. This town in Chile suffers earthquakes every day, yet a nearby mine is still one of the world's biggest producers of copper.

B

Why do we still live in the danger zones?

We are prepared to live with the danger because we love it here. It is our home and always has been. There's nowhere else we want to go.

We think that disasters only affect other people and it will never happen to us. We just ignore the dangers and get on with life.

People lived in earthquake zones long before they understood the risks. We did not know about plates and zones of activity until the 1960s.

We simply can't afford to move anywhere else. Even after a disaster we would rather return to the life we are used to than start again elsewhere.

Scientists now know everything about earthquakes and volcanoes and we are all well prepared for any disaster. We feel safe enough.

Some settlements in danger zones, like Mexico City, Tokyo and San Francisco have grown into enormous cities. Where could all the people move to?

For many people it would seem that the advantages of living in a danger zone far outweigh the risk of coping with the remote possibility of an earthquake or volcanic eruption. As scientists begin to develop more accurate methods of predicting these events, the risk of living in danger areas is further reduced.

Perhaps for this reason, the number of people populating the sides of active volcanoes and valley areas has increased in recent years. Even areas liable to earthquakes, a still far from predictable hazard, have seen a steady growth in population. Measuring the advantages against the disadvantages of living in a danger zone is called **risk assessment**.

Activities

1 a) Make a larger copy of drawing **C** and complete it using information from pages 65, 68, 69 and this double page.
   b) Are there more disadvantages or more advantages?
   c) Do you think the disadvantages are more important than the advantages? Give reasons for your answer.

C

Living in areas of earthquakes and volcanic eruptions

Disadvantages
  •
  •
  •

Advantages
  •
  •

2 Name at least six places that are well populated and lie in areas of tectonic activity.

3 Imagine that you and your family are living in Kobe, Japan, despite the recent earthquake. Make a list of reasons why you still live there and do not want to move away. Think about your way of life, jobs, schools, friends, and the points shown in drawing **B**.

Summary

Areas affected by tectonic activity present both advantages and disadvantages. People living in these danger zones have to balance the threat of earthquakes and eruptions against the benefits that the area may provide.

What factors affect temperature?

We all know what **weather** is. It is the day-to-day condition of the atmosphere in terms of temperature, rainfall, sunshine and wind. Climate is different. **Climate** is the average weather over a long period of time. There are several different climates across the world and each has its own distinctive pattern of temperature and rainfall. These two pages look at the factors that affect temperature, whilst pages 76 and 77 consider variations in rainfall.

Temperatures vary not only from place to place but also from time to time. The four main reasons for this are **latitude**, **distance from the sea**, **prevailing winds** and height above sea-level or **altitude**.

Latitude

Places that are near to the Equator are much warmer than places that are near to the poles. This is due to a combination of the curvature of the Earth, the angle of the sun in the sky, and the layer of atmosphere that surrounds the Earth.

At the Equator the sun is always at a high angle in the sky. When it is overhead it shines vertically downwards. Its heat is concentrated upon a small area which, as a result, warms up rapidly and becomes very hot. Going towards the poles, the sun's angle in the sky decreases. As the rays now have a greater area to heat, temperatures remain much lower than at the Equator.

A

North Pole

Britain

Greater curvature – greater area of land to heat up

Equal amounts of heat from the sun

Atmosphere

Equator

Directly above Earth. Small area to heat up

Less atmosphere to pass through

More atmosphere to pass through

Greatest curvature – very large area to heat up

South Pole

The atmosphere surrounding the Earth contains dust, smoke and other solid particles. These particles absorb heat. As the sun's rays pass through the atmosphere at a more direct angle (and therefore more quickly) at the Equator than nearer the poles, then less heat is lost.

Distance from the sea

Land and sea respond differently to temperature. The land heats up quickly, but also loses heat quickly. In contrast, the sea heats up and cools down much more slowly. This has two main effects on climate.

1 Places far from the sea have a **great range of temperature**. In summer they heat up quickly and become warmer than expected whilst in winter they rapidly lose their heat and become very cold.

2 Places near to the sea have a **small range of temperature**. This is because in summer the sea warms up very slowly and temperatures remain lower than might be expected. In winter the sea retains much of its heat and warms the coastal areas so that they enjoy relatively mild temperatures.

Diagram **B** shows temperatures across Europe at about the same latitude. Each place should have similar temperatures, but look at how different they are.

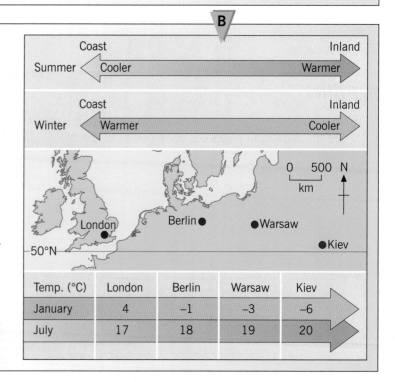

B

	Coast		Inland
Summer	Cooler		Warmer

	Coast		Inland
Winter	Warmer		Cooler

Temp. (°C)	London	Berlin	Warsaw	Kiev
January	4	–1	–3	–6
July	17	18	19	20

Prevailing winds

The prevailing wind is the direction from which the wind is most likely to come. In Britain the prevailing wind is from the west or south-west.

The temperature of the wind depends on where it comes from and the type of surface over which it passes. If the wind blows from the land, it will be warm in summer but cold in winter. If, as in Britain, the prevailing wind comes from the sea, it will lower temperatures in summer but raise them in winter. This also helps to explain why places in Britain have cooler summers and milder winters than places further east into Europe.

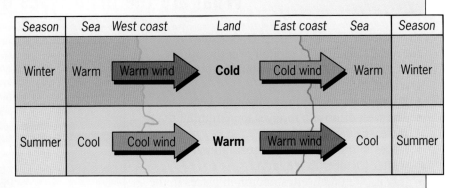

Season	Sea	West coast	Land	East coast	Sea	Season
Winter	Warm	Warm wind →	**Cold**	Cold wind →	Warm	Winter
Summer	Cool	Cool wind →	**Warm**	Warm wind →	Cool	Summer

Altitude (height of the land)

Mountains have much lower temperatures than the lowlands around them. Temperatures decrease, on average, by 10°C for every 1000 metres in height. This is because as height increases, the air becomes less dense and so is less able to retain the heat it receives from the ground.

This explains why snow lies for several months each winter in the Scottish Highlands, and throughout the year on several high mountains on or near the Equator. Mount Kenya and Kilimanjaro in Africa, and Chimborazo and Cotapaxi in South America, are some examples.

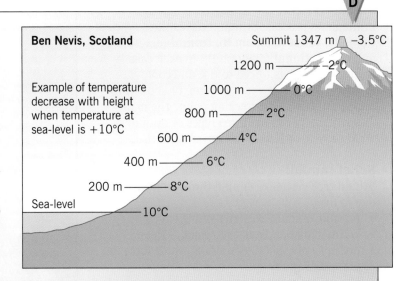

Ben Nevis, Scotland

Example of temperature decrease with height when temperature at sea-level is +10°C

Summit 1347 m −3.5°C
1200 m — −2°C
1000 m — 0°C
800 m — 2°C
600 m — 4°C
400 m — 6°C
200 m — 8°C
Sea-level — 10°C

Activities

1 a) What is the difference between *weather* and *climate*?
   b) What is meant by *range of temperature*?

2 a) What is meant by the term *prevailing wind*?
   b) Describe two ways that a prevailing wind can bring cold weather.

3 Look at diagram **E** and explain, with the help of diagrams, why:
   a) A is warmer than B in summer
   b) C is warmer than E in winter
   c) E has a greater annual range of temperature than C
   d) C is warmer than D in winter
   e) F is colder than D.

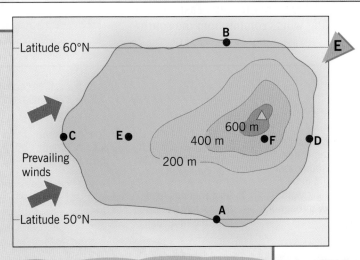

Latitude 60°N
Latitude 50°N
Prevailing winds
600 m
400 m
200 m
B E C E F D A

Summary

Temperatures vary from place to place and from time to time. The main factors that affect temperature are latitude, distance from the sea, direction of the prevailing wind and altitude.

What are the main types of rainfall?

The map and graphs on page 79 describe the uneven distribution of rainfall over the British Isles. The main features of Britain's rainfall can be described as follows:

- Rainfall may occur throughout the year.
- The west of Britain receives more rainfall than the east.
- Places in the west receive most rainfall during winter (October to January).
- July is often the wettest month in places in the east.

So how does it rain and why does Britain have this particular pattern of rainfall?

Diagram **A** shows the several stages in the rainmaking process. Britain receives three types of rainfall – **relief** (diagram **B**), **frontal** (diagram **C**) and **convectional** (diagram **D**). In each case warm, moist air is forced to rise and cool. Condensation causes rain to form, or snow if the temperature is below freezing point. The main difference between the three types of rainfall is what causes, or forces, the warm air to rise.

Relief rainfall

Britain receives relief rainfall throughout the year due to the:

- prevailing south-westerly winds which bring warm, moist air from the Atlantic Ocean
- presence of coastal mountains which force the air to rise and cool.

West coasts receive more rain than east coasts due to the prevailing winds coming from the south-west.

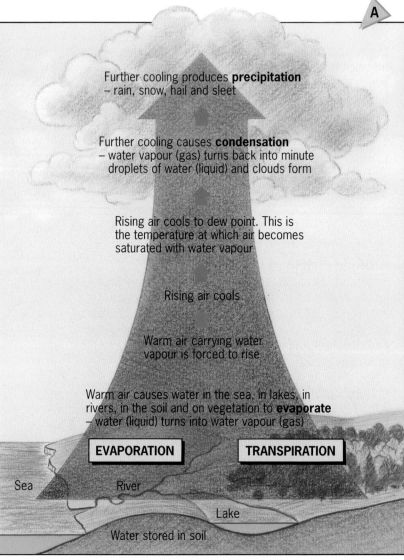

A

Further cooling produces **precipitation** – rain, snow, hail and sleet

Further cooling causes **condensation** – water vapour (gas) turns back into minute droplets of water (liquid) and clouds form

Rising air cools to dew point. This is the temperature at which air becomes saturated with water vapour

Rising air cools

Warm air carrying water vapour is forced to rise

Warm air causes water in the sea, in lakes, in rivers, in the soil and on vegetation to **evaporate** – water (liquid) turns into water vapour (gas)

EVAPORATION   **TRANSPIRATION**

Sea   River

Lake

Water stored in soil

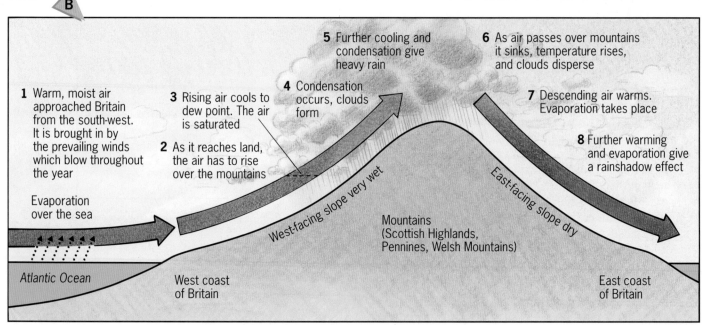

B

5 Further cooling and condensation give heavy rain

6 As air passes over mountains it sinks, temperature rises, and clouds disperse

4 Condensation occurs, clouds form

1 Warm, moist air approached Britain from the south-west. It is brought in by the prevailing winds which blow throughout the year

3 Rising air cools to dew point. The air is saturated

7 Descending air warms. Evaporation takes place

8 Further warming and evaporation give a rainshadow effect

2 As it reaches land, the air has to rise over the mountains

Evaporation over the sea

West-facing slope very wet

East-facing slope dry

Mountains (Scottish Highlands, Pennines, Welsh Mountains)

Atlantic Ocean

West coast of Britain

East coast of Britain

Frontal rainfall

Frontal rain results from the meeting of a warm mass of air and a cold mass of air. As the two air masses have different temperatures, they will have different densities and so do not mix easily. The boundary between warm and cold air is called a **front**. As warm air is lighter than cold air, it is forced to rise over the cold air. Frontal rain occurs in depressions which form to the west of the country, over the Atlantic Ocean (page 80). In a depression warm, moist air from the tropics is forced to rise over colder, drier air from polar regions. Depressions:

- usually approach the British Isles from the south-west and so give more rain to western parts of Britain
- are more frequent in winter which explains why western parts of Britain receive most of their rainfall during that season.

Convectional rainfall

Convectional rainfall is caused by the sun heating the ground. The heated ground will, in turn, warm the air which is in contact with it. As the air warms, it gets lighter and is forced to rise in strong upward convection currents. Water on the ground's surface will evaporate and will also rise. As the warm, moist air rises in convection currents, it cools and often gives thunderstorms. Convectional rainfall, which occurs most afternoons on the Equator (page 92), is most likely when the sun is at a high angle in the sky. Britain is usually too cool for this type of rainfall apart from places in the east and south-east which have the highest summer temperatures (map **A** page 78). This explains why July is often the wettest month in the east of Britain.

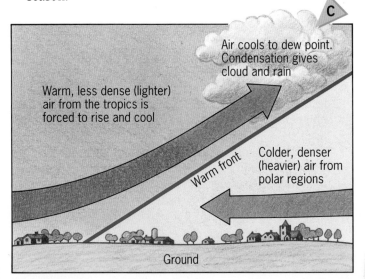

C

Air cools to dew point. Condensation gives cloud and rain

Warm, less dense (lighter) air from the tropics is forced to rise and cool

Warm front

Colder, denser (heavier) air from polar regions

Ground

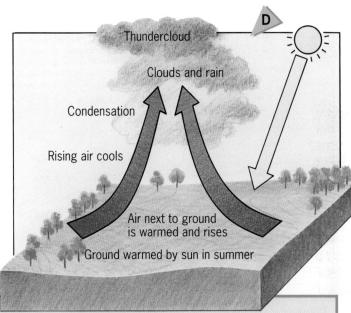

D

Thundercloud

Clouds and rain

Condensation

Rising air cools

Air next to ground is warmed and rises

Ground warmed by sun in summer

Activities

1 a) Put the following phrases in their correct order to answer the question, 'Why does it rain?'
- *rising air cools* • *further cooling gives condensation* • *warm air and water are forced to rise* • *warm air causes water in the sea and lakes to evaporate* • *further cooling causes precipitation* • *air is cooled to dew point.*

b) Give three ways by which warm, moist air is forced to rise.

2 Along the top of table **E** are several factors which affect the distribution of rainfall in the British Isles. Copy the table and tick the appropriate boxes to show which of these factors affect the three distributions listed in the left-hand column.

E

	Prevailing south-west wind	Mountains on western side of Britain	Depressions come from the Atlantic Ocean	Depressions are most frequent in winter	Convectional thunderstorms
West of Britain gets more rain than the east					
Places in the west get most of their rain in winter					
Places in the east often have July as their wettest month					

Summary

Rain is caused by warm, moist air being forced to rise and cool. The three main types of rainfall in the British Isles are relief, frontal and convectional.

What is Britain's climate?

Britain has a variable climate. This means that the weather changes from day to day, which makes it difficult to forecast accurately. Britain's average climate is cool summers, mild winters and rain spread evenly throughout the year. This type of climate can be described as **equable** or **temperate**, which means that it never has any extremes. In other words it is never too hot or too cold and rarely too wet or too dry. Not all of Britain has the same climate, however. There are variations between the north and south, and east and west. These are explained on page 90.

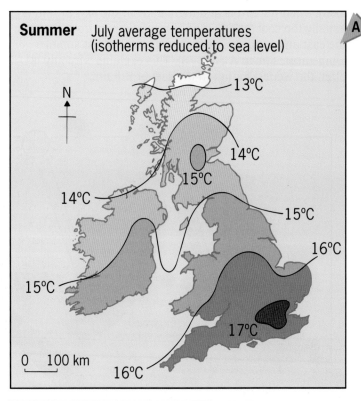

Summer July average temperatures (isotherms reduced to sea level) **A**

1 Temperature

Maps **A** and **B** show the average temperatures across Britain in January and July. The isotherms drawn on the maps are lines of equal temperature. On these maps, the isotherms ignore height and are reduced to sea-level. This makes it easier to both recognise patterns and make comparisons between places.

Summer The isotherms on map **A** show that:
- Temperatures in the summer are highest in the south and decrease northwards. This is mainly because the sun is at a higher angle in the sky in the south and provides more heat.
- Temperatures inland are higher than those near the coasts. The main reason for this is that the land warms up quickly in summer. The sea remains cool and keeps temperatures in coastal areas relatively low.

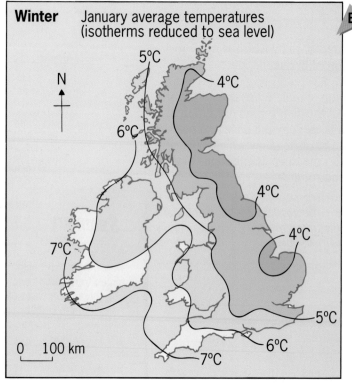

Winter January average temperatures (isotherms reduced to sea level) **B**

Winter The isotherms on map **B** show that:
- Temperatures in winter are highest in the west and decrease eastwards. The main reason for this is that the west coast is warmed by an ocean current called the North Atlantic Drift. Another reason is that the prevailing south-westerly winds blow across the relatively warm waters of the Atlantic Ocean and raise west coast temperatures.
- Temperatures are highest in south-west England. This is mainly because the area is almost surrounded by the sea which in winter is warmer than the land.

Some other reasons for the seasonal and regional differences in temperature in the British Isles are explained on pages 74 and 75.

2 Rainfall

Map **C** shows the expected amounts of rainfall over the British Isles in an average year. It is accompanied by six graphs. Three are for places in the west of Britain and three for places in the east.

The map and graphs show that annual rainfall totals:
- are highest in the north-west of Scotland
- decrease rapidly from the north-west of Scotland to the south-east of England.

This is mainly because the prevailing winds are from the west and are laden with moisture when they blow ashore from the Atlantic Ocean. Another reason is that most rain-bearing **depressions** (pages 80 and 81) approach from the west, so western areas receive more rain. North-west Scotland is particularly wet because much of the area is high land and so receives relief rain.

The graphs also show a contrast in the seasonal distribution of rainfall. Whereas places in the west receive most rain in winter (October to February), places in the east tend to get most rain in July (summer). The reasons for the annual and seasonal differences in rainfall are explained on pages 76 and 77.

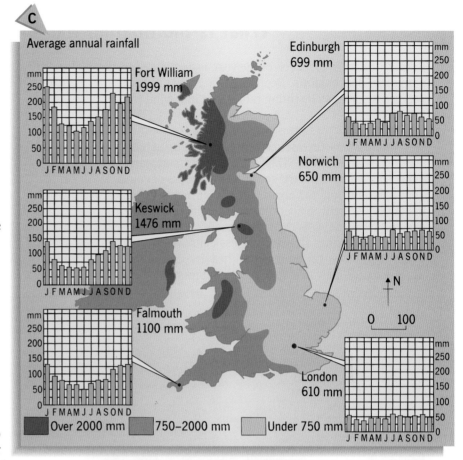

C Average annual rainfall

Fort William 1999 mm
Keswick 1476 mm
Falmouth 1100 mm
Edinburgh 699 mm
Norwich 650 mm
London 610 mm

N

0   100

Over 2000 mm   750–2000 mm   Under 750 mm

Activities

1 Complete the following sentences.
   - The climate of Britain is one of ...
   - It is described as temperate or equable because ...

2 Copy and complete table **D** using information from maps **A** and **B** and the graphs in **C**.

D

Place	Temperature (°C)		Rainfall (mm)		
	January	July	January	July	Total
Fort William					
Falmouth					
London					
Edinburgh					

3 Look at photos **D** and **E** and their locations on map **C**.
   a) Describe the weather that you can see in each photo.
   b) Is the weather typical for that area at the time of year?
   c) Suggest reasons for the weather in each photo.

E Winter in north-west Scotland

F Summer on a beach in south-west England

Summary Britain usually has cool summers and mild winters. Rain may fall throughout the year and winters are usually slightly wetter than summers.

What is the weather like in a depression?

Britain's weather, for much of the year, is dominated by the passing of **depressions**. Depressions are areas of low pressure which usually bring rain, cloud and wind to the British Isles.

Most depressions develop to the west of the British Isles over the Atlantic Ocean. This is where a mass of warm, moist tropical air from the south meets a mass of colder, drier polar air from the north. The two air masses, because they have different temperatures, have different densities (weight). This prevents them from easily mixing (page 77). Instead, the warmer air, which is less dense (lighter) is forced to rise over the more dense (heavier) colder air. As the warm air rises it creates an area of low pressure at ground level. The boundary between two air masses is called a **front**. There are two fronts in a typical depression:

1 The warm front, which passes first, is where the advancing warm air is forced to rise over the cold air.
2 The cold front, which follows, is where the advancing cold air undercuts the warm air in front of it.

In both cases the warm, moist air is forced to rise. If it cools to dew point it condenses to form cloud and to give frontal rain (page 77). Although each depression is unique, the usual weather which they bring on their eastward journey across the British Isles has an easily identifiable pattern (diagram **A**).

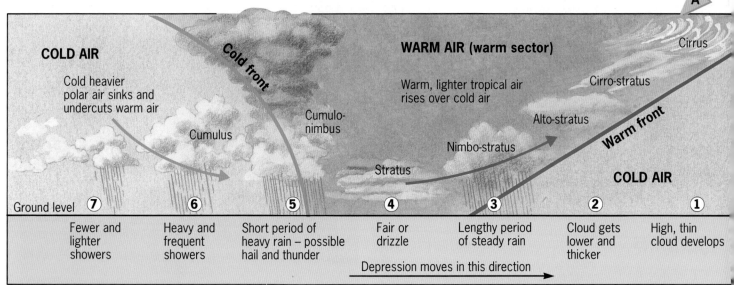

The first sign of an approaching warm front of a depression is the formation of high, thin clouds (cirrus). In time the clouds get lower and thicker (stratus). Winds slowly begin to increase in strength and blow in an anti-clockwise direction from the south-east. As warm air rises there is a rapid fall in atmospheric pressure. As the warm front passes, temperatures rise and winds become stronger, blowing from a south-westerly direction. Steady rain falls for a lengthy period from the low, thick clouds (nimbo-stratus).

The weather within the warm sector is a little less predictable. Light rain or drizzle may continue, or the clouds may break to give weak sunshine. Winds usually decrease in strength.

The most extreme conditions occur as the cold front passes. Winds often reach gale force and swing round to the north-west. Rainfall is very heavy, and can at times be accompanied by hail and even thunder (cumulo-nimbus clouds). The rain, however, is of shorter duration than that at the warm front. As the cold air replaces the warm air,

temperatures fall and atmospheric pressure rises. In time the heavy rain gives way to frequent and heavy showers and winds slowly begin to decrease in strength. Temperatures remain low due to the winds which continue to come from the north-west. Eventually the showers die out and the clouds disperse to give increasingly longer sunny intervals. Usually, however, this is just a short lull before the approach of the next depression and the repetition of a similar weather sequence. An average depression can take between one and three days to pass over the British Isles.

Diagram **A** is a section, viewed from ground level, through a typical depression. Diagram **B** shows three weather maps, viewed from the air, drawn at different times during the passing of a depression. Weather maps have their own weather symbols (page 83). In diagram **B** the warm front is shown by the red line and the cold front by the blue line. The black 'circular' lines are **isobars**. Isobars are lines that join up places of equal pressure. The closer together the isobars are on a weather map, the stronger the wind will be.

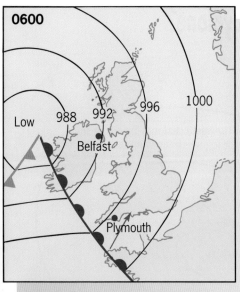

0600

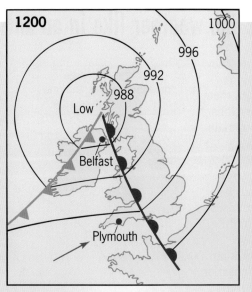

1200

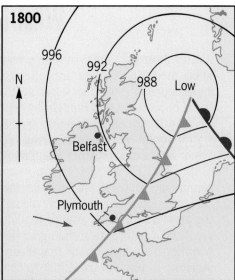

1800

Plymouth
Dry and sunny. Light clouds and rain approaching from west. Winds gentle but increasing from the south-east. Temperatures expected to rise.

Plymouth
Rain clearing to give patchy cloud and clear spells.
Strong winds from the south-west.
Warm.

Plymouth
Stormy weather.
Gale force winds from the north-west.
Heavy rain giving way to showers.
Colder, becoming brighter later.

B

Activities

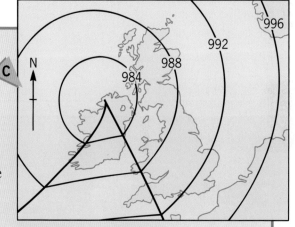

C

1 **a)** Make a copy of diagram **C**.
   b) Mark and label the following:
   - warm front
   - cold front
   - warm sector (warm air)
   - area of low pressure
   - lengthy period of steady rain
   - short period of heavy rain.

2 Use diagrams **A** and **B**, together with the written information on these two pages, to complete table **D** to describe the weather associated with the passing of a typical depression.

⑤ After the cold front passes	④ As cold front passes	③ During the warm sector	② As warm front passes	① As the warm front approaches	Weather conditions	**D**
					Temperature	
					Pressure	
					Cloud amount and type	
					Precipitation	
					Wind direction	
					Wind speed	

3 Use diagram **B** to:
   a) Describe the weather conditions at Belfast for 0600 hours, 1200 hours (noon) and 1800 hours.
   b) Explain why the weather at Belfast changed during this period of time.
   c) Give a short forecast for the expected weather at Belfast 0000 hours (midnight).

Summary

Britain's weather is dominated by the passing of depressions. These form when warm, moist air meets colder, drier air. They are areas of low pressure which bring cloud, rain and wind to the British Isles.

81

What is the weather like in an anticyclone?

Anticyclones are areas of high pressure. They affect the British Isles far less often than do depressions. The weather associated with them is the opposite to that brought by depressions. Once anticyclones develop they tend to remain stationary for several days, giving very dry, bright and settled weather (diagram **A**).

Anticyclones form in places where the air is descending. As more and more air descends, so the pressure increases and an area of 'high pressure' develops. Descending air is also warming air, which means that it can pick up moisture through evaporation. As condensation is unlikely in these conditions, clouds rarely form and the weather remains fine and dry. Winds are very gentle and at times may die away completely to give a period of calm.

However, there are important seasonal differences in the weather associated with anticyclones. These are shown in photos **B** and **C** below.

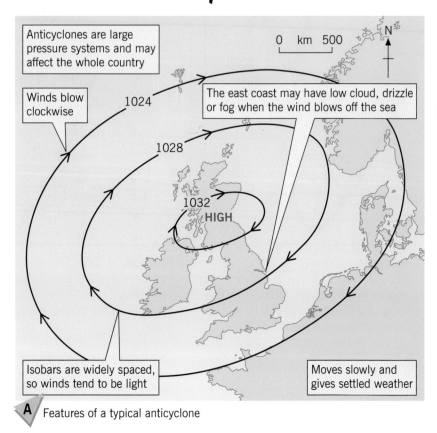

Anticyclones are large pressure systems and may affect the whole country

Winds blow clockwise

The east coast may have low cloud, drizzle or fog when the wind blows off the sea

1024

1028

1032 HIGH

Isobars are widely spaced, so winds tend to be light

Moves slowly and gives settled weather

0   km   500

A Features of a typical anticyclone

Summer conditions
- Very little cloud
- Dry with light winds
- Sun high in the sky, so hot and sunny
- Cloudless evening skies allow heat to escape, so nights can be cool
- Early morning dew and mist
- Risk of thunderstorms at end of 'heatwave' conditions.

Winter conditions
- Dry and bright with very little cloud
- Sun low in the sky, so cold conditions
- Clear evening skies mean that nights can be very cold
- Early morning frost and fog may last all day
- Extensive low cloud or fog may produce overcast or 'gloomy' conditions.

Synoptic charts

Synoptic charts are maps that show several weather conditions at a particular time. Diagram **D** is a daily weather map as you might see on television or in a newspaper. It aims to give a clear, visual and simplified weather forecast. Synoptic charts issued by the Meteorological Office are more complicated. They use official symbols which give weather conditions at specific weather stations. These are shown in diagrams **E** and **F**.

D

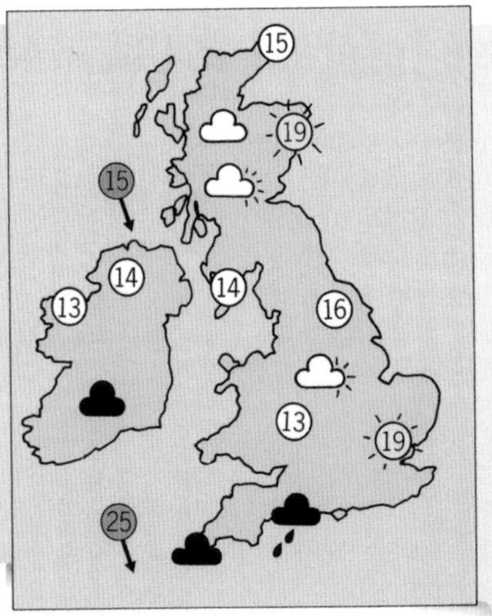
15
19
15
14
14
16
13
13
19
25

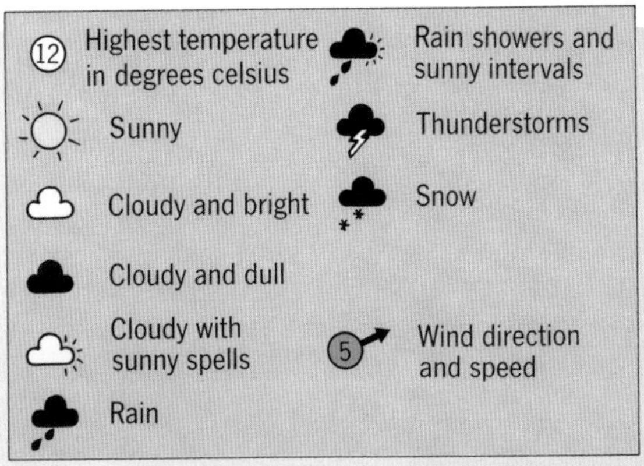
12 Highest temperature in degrees celsius
Sunny
Cloudy and bright
Cloudy and dull
Cloudy with sunny spells
Rain
Rain showers and sunny intervals
Thunderstorms
Snow
5 Wind direction and speed

Britain today: The south-west of England will be cloudy, wet and windy. Western Britain will be mainly cloudy and bright. Eastern Britain will be sunny and warm.

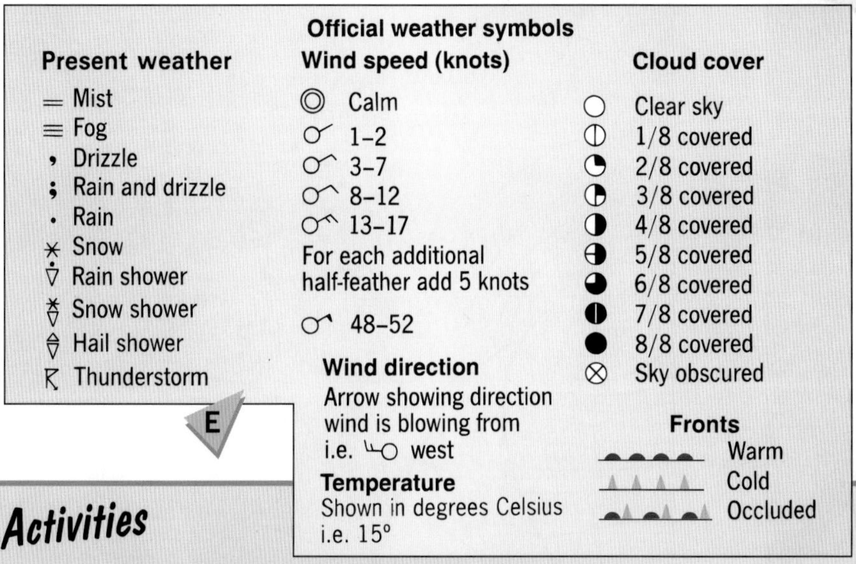
Official weather symbols
Present weather
Mist
Fog
Drizzle
Rain and drizzle
Rain
Snow
Rain shower
Snow shower
Hail shower
Thunderstorm
Wind speed (knots)
Calm
1–2
3–7
8–12
13–17
For each additional half-feather add 5 knots
48–52
Wind direction
Arrow showing direction wind is blowing from
i.e. west
Temperature
Shown in degrees Celsius
i.e. 15°
Cloud cover
Clear sky
1/8 covered
2/8 covered
3/8 covered
4/8 covered
5/8 covered
6/8 covered
7/8 covered
8/8 covered
Sky obscured
Fronts
Warm
Cold
Occluded

E

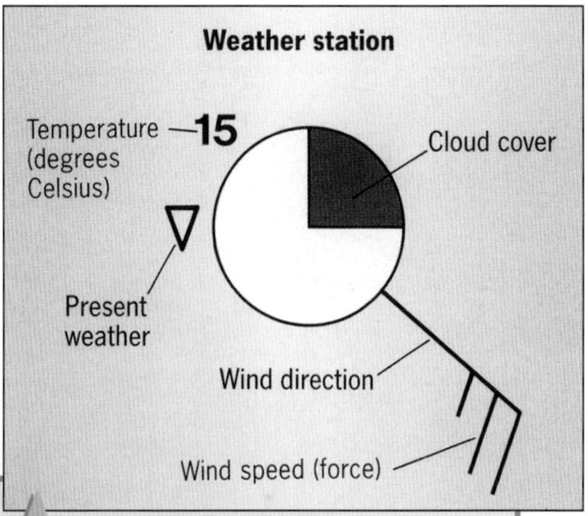
Weather station
Temperature (degrees Celsius)
15
Cloud cover
Present weather
Wind direction
Wind speed (force)

F

Activities

1 **a)** Draw a star diagram to show at least ten features of an anticyclone.
  b) In what ways is the weather of a summer anticyclone different from that of a winter anticyclone?

2 The weather conditions in a depression and an anticyclone are very different. Make an enlarged copy of table **G** and complete it to show the main differences.

G

	Depression	*Anticyclone*	
Pressure			
Wind direction			
Wind speed			
Cloud cover			
Precipitation		*Summer*	*Winter*
Temperature			

3 **a)** Describe five weather conditions at the two weather stations in diagram **H**.
  b) Draw two weather stations and add the following weather information:
    i) *Place X:* temperature = 26°C; present weather = mist; wind direction = NE; wind speed = force 1; cloud cover = 1/8 covered.
    ii) *Place Y:* temperature = 12°C; present weather = rain; wind direction = SW; wind speed = force 8; cloud cover = 8/8 covered.

H

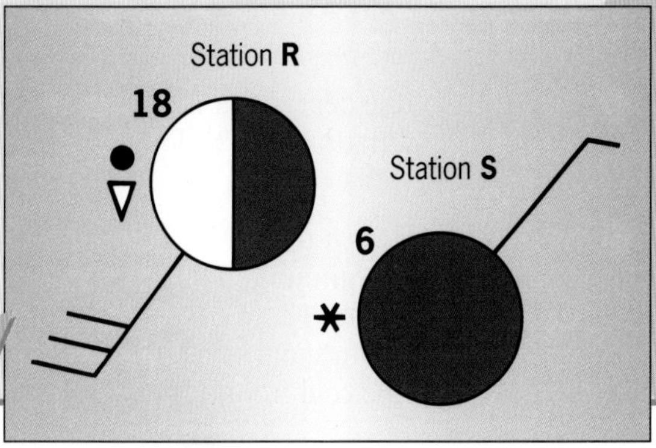
Station R
18
Station S
6

Summary

Anticyclones are areas of high pressure which usually give lengthy periods of fine, dry and settled weather. Synoptic charts are used to show the weather conditions for a particular place at a given moment of time.

How can we forecast the weather?

Synoptic charts are often issued along with **satellite images** (photos **A** and **C**; maps **B** and **D**). Satellite images are photos taken from space and sent back to Earth. They are essential when trying to produce a weather forecast or for making short-term predictions about likely changes in the weather. Satellite photos, which usually have lines of latitude and longitude superimposed upon them, show images of cloud patterns (photo **A**).

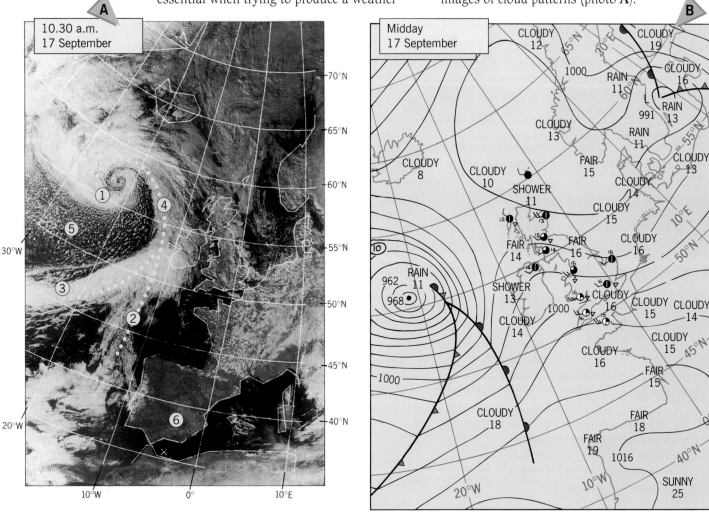

Photo **A** shows a depression approaching the British Isles. It was taken $1\frac{1}{2}$ hours before synoptic chart **B**. The cloud patterns are shown in white. Fronts have been added as yellow dotted lines.

Look at the following features, which are labelled 1 to 6.

① The centre of the depression is a mass of swirling cloud (latitude 57°N, longitude 22°W).

② The warm front is a thickening band of cloud which is beginning to obscure the coastline of Ireland. The warm front, marking the advance of the depression, lies to the east of the cold front.

③ The cold front is a long tail of cloud extending south-westwards back into the Atlantic Ocean.

④ The occluded front is a very thick band of cloud resulting from the cold front having caught up with the warm front. As there is no warm sector in an occlusion, there is no chance of any clearing skies as is possible between the fronts of some depressions.

⑤ A band of heavy showers, shown as patches of cloud, follow behind the cold front (between latitude 50° to 55°N and longitude 15° to 30°W).

⑥ Clear skies over Spain and the western Mediterranean indicate the existence of an anticyclone.

Photo **C** is part of a sequence of satellite images which were taken to show the passage of the depression seen in photo **A**. It was taken 24 hours after photo **A**. The full sequence of images indicated the direction and the speed at which the cloud, and therefore the depression, moved. Forecasters use information from satellite images together with data collected from weather stations on land and weather ships at sea, to draw synoptic charts (maps **B** and **D**). The use of satellite images along with synoptic charts helps meteorologists to forecast the weather.

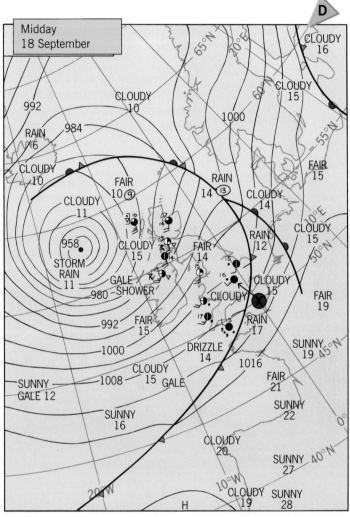

Activities

1 Give the meaning of the following terms (pages 80 to 83 and the Glossary will help you):

   a) *anticyclone*   **b)** *depression*   **c)** *warm front*
   d) *cold front*   **e)** *occlusion*   **f)** *synoptic chart*
   g) *satellite image*.

2 Match the following weather features with a letter from photo **C** (map **D** will help you):

   a) *anticyclone*   **b)** *depression*   **c)** *warm front*
   d) *cold front*   **e)** *occlusion*.

3 **a)** Using the weather symbols from page 83, describe the weather conditions at station Ⓧ on map **D**.

   b) Suggest reasons for the station's present weather, cloud cover and wind direction and force.

4 How did the weather over Spain differ from that over England during the period of the forecast?

Summary

Satellite images are photos taken from space and relayed back to Earth. They show cloud patterns which, together with synoptic charts, help forecasters describe and predict the weather.

How does climate affect our lives?

Weather and climate affect our lives in all kinds of ways. For example, we have to take account of the weather when deciding what clothes to wear, where to go on holiday, what to eat and what to do in our free time. It also affects sporting activities like tennis, cricket and skiing which need, and rely on, certain types of weather.

The climate of an area influences the environment and human activity to an even greater extent. It determines the vegetation and affects, for example, agriculture, water provision, the design of buildings and the overall way of life that people are able to lead.

Over the years, people have learned how to cope with the weather and climate. Sometimes, however, the weather can be more severe or different from what is usually expected. The effects on people are then greater because they are less likely to have been planned for. Some ways that people are affected by, and respond to, weather and climate in Britain are shown below.

Summer tourism In summer, the south is sunnier and warmer than most other parts of Britain and attracts large numbers of tourists. Tourism can bring jobs and wealth to an area. In 2000, visitors to the south-west region spent £6.4 billion and helped support 235 000 jobs.

Winter tourism Scotland can be cold and wet in winter but is the only location in the UK to receive sufficient snowfall to support skiing. There are five main centres attracting half a million skiers a year, providing 3000 jobs and an income of £30 million.

Crop farming Climate is an important factor in determining the best type of farming for an area. Crops such as wheat and barley grow best in warm, sunny weather with frequent light showers. These conditions are most common in the south and east of Britain.

Hill sheep farming The highland areas of Britain tend to be colder, wetter and less sunny than lowland areas. This makes farming difficult. Farmers cope with the problem by raising breeds of sheep like Herdwick and Swaledale that can survive harsh climates.

Activities

1 Describe the climatic reasons in the UK for:
   a) crop growing in the south and east
   b) sheep farming in the upland areas
   c) water storage in the west
   d) seaside resorts in the south.

2 a) Which areas of the UK are most affected by snow and ice?
   b) Make a list of problems caused by snowstorms. Use the headings **During the storm** and **Following the storm**.
   c) What can be done to reduce the worst effects of snow and ice?

3 a) Explain why recent floods in the UK have caused more problems than in the past.
   b) Suggest what can be done to help people cope with floods and reduce the damage that they cause. Use the headings **Before the flood**, **During the flood** and **After the flood**.

Summary

Weather and climate affect environments and human activity in many different ways. People in Britain have learned to cope with the weather but severe conditions can still cause problems.

Water supply Water is an essential need for people and industry but most of the UK's population live in the south and east which is relatively dry. Water has to be stored in reservoirs in the wetter west, then transferred by pipeline to the areas in need.

Shopping malls Shopping in cold, wet and windy conditions is not pleasant. Covered shopping centres have solved the problem by providing a warm and dry environment. Many people now use these centres as much for recreation as for shopping.

Flooding Floods have always been a problem for some areas of the UK but recent changes in weather patterns have made the problem much greater. Early warning systems and flood protection schemes have helped reduce the more damaging effects of flooding.

Snow and ice Heavy falls of snow and the formation of ice can disrupt people's normal way of life. The areas most prone to snow are the Scottish Highlands and the east coast of England. Bad-weather warnings help reduce the worst effects and salt is used for de-icing roads.

What are the world's main climates?

As we have seen on page 74, weather and climate are about the same things – temperature, rainfall, wind speed, sunshine and the amount of cloud. Yet weather and climate are not the same. The difference is one of time scale. Weather is the condition of the atmosphere today or tomorrow, whilst climate is the average weather over a long period of time.

There are several different climates around the world. Some are the same throughout the year whilst others have **seasonal** differences. Near the Equator, for example, almost every day is hot and sunny, with heavy rain in the afternoon. The Mediterranean, on the other hand, has seasonal differences, with dry summers and wet winters, the opposite to parts of Africa and South-east Asia where summers are wet and winters dry.

Although climates vary a great deal around the world, large areas have similar characteristics, which enables climatic regions to be identified. Map **A** shows six different types of climate. Maps like this should be used with care, however. They are always simplified and do not show local variations or the gradual changes in climate between regions.

A

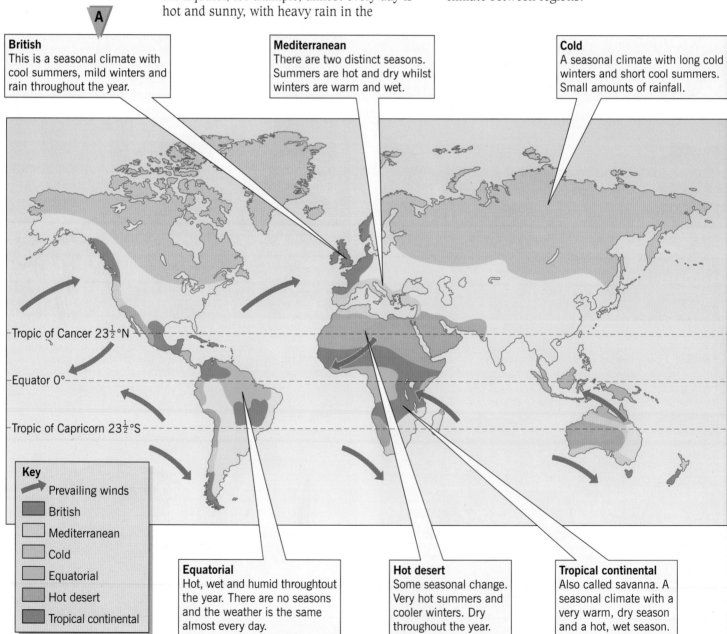

British
This is a seasonal climate with cool summers, mild winters and rain throughout the year.

Mediterranean
There are two distinct seasons. Summers are hot and dry whilst winters are warm and wet.

Cold
A seasonal climate with long cold winters and short cool summers. Small amounts of rainfall.

Tropic of Cancer 23½°N

Equator 0°

Tropic of Capricorn 23½°S

Key
- ➤ Prevailing winds
- British
- Mediterranean
- Cold
- Equatorial
- Hot desert
- Tropical continental

Equatorial
Hot, wet and humid throughtout the year. There are no seasons and the weather is the same almost every day.

Hot desert
Some seasonal change. Very hot summers and cooler winters. Dry throughout the year.

Tropical continental
Also called savanna. A seasonal climate with a very warm, dry season and a hot, wet season.

We have already looked in the previous unit at some of the factors that affect climate. Those that mainly affect temperature include latitude, distance from the sea, prevailing winds and altitude (pages 74–75). Relief, frontal and convectional are three types of rainfall and are explained on pages 76–77.

Differences in pressure also affect climate.
- Low pressure is rising air which usually brings cloud, rain and wind. Low pressure systems are called depressions (pages 80–81).
- High pressure is descending air and usually gives fine, dry and settled weather. High pressure systems are called anticyclones (page 82).

Drawing **B** shows the pressure belts that circle the globe. These are caused by warm air rising and cool air descending. In summer the belts move northwards whilst in winter they move south. This movement is the main cause of seasonal changes in climate.

Global pressure belts **B**

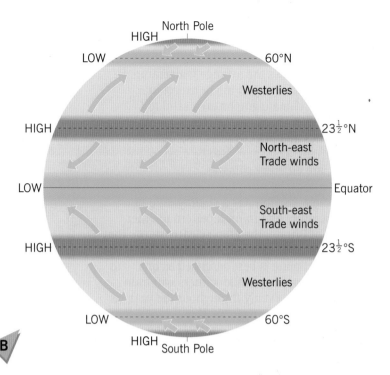

C Some reasons for climate types

Climate type	Pressure	Prevailing winds	Rainfall
British	Low all year	South-westerly all year	Frontal and relief all year
Mediterranean	Low in winter	South-westerly in winter	Frontal and relief in winter
	High in summer	North-east trades in summer	Very little
Cold	High most of year	North-east all year	Very little
Equatorial	Low all year	Light and variable	Convectional all year
Hot desert	High all year	North-east trades all year	Very little
Tropical continental	High in 'winter'	North-east trades in 'winter'	Very little
	Low in 'summer'	Light and variable in 'summer'	Convectional rain

Activities

1 a) What is meant by a *seasonal* climate?
  b) What is the difference between Mediterranean and tropical continental climates?

2 Diagram **E** shows one of the reasons why it does not rain very often in hot deserts. Copy and complete the diagram to show why it rains a lot in equatorial regions.

3 Write *increase* or *decrease* to complete these sentences.
  a) Temperatures ...... with distance from the Equator.
  b) Temperatures ...... with altitude.
  c) Summer temperatures ...... with distance inland.

4 a) Describe the main features of a hot desert climate.
  b) Name four places in the world where this type of climate may be found.
  c) Explain why North Africa has a hot desert climate. Map **A**, diagram **B** and table **C** will help you.

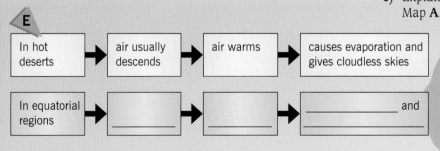

E

| In hot deserts | → | air usually descends | → | air warms | → | causes evaporation and gives cloudless skies |

| In equatorial regions | → | | → | | → | _____ and |

Summary

Climatic regions can be defined by looking at variations in temperature and rainfall across the world. These variations depend on several different factors.

89

What is the British climate?

Graph **A** shows the average temperatures and rainfall that you might expect in London. Climate graphs are useful to geographers because they show monthly statistics and help us recognise seasonal variations. They may also be used to compare different climates.

Although London has a slightly different temperature and rainfall pattern to other parts of Britain, the graph clearly shows that the climate is seasonal. In general, the British climate is best described as one with cool summers and mild winters with rain spread fairly evenly throughout the year. As we have seen on page 78, this type of climate is described as equable, or temperate, because it rarely has any extremes. The British climate is found not only in Britain but in several other places around the world. As map **A** on page 88 shows, these areas include south-west Canada, southern Chile and the South Island of New Zealand.

Temperatures

Most of the British Isles lies between latitudes 50°N and 60°N. This is far from the Equator, so the sun is only powerful enough to give moderate average temperatures. Winters, though, are much warmer than might be expected for such a high latitude. This is due mainly to the effect of the prevailing south-westerly winds which cross the relatively warm Atlantic Ocean and raise temperatures by several degrees.

Rainfall

The British Isles is relatively wet and receives rainfall throughout the year. The main reason for this is that the prevailing south-westerly winds bring in moist air from the Atlantic Ocean and cause relief rainfall as they cross the mountains of western Britain. Britain is also in the path of depressions which funnel in from the west giving frontal rain.

Climate variations

The climate is not exactly the same all over Britain. Map **B** divides the country into four different regions and shows both the regional and seasonal differences in climate. Three main features stand out:
- In summer the south is warmer than the north.
- In winter the west is warmer than the east.
- The west is wetter than the east.

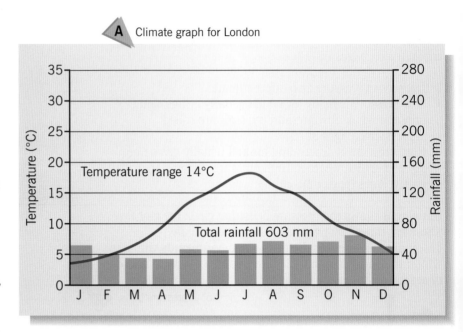

A Climate graph for London

Temperature range 14°C

Total rainfall 603 mm

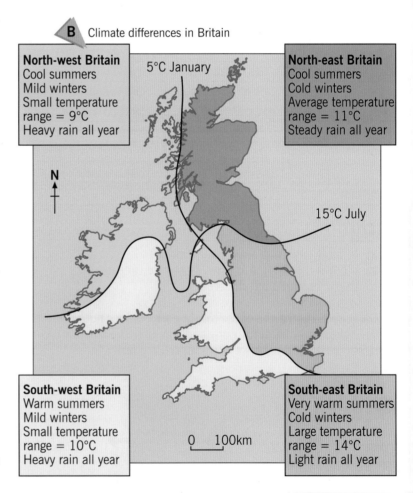

B Climate differences in Britain

5°C January

15°C July

North-west Britain
Cool summers
Mild winters
Small temperature range = 9°C
Heavy rain all year

North-east Britain
Cool summers
Cold winters
Average temperature range = 11°C
Steady rain all year

South-west Britain
Warm summers
Mild winters
Small temperature range = 10°C
Heavy rain all year

South-east Britain
Very warm summers
Cold winters
Large temperature range = 14°C
Light rain all year

N

0   100km

Climate differences in Europe

As might be expected, the climate is not the same all across Europe. Indeed, map **C** shows that there are four very different types of climate. The Mediterranean climate is explained on page 94, cold climates on page 95, and the British (sometimes called temperate maritime) on page 90 opposite.

The fourth European climate is called continental interior. This stretches from central Europe westwards into Russia. The climate here is one of hot damp summers and very cold dry winters. The main reason for this climate's characteristics is distance from the sea. As we saw on page 74, inland areas are warmer in summer and colder in winter than coastal places. Being far from the sea they also tend to be drier.

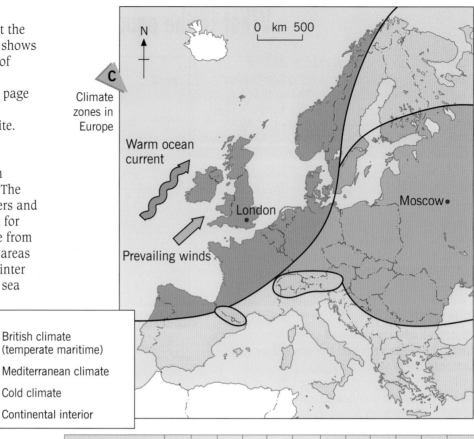

C Climate zones in Europe

Warm ocean current

Prevailing winds

Key

- British climate (temperate maritime)
- Mediterranean climate
- Cold climate
- Continental interior

Activities

1 a) Look at the climate graph for London and give the month and the figure for each of the following:
- maximum monthly temperature
- minimum monthly temperature
- most rainfall
- least rainfall.

b) Describe the climate shown by the graph.

2 a) Copy and complete table **D** by ranking the four regions named on map **B**.

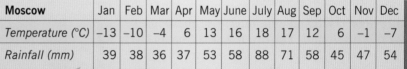

E

Moscow	Jan	Feb	Mar	Apr	May	June	July	Aug	Sep	Oct	Nov	Dec
Temperature (°C)	–13	–10	–4	6	13	16	18	17	12	6	–1	–7
Rainfall (mm)	39	38	36	37	53	58	88	71	58	45	47	54

D

Britain	North-west	North-east	South-west	South-east
Warmest in summer				
Mildest in winter				
Highest range in temperature				
Most rainfall				

3 a) Make a larger copy of graph outline **F** and complete the graph using information in table **E**.

b) Use your completed graph to answer the same questions as in Activity **1a)** and **1b)**.

c) What type of climate is shown by the graph?

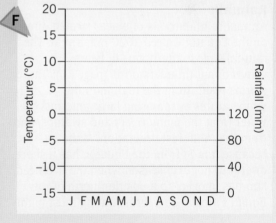

F

4 With help from map **C** and pages 74–75, explain why the continental interior climate is different from the British climate. There are three main reasons.

Summary

The British climate has cool summers, mild winters and rain spread throughout the year. However, the seasonal patterns of climate vary from year to year and from region to region. There are four main types of climate in Europe.

What is the equatorial climate?

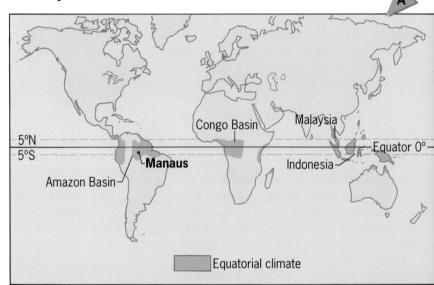

Places with an equatorial climate lie in a narrow zone which extends roughly 5° either side of the Equator (map **A**). The zone is not, however, continuous. It is broken by the Andes Mountains in South America and the East African Plateau in Africa.

Graph **B** is a climate graph for Manaus. Manaus is located 3° south of the Equator in the centre of the Amazon Basin in Brazil. It is typical of an equatorial climate in that it is hot, wet and humid throughout the year. The climate is unique in that it has no seasons, and a daily weather pattern that is repeated virtually every day of the year.

Temperatures
Temperatures are high and constant throughout the year and the annual range of temperature is very small (2°C). The major influence on temperature is the position of the sun. Even when it is not directly overhead, it always shines from a very high angle in the sky (page 74). Evening temperatures rarely fall below 22°C while daytime temperatures, due to afternoon cloud and rain, rarely rise above 32°C. Places on the Equator receive 12 hours of daylight and 12 hours of darkness every day of the year.

Rainfall
Annual rainfall totals of places located directly on the Equator exceed 2000 mm a year. The rain falls most afternoons in heavy convectional thunderstorms (page 77). These storms result from the high morning temperatures evaporating large amounts of water from the many rivers and swamps, and from the rainforest vegetation. Manaus, because it is 3° from the Equator, has a short, drier (but not dry) season when the overhead sun has 'moved' to a position north of the Equator. Heavy dew forms during most nights.

Although equatorial areas experience strong vertical air movements, surface winds are light and variable. There are no prevailing winds.

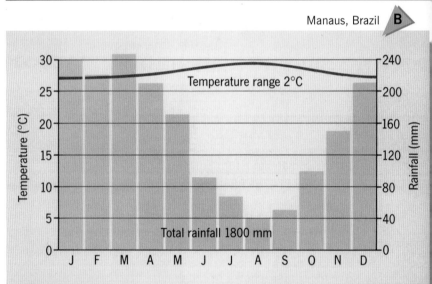

Manaus, Brazil **B**

Activities

1 **a)** Describe the main features of the equatorial type of climate.
   b) Describe the location of those places with an equatorial climate
   c) With the help of an atlas, name four countries that have an equatorial climate.

2 Give four main differences between the climate graph for Manaus and the climate graph for London shown on page 90.

3 Why do equatorial climates have:
   a) high temperatures throughout the year and a low annual temperature range
   b) heavy convectional thunderstorms during most afternoons?

Summary

Equatorial climates, which are hot, wet and humid throughout the year, are usually located within 5° north or south of the Equator and where there is low relief.

What is the tropical continental (savanna) climate?

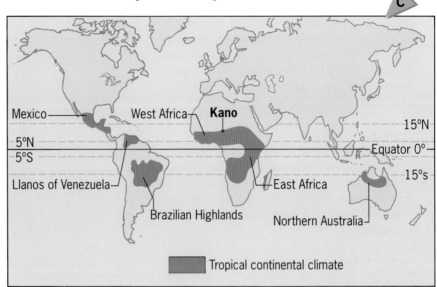

C

Places with this type of climate are located in the centre of continents, approximately between latitudes 5° and 15° north and south of the Equator (map **C**). It is also found on the higher land of the East African Plateau which straddles the Equator. Graph **D** is a climate graph for Kano in northern Nigeria. The climate has two distinct seasons.

1 A very warm, dry season when conditions are similar to those of the hot desert (page 88).

2 A hot, wet season when the weather more resembles that of equatorial areas (page 92).

Temperatures

Temperatures are high throughout the year and there is a relatively small annual range. The cooler (though not by British standards) season occurs when the sun is overhead in the opposite hemisphere. Temperatures rise as the angle of the sun in the sky increases (page 74), only to fall slightly at the time when there is most cloud and rainfall. Most areas are too far inland to be influenced by any moderating effect of the sea. Several places with this climate are in upland areas where temperatures are slightly reduced due to the increase in relief (altitude, page 75).

Rainfall

During the dry season the prevailing trade winds blow from the east. Any moisture that they carried will have been shed long before they reach the central parts of continents. The dry season is shorter towards the Equator and longer away from the Equator. The rainy season coincides with the time when the sun is overhead and the dry prevailing winds die away. The higher temperatures result in warm air being forced to rise to give frequent afternoon convectional thunderstorms (page 77). Unfortunately the length of the rainy season and the total amounts of rain are unreliable, while the heavy nature of the rain can do more damage than good (pages 96–97).

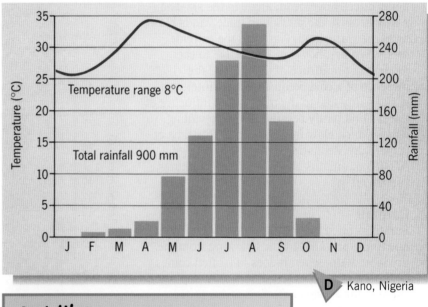

D Kano, Nigeria

Activities

1 a) Describe the main features of the tropical continental type of climate

  b) Describe the location of those places with a tropical continental climate.

  c) With the help of an atlas, name four countries that have a tropical continental climate.

2 Give four main differences between the climate graph for Kano and the climate graph for London shown on page 90.

3 Why do tropical continental climates have:
  a) a hot wet season
  b) a warm dry season?

Summary

Tropical continental climates are usually located in the centre of continents between 5° and 15° north and south of the Equator. They have a very warm, dry season and a hot, wet season.

What is the Mediterranean climate?

Places with a Mediterranean type of climate are located on west coasts of continents between latitudes 30° and 40° north and south of the Equator (map **A**). The exception is the area surrounding the 'inland' Mediterranean Sea. Graph **B** is a climate graph for Athens in Greece. The Mediterranean climate has two distinct seasons.

1 Hot, dry summers when the weather has similarities with that of the hot deserts (page 88).
2 Warm, wet winters when the weather more resembles that of the British Isles (pages 78 and 90).

Temperatures

Temperatures can be very high in summer. This is partly due to the sun being at a high, though never directly overhead, angle in the sky and partly because the prevailing trade winds blow from the warm land. Places on the extreme west coast are, however, cooler due to the moderating influence of the sea and the presence of a cold ocean current. Winters are warm partly due to the moderating influence of the sea and partly because of the prevailing winds which blow from the sea at this time of year. Many Mediterranean areas have high coastal mountains which lower temperatures considerably.

Rainfall

Summers often experience drought conditions due to the prevailing winds blowing from the dry land. Rain, when it does fall, often comes in short but heavy convectional thunderstorms. Winters can be very wet. This is due to the prevailing winds blowing from the sea and depressions moving eastwards which, together, give relief and frontal rain. Snow falls at higher altitudes.

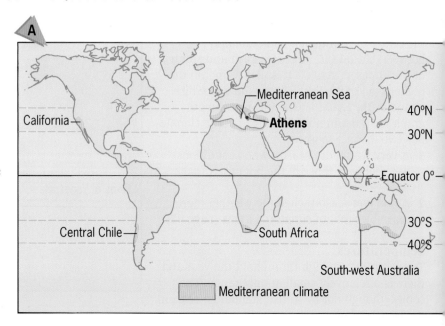

A

Mediterranean Sea
California
Athens
40°N
30°N
Equator 0°
Central Chile
South Africa
30°S
40°S
South-west Australia

Mediterranean climate

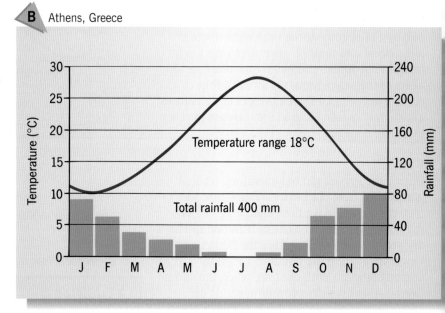

B Athens, Greece

Temperature range 18°C

Total rainfall 400 mm

Temperature (°C)

Rainfall (mm)

J F M A M J J A S O N D

Activities

1 **a)** Describe the main features of the Mediterranean type of climate.
   b) Describe the location of those places with a Mediterranean climate.
   c) With the help of an atlas, name six countries that have a Mediterranean climate.

2 Give four main differences between the climate graph for Athens and the climate graph for London on page 90.

3 Why do Mediterranean climates have:
   a) hot, dry summers
   b) warm, wet winters?

Summary

Mediterranean climates are usually located on west coasts of continents between 30° and 40° north and south of the Equator. Most places have hot, dry summers and warm, wet winters.

What are cold climates?

In high latitudes between 60° and 75° both north and south of the Equator, temperatures can be very low indeed. As map **C** shows, this cold climate stretches across the north of America, northern Europe and northern Russia in a wide belt. Mountain areas such as the Himalayas, Alps and Rockies also have a similar climate due to their extremes of altitude.

Temperatures

Cold climates have short and cool summers. Although the hours of daylight are long, temperatures remain low owing to the low angle of the sun in the sky and, for inland areas, the distance from the sea. Winters are both very long and very cold. Places north of the Arctic Circle have a period when the sun never rises above the horizon. Strong winds from the continental interior mean there is a high **wind-chill factor**.

Rainfall

Rainfall in these areas is very low. There are three main reasons for this. The first is that the air is cold and cannot hold much moisture. The second is that the distance from the sea further reduces the amount of moisture in the air. The third is that amounts of relief and frontal rain are reduced by the rainshadow effects caused by mountain ranges to the west. In America these are the Rockies, and in Europe the Norwegian mountains. Precipitation falls mainly as short convectional showers in the summer and as snow in the winter.

Further towards the poles, conditions are more severe with even lower temperatures. The climate in these areas is called **Arctic** or **Polar**.

C

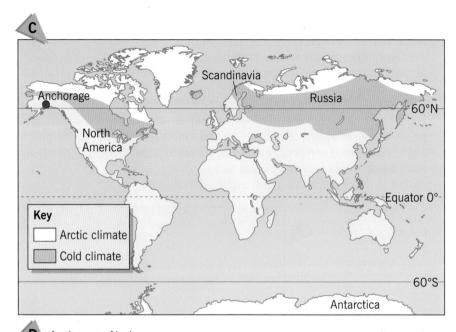

D Anchorage, Alaska

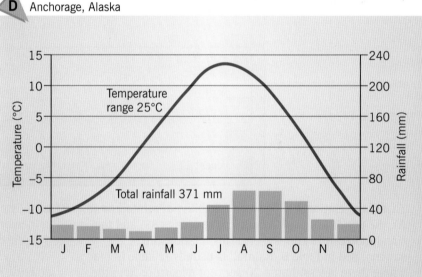

Activities

1 **a)** Describe the main features of the cold climate.
   b) Describe the location of those places with a cold climate.
   c) With the help of an atlas, name five places that have a cold climate.

2 Give four main differences between the climate graph for Anchorage and the climate graph of London on page 90.

3 Why do cold climates have:
   a) low temperatures in both summer and winter
   b) only small amounts of rainfall throughout the year?

Summary

Cold climates are found at latitudes above 60° in both the northern and southern hemispheres. Most places have cool summers and very cold winters. Rainfall amounts are low.

Why is rainfall unreliable in some parts of the world?

Rainfall in the British Isles is reliable. We know that in a usual year it will rain every few days, rainfall will be spread evenly throughout the twelve months, and that the total amount for each place can be fairly accurately predicted. Occasionally some seasons are wetter and others drier than is expected, but annual totals are nearly always within 10 per cent of the expected average total (map **A**). One of the driest summers ever recorded in England and Wales was in 1976. It caused a serious drought. Yet it was followed by the third wettest winter ever recorded and so, over a twelve-month period, rainfall virtually averaged itself out.

Not only is rainfall reliable in Britain, but so too is water supply. When it does rain it usually falls steadily for a period of several hours. This allows time for the water to infiltrate into the ground where it can be stored for use in drier periods. The British Isles rarely gets the severe storms that cause rapid surface run-off and flash floods. Britain also has the money and technology to build dams so that surplus water can be stored in reservoirs and then piped where and when it is needed. On a global scale, many countries still do not have a guaranteed supply 'at the turn of a tap'.

Reliability of rainfall 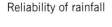 **A**

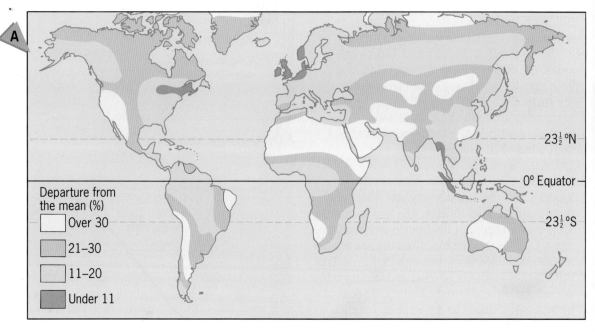

There is a relationship between reliability of rainfall (map **A**), the pattern of world climatic types (page 88) and the circulation of the atmosphere. This relationship is summarised in table **B**. Rainfall is usually reliable in places where there is less than a 20 per cent departure from the mean (average). Rainfall is least reliable in areas with over a 30 per cent departure from the mean, areas which are also

usually either too dry or too cold for permanent human settlement. It is those climates with a pronounced wet and dry season, especially in tropical continental areas (page 93), where rainfall is most unreliable in relation to human activity. Here, in some years, the rains may fail to arrive. In other years, when it does rain, total amounts can come from just a few heavy storms. The rain will then be too heavy to

B

Departure from the mean (%)	Atmospheric pressure	Vertical air movement	Surface prevailing winds	Rainfall distribution	Effectiveness of rain
Under 20	Low	Rising air cools, condenses and gives rain	• Equatorial – variable • British type – south-west from the Atlantic Ocean	Rain throughout the year	Equatorial – heavy but intercepted by vegetation British type – effective
20–30	Seasonal – tropical continental	Rising air gives rain for part of the year. Descending air gives no rain for rest of year	• Dry North East Trades in winter • Light and variable in summer	Wet season and a dry season	Heavy storms; little interception; less effective and much surface run-off
Over 30	High (i) Hot desert (ii) Polar	Descending air warms, so no condensation	Trade winds from dry land	Very little	Ineffective – short, intense storms

allow infiltration and so the precious water is lost through surface run-off and flash floods. As the rainy season coincides with the time of highest temperatures then evaporation rates, and therefore water loss, will also be at their greatest. Drought frequently occurs in the Sahel countries along the southern fringes of the Sahara Desert when annual rainfall totals fall below average for several successive years. Unreliable rainfall is one of several important reasons why so many countries with a tropical continental climate, especially in Africa, have remained economically less developed. This creates a vicious circle since, because they are economically less developed, they have less capital and technology to improve their water supply.

Diagram **C** gives some of the consequences of unreliable rainfall upon human activity in Kenya. In a normal year it is estimated that Kenya receives enough rainfall, if it was spread out evenly, to sustain a population several times greater than it has at present. Unfortunately the distribution of rainfall and population does not always match, and rainfall is not always reliable.

C

Massai herder
We depend upon our cattle, sheep and goats. When the rains fail, there is overgrazing and eventually our amimals will die.

Industrialist
We need energy to manufacture goods. We have to rely upon hydro-electricity. If the rains fail, reservoirs dry up and the power stations have to close down. We had big power cuts in Nairobi in 1992.

Wildlife worker
Drought also kills wildlife. In the 1970s thousands of elephants died.

Villager
We have no piped water. If the rains fail, rivers and waterholes dry up and we may have to walk many miles to get water.

Farmer
We need water for our crops. We need more crops since our population is growing so rapidly. If the rains do not come at the right time, and in sufficient quantity, our crops will fail and we will be short of food.

Government official
Kenya is an economically developing country. We do not have enough money to build dams to store water for times of drought. We have several dams but not enough. One big dam was built for us using French money but that accounts fot two-thirds of our national debt.

Activities

1 Give three reasons why rainfall in the British Isles is reliable.

2 **a)** Write out the paragraph below using the correct word from each pair in brackets.

> Rainfall is most unreliable in areas with (high/low) pressure, when air is (descending/rising) and (cooling/warming), where winds blow from the (land/sea), and when it falls in (long/short) (storms/periods).

b) Which two of the following four climates have the least reliable rainfall:
- equatorial
- tropical continental
- hot deserts
- British type?

3 **a)** How does an unreliable water supply affect a country such as Kenya?

b) Do you agree or disagree with the statement made in diagram **D**? Give reasons for your answer.

D

The most serious single climatic hazard in an economically less developed country is the variability and uncertainty of rainfall.

Summary

The distribution of areas with a markedly unreliable rainfall can be related to the general pattern of atmospheric circulation and to specific climatic types.

What are the effects of drought?

A **drought** is a long period of weather that is drier than usual. When a drought lasts for a particularly long time or rainfall amounts are considerably lower than average, the drought becomes a climatic hazard and can cause severe problems for people. As always, the problems are greatest in the poorer, developing countries where a lack of resources and a shortage of money make coping with disasters much more difficult.

In recent times the worst droughts have been in the Sahel region of Africa just south of the Sahara Desert (pages 126–127). From the 1920s to the 1960s, rain was reliable here and people were encouraged to move into the area. In the early 1970s, however, a series of droughts occurred, badly affecting Ethiopia, Sudan, Chad, Niger, Mali and Mauritania. Millions of nomadic farmers were forced to leave the area which soon became a barren desert. The 1980s saw a repeat of the problem, particularly in Ethiopia and Sudan where thousands of people died and millions were forced off their land. Despite huge aid programmes which have helped generate long-term agricultural improvement, the problems have continued through the 1990s and into the 21st century.

Drought in Ethiopia, 1983–84

Ethiopia is one of the world's poorest countries and in 1983–84 suffered its worst drought in living memory. Television pictures of the drought showing starving people helped to attract large amounts of aid from all around the world. Aid came from charities like Oxfam and Band Aid as well as from governments and other organisations (pages 264–265). Sadly, civil war and poor transport left much of the donated foodstuffs rotting in ports, never to reach the people most in need.

A Children waiting for a feeding centre to open

B **MAINLY DEVELOPING COUNTRIES**

```
              Drought
                 │
                 ▼
       Farming land dries out
         ╱              ╲
        ▼                ▼
  Animals die         Crops fail
         ╲              ╱
          ▼            ▼
    People starve, farmers lose income
                 │
                 ▼
   People die, migrate, or are given food aid
                 │
                 ▼
      Aid given to dig new wells and plant
           drought-resistant crops
```

Short-term effects
- A serious shortage of food and water caused widespread starvation and illness.
- About 500 000 people died. The young and elderly were especially affected.
- Millions more needed food and medical aid which was organised mainly by charities in developed countries.
- People migrated from areas that were too remote to receive food aid. Many ended up in refugee camps (page 143).

Long-term effects
- Up to a million people who were malnourished and poverty stricken continued to need aid (page 274).
- Regular aid from developed countries has improved agricultural output and provided people with security.

Activities

1 a) What exactly is a *drought*?
  b) What is meant by the term *climatic hazard*?

2 Write newspaper articles for each of the headlines below. For each one, include information on the problem and its causes and effects. Also say what was done to limit the effects of drought in both the short term and the long term.

C **Ethiopia devastated by drought**

Water shortages cause problems in the UK

3 Explain why drought in less developed countries like Ethiopia usually causes more problems than in richer developed countries like the UK. Give examples to support your answer.

Summary

Drought is a climatic hazard and can cause serious problems for people. Less developed countries suffer most because they are too poor to cope with drought and the resultant food shortages and famine that it may lead to.

Drought in the UK, 1995–96

The UK has a temperate climate and can expect to have rain throughout the year – not in any great amounts, but certainly sufficient to provide for the country's needs. In the last part of the 20th century, however, the country suffered two major droughts and both caused unexpected problems for the country. In the north of England water was so scarce that a fleet of 200 tankers working 24 hours a day, 7 days a week, was used to transfer water to empty reservoirs.

D Ladybower reservoir, Derbyshire

E **MAINLY DEVELOPED COUNTRIES**

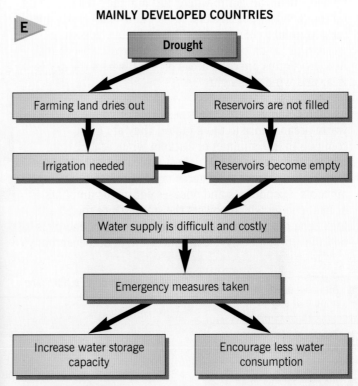

Drought

Farming land dries out → Reservoirs are not filled

Irrigation needed → Reservoirs become empty

Water supply is difficult and costly

Emergency measures taken

Increase water storage capacity / Encourage less water consumption

Short-term effects
- Garden hosepipes were banned and water rationing was introduced in some places.
- Clay soils dried out and buildings were damaged as their foundations moved.
- Grass stopped growing, leaving a shortage of cattle feed. Crops died in the hot, dry conditions.
- Heathland and forest areas became tinder-dry. Large areas were destroyed by fire.

Long-term effects
- Legislation was introduced to try to reduce water consumption by both private and industrial users.
- Water authorities planned to increase water storage capacity and link reservoirs to make transfers easier.

What are tropical storms (hurricanes)?

The tropics experience some of the fiercest and most destructive storms on Earth. The **hurricanes** that occur in that region claim an average 20 000 lives each year and cause immense damage to property, vegetation and shipping.

A hurricane (also called a cyclone, typhoon or willy-willy) is a particularly powerful **tropical storm**. It rotates around an area of intense low pressure and produces very high winds and torrential rain. Wind speeds commonly exceed 118 km/h (73 mph) and in the most powerful storms have been known to reach 300 km/h (186 mph) in gusts. Rainfall is almost continuous throughout the storm but is heaviest near the centre where 300 mm may fall in just 24 hours. (London can expect 610 mm in a year.) Hurricanes vary from 80 to 650 km in width and move generally westwards, often on an erratic and unpredictable course. At their centres are calm areas or 'eyes' where the sky is clear and winds are light.

As map **C** shows, most tropical storms originate over warm oceans close to the Equator. They are most common in late summer or autumn when sea temperatures are at their highest (at least 26°C). At these temperatures water evaporates rapidly and as the rising air cools it condenses and releases enormous amounts of heat energy which powers the storm. Once the hurricane reaches land, however, the source of energy is lost and the storm quickly declines in strength and eventually

A A hurricane viewed from space: notice the swirling shape, thick cloud and the 'eye' of the storm

blows itself out. Diagram **B** shows the development and structure of a typical hurricane.

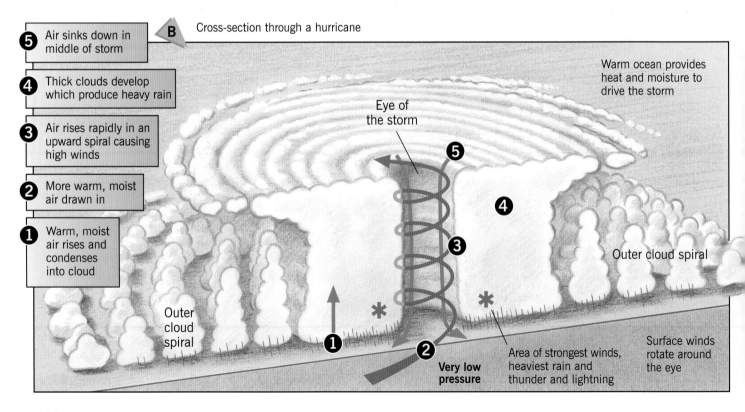

B Cross-section through a hurricane

5 Air sinks down in middle of storm

4 Thick clouds develop which produce heavy rain

3 Air rises rapidly in an upward spiral causing high winds

2 More warm, moist air drawn in

1 Warm, moist air rises and condenses into cloud

Warm ocean provides heat and moisture to drive the storm

Eye of the storm

Outer cloud spiral

Outer cloud spiral

Very low pressure

Area of strongest winds, heaviest rain and thunder and lightning

Surface winds rotate around the eye

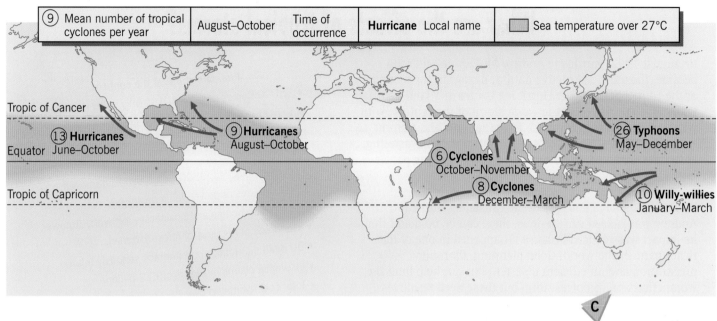

| ⑨ Mean number of tropical cyclones per year | August–October | Time of occurrence | **Hurricane** Local name | Sea temperature over 27°C |

C

Hurricanes are a major natural hazard. Their effects can be catastrophic, causing widespread damage to property, disrupting communications, destroying crops and resulting in considerable loss of life. They have probably caused more deaths worldwide than any other form of natural hazard except perhaps droughts. Although the extremely strong winds and intense rainfall cause a huge amount of damage, by far the most dangerous feature of a hurricane is the **storm surge**. This occurs at the **eye of the storm** where very low pressure causes the sea-level to rise by up to 10 metres. The resulting high seas and tidal waves lead to serious flooding of low-lying coastal areas close to the hurricane track. The storm surge that swept up the Bay of Bengal in 1970 killed 300 000 people in north-east India and Bangladesh and made a further 2.5 million people homeless (pages 36 and 103).

The extent of the damage and loss of life caused by tropical storms varies and is closely related to the stage of development of the affected area. Poorer countries suffer most because their buildings, warning systems, defences and emergency services may be inadequate. Wealthier countries, however, can afford to prepare for such disasters and so minimise the potential destruction and loss of life.

Activities

1 a) Draw a sketch of photo **A** and label the thick cloud, swirling shape and the eye of the storm.
   b) Describe the main features of a hurricane. Include the following in your description:
      • *pressure* • *winds* • *rainfall* • *size*
      • *movement* • *the 'eye' of the storm*.

2 With the help of a simple diagram, describe how hurricanes develop.

3 a) From map **C**, describe and explain the distribution of tropical storms.
   b) Why do these storms weaken when they move over land or cross cooler areas of ocean?

4 With help from diagram **D**:
   a) Explain how a storm surge happens.
   b) Suggest why its effects can be devastating.

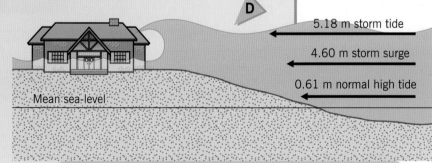

D

5.18 m storm tide

4.60 m storm surge

0.61 m normal high tide

Mean sea-level

Summary Hurricanes are tremendously powerful tropical storms that can cause widespread destruction. They develop over warm tropical oceans but soon decline on reaching land where they lose their source of energy.

What are the effects of tropical storms?

In August 1992, Hurricane Andrew hit Florida and soon became 'the most destructive natural hazard in the history of the United States'. The Hurricane Center in Miami had warned people to expect the worst storm of the century, with winds of 240 km/h (150 mph), torrential rain and a storm surge of 5 metres. The effects of the storm were devastating, with trees uprooted, buildings destroyed and many people injured or killed (figure **A**).

Whilst the damage and loss of life caused by Hurricane Andrew was indeed very serious, there can be no doubt that its impact was reduced because it happened in one of the richer parts of the world. Good planning, thorough preparation and an efficient disaster plan can help limit the worst effects of a tropical storm but these need resources and large sums of money to implement. Florida is fortunate in that it is a very wealthy region, and as figure **B** shows, is well prepared for these storms which affect the state almost every year.

Florida's storm of the century
* 25000 homes destroyed: 100000 badly damaged.
* 175000 people homeless in South Florida alone.
* 82000 businesses destroyed or closed down.
* 120300 job losses.
* 52 roads blocked, 9500 traffic signals damaged.
* 5311 metres of power cable destroyed.
* 1.3 million homes and businesses left without power.
* Hundreds of hectares of forest flattened.
* 30 deaths in total, and hundreds seriously injured.
* Insurance claims in excess of £12 billion.
* Total cost estimated over £50 billion.

Florida, USA: August 1992

* Disaster plan prepared to cover time before, during and after the storm.

* Evacuation plan prepared with accommodation, food and medical care made available.

* Training provided for emergency services such as police, fire and ambulance crews.

* Efficient warning and information system set up using television, radio and home visits.

* Research organisations follow storm development and predict its track.

* Help given after the storm in the form of financial support and rebuilding programmes.

Activities

1 a) Describe the main features of Hurricane Andrew.
   b) Draw a star diagram to show the hurricane's main effects.

2 a) Describe the main features of the Bangladesh cyclone.
   b) Draw a star diagram to show the cyclone's main effects.

3 Compare the disaster preparations of Florida and Bangladesh by completing table **C**.

	Florida, USA	Bangladesh
Warning and information provision		
Evacuation plans		
Emergency services		
Provision of aftercare		
Other features		

Bangladesh is one of the poorest and most densely populated nations in the world. Most of the country is an extensive, flat, low-lying river delta with rich soils and a hot, wet climate which supports millions of rice farmers. Unfortunately, Bangladesh lies in an area where tropical storms are common. In many years, thousands of people die as these violent storms (called **cyclones** in this part of the world) sweep in from the Bay of Bengal destroying everything before them.

In April 1991 a particularly severe storm hit Bangladesh. Winds of 225 km/h (140 mph) and waves 7 metres high swept through the coastal districts flattening houses and causing serious flooding. The death toll was enormous and most people living in the area lost everything they owned, including their livelihood (figure **D**).

The problem in Bangladesh is one of poverty. People simply cannot afford to leave the danger zone to live elsewhere and the government is unable to fund adequate disaster plans. Without help from richer countries in the form of aid and support, there is little hope of reducing the impact of such storms in the future.

D

Storm disaster hits Bangladesh again
- Poorly constructed buildings blown away.
- Hundreds of villages entirely wiped out.
- An estimated 130 000 people killed, or washed away and never seen again.
- All food supplies lost, and drinking water contaminated with raw sewage and dead bodies.
- Half a million head of cattle drowned and 63 000 hectares of crops destroyed.
- Most roads and bridges in the area damaged.
- Electricity supplies cut off for months.
- More than 4 million people faced with starvation and disease during the weeks after the storm.

Bangladesh, April 1991
- Badly designed and poorly constructed buildings unable to withstand high winds.
- Emergency electricity supplies and telephone links failed to work.
- Megaphone warning system ineffective.
- Evacuation difficult because of a lack of transport and a shortage of safe places to go.
- Government failed to provide enough cyclone shelters for everyone affected.
- Lack of emergency food and shortage of clean water caused further deaths by starvation and disease.
- Local rescue workers poorly prepared and unable to reach stricken areas because of inadequate transport.

E

4 The following is a list of problems facing the authorities. List them in order of urgency, putting the most important first. Give reasons for your choice of the four most urgent points.
- Help rebuild homes
- Search for missing people
- Rescue stranded people
- Supply safe drinking water
- Open up all roads
- Provide first aid
- Supply food
- Re-open businesses

5 Explain why tropical storms in less developed countries like Bangladesh usually do more damage than similar storms in richer developed countries like the USA. Give examples to support your answer.

Summary Tropical storms can cause considerable damage and loss of life. Richer countries can afford more effective disaster plans and are better able to cope with these storms than poorer countries.

What are the causes and effects of global warming?

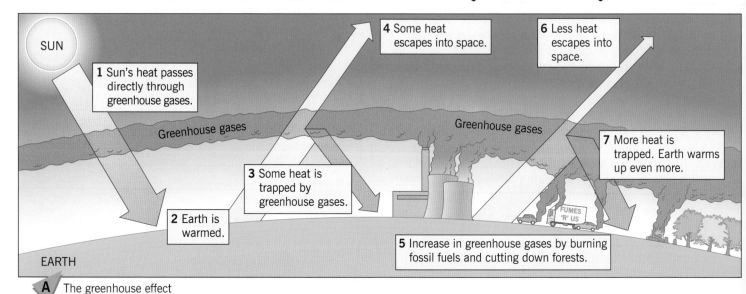

SUN

1 Sun's heat passes directly through greenhouse gases.

4 Some heat escapes into space.

6 Less heat escapes into space.

Greenhouse gases

Greenhouse gases

3 Some heat is trapped by greenhouse gases.

7 More heat is trapped. Earth warms up even more.

2 Earth is warmed.

FUMES 'R' US

5 Increase in greenhouse gases by burning fossil fuels and cutting down forests.

EARTH

A The greenhouse effect

The South Pacific island of Kiribati was shown on television celebrating the new millennium, and the rest of the world linked to it by satellite saw a sight that could soon be only a memory. For it is predicted that by the start of the next century, Kiribati and several other low-lying islands will have all but disappeared under the sea. Indeed by the year 2100, some 50 million people world-wide could be affected by rising sea-levels.

Rising sea-levels are a result of rising temperatures. The world is now warmer than it has been for many thousands of years. In the 20th century, average temperatures rose by 0.6°C, with most of this increase in the last 40 years. The warmest years ever recorded were in 1990 and 1995. Some scientists believe that the Earth's climate is going to get even warmer. They predict that by 2100 the average global temperature could be between 1.6°C and 4.2°C higher than it is today. This heating up of our planet is called **global warming**, and it is likely to cause serious problems in the future.

Global warming is thought to be due to the **greenhouse effect**. As diagram **A** shows, the Earth is surrounded by a layer of gases, including carbon dioxide. This keeps the Earth warm by preventing the escape of heat that would normally be lost from the atmosphere. The gases act rather like the glass in a greenhouse. They let heat in but prevent most of it from getting out. The burning of fossil fuels such as oil, coal and natural gas produces large amounts of carbon dioxide. As the amount of this gas increases, the Earth becomes warmer.

Greenhouse gases **B**

CFCs (chlorofluorocarbons) from aerosols, plastic foam and fridges, are the most damaging of greenhouse gases.

CFCs 13%

Carbon dioxide is produced by power stations, factories and road vehicles that burn fossil fuels.

Nitrous oxide 5%

Carbon dioxide 72%

Nitrous oxide is emitted from car exhausts, power stations and fertilisers.

Deforestation and the burning of rainforests also produces carbon dioxide.

Methane 10%

Methane is released from waste dumps, farms and rice fields.

Nobody knows exactly what the effects of global warming will be. Scientists are in general agreement that changes in world climates and sea-levels are almost inevitable, but the exact nature of these changes are difficult to predict. No doubt some of the effects will be harmful whilst others may bring benefits. Some of the effects predicted by scientists are shown on map **C** and in the list below.

- Sea temperatures would rise, the water would expand and sea-levels could rise by 0.25 to 1.5 metres.
- Ice caps and glaciers would start to melt, causing sea-levels to rise even further.
- Low-lying areas would be flooded. Some islands like Kiribati would disappear altogether.
- There might be more violent storms, and extremes of hot weather.
- Hot regions would become hotter and deserts would spread.
- Climatic belts and vegetation belts would move.
- Some plants and animals would become extinct.
- There could be an increase in insect pests like aphids and fleas.
- Tropical diseases may spread to temperate regions such as the UK.

Gaining international agreement to reduce the releases of greenhouse gases is difficult. Industrialised countries are reluctant, claiming that the high economic costs involved could cause job losses and a lowering of standards of living. Developing countries are reluctant, believing that they need to increase energy consumption if they are to create new jobs and

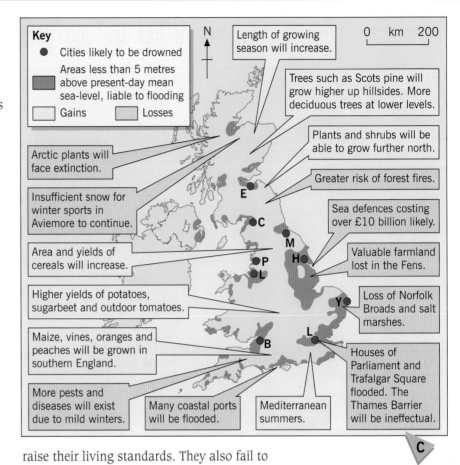

raise their living standards. They also fail to see why they should help solve a problem that they did not create.

In 1997, the Kyoto Protocol brought agreement between countries that greenhouse gases should be reduced. So far, many European countries, including Britain, have admitted difficulty in achieving the target. In 2001, despite producing 25 per cent of the world's greenhouse gases already, President Bush stated that the USA would not be reducing emissions to the agreed levels.

Activities

1 a) Give the meaning of the term *global warming*.
   b) Draw a labelled diagram to explain the greenhouse effect.
   c) What are the four main greenhouse gases?

2 a) How is global warming predicted to affect the world's temperatures and sea-levels? Give facts, figures and examples to illustrate your answer.
   b) Name six UK cities that may be drowned.
   c) Describe how farming in the UK may benefit from global warming.

3 a) Why is **international** agreement necessary for the reduction of greenhouse gases ?
   b) Why is it difficult to get this international agreement?

Summary Global warming is caused by an increase in greenhouse gases. As its consequences affect the whole world, there is a need for international agreement on how to reduce the problem.

What are ecosystems?

Chaffinch | Beech | Sycamore | Squirrel | Owl
Hazel | Insects | Oak
Fox | Brambles | Rabbits | Deer
Voles and mice | Weasel | Wild flowers | Leaf litter

A A woodland in northern England

Look at diagram **A** which shows part of Wylam Wood, a typical area of deciduous woodland in northern England. Although it looks a very ordinary and rather tranquil scene, the woods are actually very complex and busy with over 1000 different species of plants and animals all living together in a carefully connected web of life. These plants and animals are part of a **community** and depend on each other for food and shelter. They also have close links with non-living elements of the environment such as the local relief, climate and soil. The relationship and interaction between plants and animals and the physical environment is called an **ecosystem**.

Diagram **B** shows the links between various elements of a simple ecosystem. Notice how the components are closely related and how each affects the other. These interactions are almost always at work although some are difficult to see either because they happen in the dark or because they are invisible to the human eye. An example of an invisible and important interaction is **photosynthesis**. In this process, plants take in carbon dioxide from the air through their leaves. In sunlight they then make sugars from carbon dioxide and water. These sugars are used by the plant to help it grow and are crucial to the **energy flows** within an ecosystem. Energy flows are explained in more detail on page 108.

B Links in a simple ecosystem

ENERGY FROM THE SUN

CLIMATE

LIVING CREATURES

VEGETATION

SOIL

Ecosystems can be any size, from the vast rainforests of the Amazon to a small woodland area or a tiny pond perhaps at the bottom of a garden. There are also many different types of ecosystem. They vary from place to place around the world as well as between and within areas such as woodland, moorland, streams, farms and even cities. These variations are due largely to the different physical conditions that exist in any particular location and they can be studied using a systems approach.

The systems diagram **C** summarises both the natural and human influences on a woodland ecosystem like the one in diagram **A**. In Britain, the influence of people is particularly important as there are few areas that have not been altered or affected by farmers, developers or industrial growth.

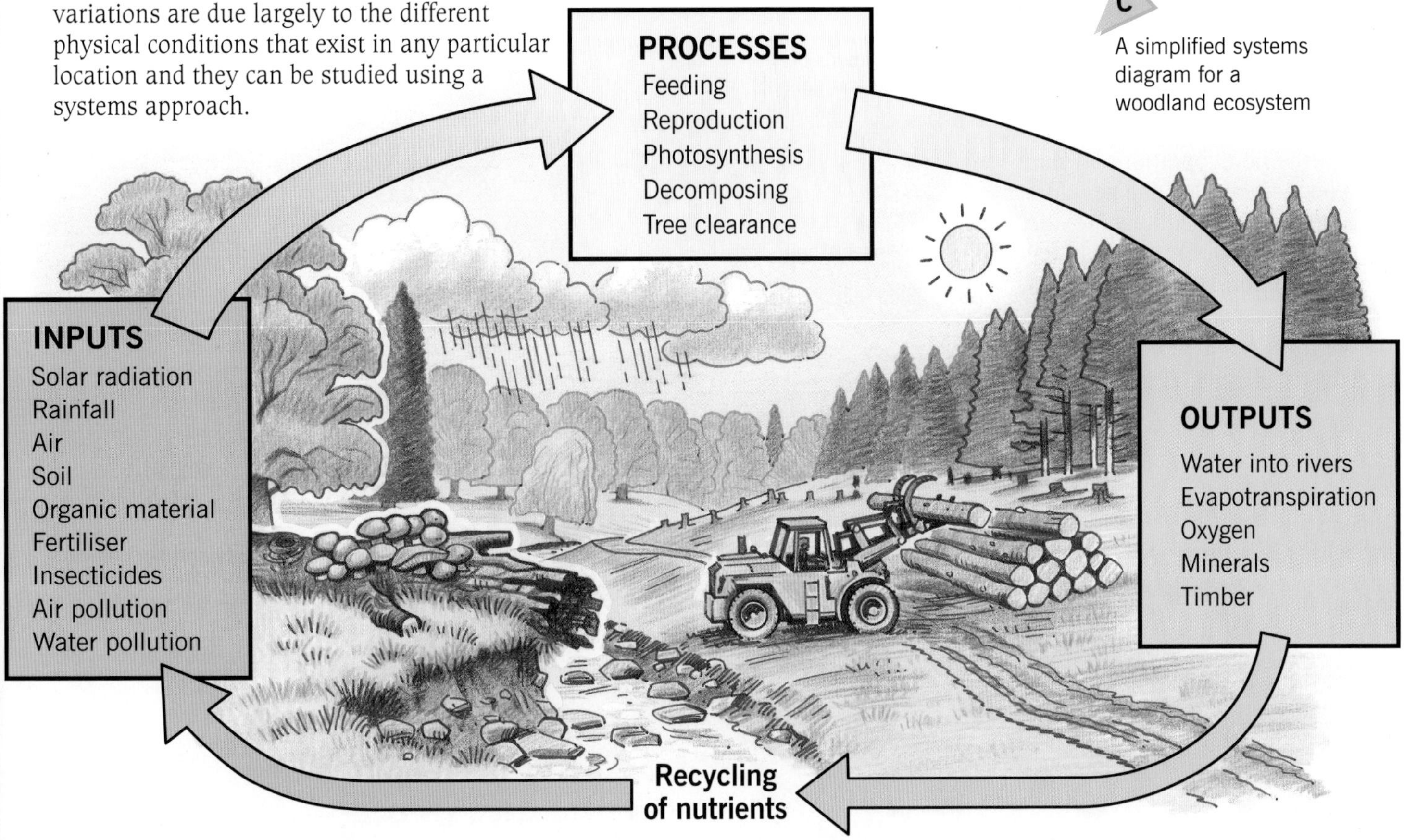

C

A simplified systems diagram for a woodland ecosystem

PROCESSES
Feeding
Reproduction
Photosynthesis
Decomposing
Tree clearance

INPUTS
Solar radiation
Rainfall
Air
Soil
Organic material
Fertiliser
Insecticides
Air pollution
Water pollution

OUTPUTS
Water into rivers
Evapotranspiration
Oxygen
Minerals
Timber

Recycling of nutrients

Activities

1 Explain the following terms and give two examples for each one.
   - *Ecosystems*
   - *Living environment*
   - *Non-living environment*.

2 Look out of the window and list:
   a) the natural features and
   b) the human features that are affecting the local environment.

3 Using diagram **B**, suggest what might be the effects on a woodland ecosystem if:
   a) half the trees were cut down
   b) chemicals were used to kill insects
   c) global warming caused climatic change.

4 **a)** Which of the influences shown in diagram **C** are human?
   b) Give three other ways in which people may affect ecosystems.

Summary

An ecosystem is a community of plants and animals which interact with each other and with the environment in order to survive. Ecosystems may be studied and compared using a systems approach of inputs, processes and outputs.

What are the features of a woodland ecosystem?

In the past much of Britain was covered in forest. The climate was colder and drier than it is today and coniferous trees such as Scots pine were most common. As the climate became gradually warmer and wetter so the type of vegetation and wildlife gradually changed to suit the new conditions. Wylam Wood is typical of the new forest cover where deciduous broadleaved trees such as oak, beech and elm are common.

Diagram **A** shows the main features of this type of ecosystem.

Rain gives water for growth

Sun provides warmth for growth and light for photosynthesis

Leaves trap energy from the sun through photosynthesis and release oxygen to the air

Tawny owls prey on smaller birds and animals such as field mice

An oak tree may contain 270 species of insect

Birds feed mainly on insects and seeds

Carbon dioxide in air combines with water and mineral nutrients to make sugars and starches

A

Animals add nutrients to the soil through their droppings

Rabbits, field mice, hedgehogs and squirrels eat nuts and berries

Weasels eat mice, voles, young rabbits and birds

Grasses and wild flowers grow in the damp shade under the tree

Soil contains minerals (rock particles), nutrients (dead organic matter) and water

Fungi and bacteria feed on dead organic matter and decompose it

Worms and insects eat dead plants and break up the soil

Roots hold the soil together and prevent soil erosion

Weathered rock releases minerals into the soil

Energy flows

Within the woodland area, trees, plants, animals and elements of the physical environment such as the weather and soils are all linked by **flows of energy**. The main source of energy is sunlight. The sun's energy is taken in by the green leaves of plants through the process of **photosynthesis**. Animals then eat the plants. These animals in turn may be eaten by other animals. This process of energy transfer through the ecosystem is called the **food chain**.

B

A woodland ecosystem

Sun

Air

Rain

Soil
Water
Rock

TOP CARNIVORES — Death

Eaten by

CARNIVORES (meat eaters) — Death

Eaten by

HERBIVORES (plant eaters) — Death

Eaten by

PRODUCERS — Death

DECOMPOSERS

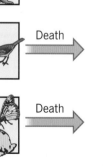
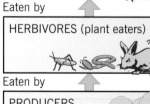

Nutrient recycling

All animals and plants require nutrients. These come from the weathering of rocks and are taken up by the roots of plants. They are then transferred to animals through the food chain. Eventually, when the plants or animals die, they rot away and decompose due to the action of fungi and bacteria. In this way the nutrients return to the soil and may be used again. Without this recycling process, the ecosystem would quickly run out of mineral nutrients (diagram **C**).

Human influences

Ecosystems are made up of several components which depend upon and affect each other. The interdependence means that changes in inputs or outputs are likely to have a knock-on effect through the system. The people of Wylam are concerned that developers might want to cut down trees to build a housing estate. This would obviously have a visual impact but would also destroy the habitat of numerous plants and animals and affect other elements of the ecosystem. Air and water pollution, the use of fertiliser, litter and even the widening of footpaths through the woods, can similarly cause severe and irreversible damage to the delicately balanced woodland ecosystem.

C Nutrient recycling

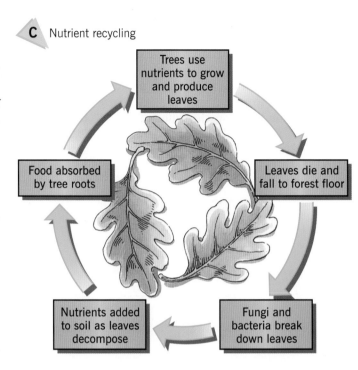

Trees use nutrients to grow and produce leaves → Leaves die and fall to forest floor → Fungi and bacteria break down leaves → Nutrients added to soil as leaves decompose → Food absorbed by tree roots → Trees use nutrients to grow and produce leaves

Activities

1 Explain the following terms.
   • *Energy flows* • *Photosynthesis*
   • *Food chain* • *Nutrient recycling*
   • *Decomposition*

2 Make a larger copy of diagram **D** and add two examples to each box from the following list.
   • *fungi* • *weasel* • *rain* • *oak tree* • *sun*
   • *rabbit* • *insect* • *flower* • *tawny owl*
   • *bacteria*

D

| Non-living environment e.g. | → | Producers e.g. | → | Herbivores e.g. | → | Carnivores e.g. | → | Decomposers e.g. |

E

SAVE OUR WOODS

3 Look at diagram **C** which shows how nutrients are recycled through leaf growth and decomposition. Draw a similar diagram to include animal life as well as plants, in the recycling process.

4 **a)** Design a poster to show the effects that tree felling would have on a woodland ecosystem. Give examples from both the living and non-living environment. Figure **E** could be your poster centrepiece.
   b) Describe two other ways that people may affect the woodland ecosystem.

Summary

Woodland environments are common in many areas of Britain and illustrate the features of complex ecosystems. Ecosystems depend on energy flows and nutrient recycling. They may be altered and damaged by human influences.

What are the characteristics of the tropical rainforests?

Tropical rainforest is the natural vegetation of places that have an equatorial climate (page 92). It provides the most luxuriant vegetation found on Earth (photos **B** and **C**). Over one-third of the world's trees grow here. There are thousands of different species with many yet to be identified and studied. As with all types of natural vegetation, the trees in the rainforest have had to **adapt** to the local environment. This means that they have had to adjust to growing in a climate that has constantly high temperatures

and heavy rainfall, and an all-year-round growing season. Some of the ways by which the trees have adapted to the climate are shown in diagram **A**. The vegetation has also had to adapt to other local conditions such as soils, flooding and competition from other plants. Different plants also survive in their own **microclimate**, perhaps as an emergent needing to reach the sunlight, perhaps as a shade-lover on the forest floor. Each plant plays an important role in the forest ecosystem.

A

How vegetation has adapted to the equatorial climate

- The trees can grow to over 40 metres in the effort to get sunlight

- The forest has an **evergreen** appearance due to the continuous growing season. This means that trees can shed leaves at any time, but always look green and in leaf

- The leaves have drip tips to shed the heavy rainfall

- Tree trunks are straight and branchless in their lower parts in their efforts to grow tall

- Lianas, which are vine-like plants, use large trees as a support to climb up to the canopy

- The forest floor is dark and damp. There is little undergrowth because the sunlight cannot reach ground level

- Dense undergrowth develops near rivers or in forest clearings where sunlight can penetrate

- Rivers flood for several months each year

- Fallen leaves soon rot in the hot, wet climate

- Large buttress roots stand above the ground to give support to the trees

Height of trees, which grow in **three layers**

Tallest trees, called **emergents**

40 m

30 m

CANOPY

20 m

UNDER-CANOPY

Lianas

10 m

Buttress roots

SHRUB LAYER

Ground level

C Ground view of the rainforest showing layers of vegetation

Until recently few parts of the rainforest had been affected by human activity. Where it had, it was usually by groups of people clearing just enough land on which to grow crops for their small community. Often, in places like the Amazon forest, the land rapidly became infertile and the people had to move and make a new clearing. This method of farming, known as shifting cultivation, allowed the forest to re-establish itself. Re-established areas often have a thicker undergrowth as the initial clearance allowed more sunlight to reach the forest floor. More recently, human activity has led to vast areas of the rainforest being totally destroyed, a process known as **deforestation** (pages 118–119). As large trees are destroyed, the habitat for other plants and wildlife is altered. Plants that cannot adapt quickly to the changed environment may die out.

Activities

1. Match up each of the eight descriptive points of the tropical rainforest in the list below with its correct number on diagram **D**.
   - *buttress roots* • *main canopy* • *lianas*
   - *under-canopy* • *branchless trunks*
   - *little undergrowth* • *emergents* • *shrub layer*

2. **a)** Why does rainforest vegetation grow so quickly?
   b) Why do some trees grow so tall?
   c) Why are buttress roots needed?
   d) Why is there so little undergrowth away from rivers?

3. **a)** Why did earlier human activity have little effect upon the tropical rainforest vegetation?
   b) Why are present-day human actions having a far greater effect?

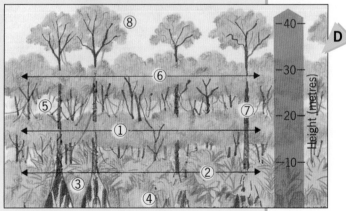

Summary

The natural vegetation of the tropical rainforest has had to adapt to the hot, wet equatorial climate. It is now being increasingly affected by human activity.

What are the characteristics of the savanna grassland vegetation?

The savanna grassland, which often includes scattered trees (photos **B** and **C**), is the natural vegetation of places with a tropical continental climate (page 93). The vegetation forms a transition between the tropical rainforests and the hot deserts (map **A**, page 88). As with all types of natural vegetation, the plants growing here have had to **adapt** to the local environment. This means that they have had to adjust to a very warm climate which has a pronounced wet season, though the rainfall is often unreliable (page 96), followed by a very long dry season. Some of the ways in which the grass and trees have adapted to the climate are shown in diagram **A**. Apart from the major problem of water supply caused by the seasonal drought, the vegetation has also had to adapt to other local conditions such as soils, relief, and competition from other plants.

A

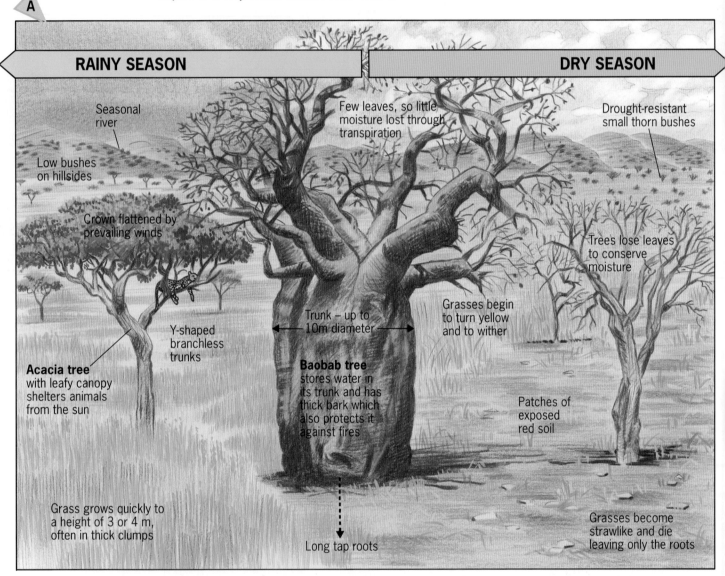

RAINY SEASON

DRY SEASON

Seasonal river

Low bushes on hillsides

Crown flattened by prevailing winds

Few leaves, so little moisture lost through transpiration

Drought-resistant small thorn bushes

Trees lose leaves to conserve moisture

Acacia tree with leafy canopy shelters animals from the sun

Y-shaped branchless trunks

Trunk – up to 10m diameter

Grasses begin to turn yellow and to wither

Baobab tree stores water in its trunk and has thick bark which also protects it against fires

Patches of exposed red soil

Grass grows quickly to a height of 3 or 4 m, often in thick clumps

Long tap roots

Grasses become strawlike and die leaving only the roots

It has been suggested that the savanna grasses are not the natural vegetation of tropical continental climates. Rather they are the result of fires started either naturally or by human activity. Fires can result from either:
- lightning associated with convectional thunderstorms in the rainy season or
- cattle herders burning off the old grass during the dry season to encourage new shoots to sprout when the rains come.

The thick bark of the baobab tree (photo **B**) acts as a protection against fires.

B

Baobab tree

Recently many parts of the savanna grasslands, especially in Africa, which are near to the desert margins have suffered from drought and **desertification** (pages 126–127). The unreliable rainfall has caused vegetation to die. With insufficient grass for the large herds of wild and domestic grazing animals, the area suffers from **overgrazing** (pages 126–127). As the human population increases, former nomadic tribes, like the Maasai in Kenya, find their traditional grazing grounds reduced in size as the land is settled permanently or is used to grow crops. This adds to overgrazing in the areas to which they are restricted. The increase in the human population has also meant that more trees and shrubs have to be cut for fuelwood or to create extra land for crops. With fewer trees and less grass to protect the land from the weather, the soil may be washed away during the rainy season or blown away during the dry season.

C

Acacia trees on the savanna grassland

Activities

1 Photo **C** was taken during the rainy season.
   a) Describe the appearance of the natural vegetation of the savanna grasslands during the rainy season.
   b) What differences will there be in the appearance of the natural vegetation during the dry season?
   c) Draw a baobab tree similar to the one shown in photo **B**. Add at least four labels to show how the tree has adapted to the climate.

2 Diagram **D** shows how the natural vegetation of parts of the savanna grasslands is being changed.
   a) Name three changes that are natural (not the result of human activity).
   b) Name four changes that are the result of human activity.
   c) Describe how these changes are altering the natural vegetation of the savanna grasslands.

Summary

The natural vegetation of the savanna grassland has had to adapt to a warm climate which has a wet and a dry season. It is being increasingly affected by human activity.

D

Drought for several years

Thunderstorm

Maasai herders

Collecting fuelwood

New houses

Herds of domestic cattle and goats

Large herds of wildebeeste, zebra and other wild herbivores

Newly planted maize

What are the characteristics of coniferous forests?

Coniferous forest is the natural vegetation of places that have a cold climate (page 95). In the northern hemisphere they stretch in a great belt from Scandinavia through Siberia and across the Bering Straits into Alaska and Canada. As with all types of natural vegetation, the trees growing here have had to adapt to the local environment. This means that they have had to adjust to a harsh climate with short cool summers and long, very cold winters. Precipitation is low and falls mainly in the summer. Snow is common in winter. Some of the ways in which the trees have adapted to the climate are shown in diagram **A**.

Compared with other forests, the coniferous forests have little variety and contain relatively few plant and animal species. This is largely due to the climate, as few species can adapt to the freezing temperatures and short growing season. The most common trees are pine, fir and spruce. These are evergreens and often grow very tall, perhaps up to 30 or 40 metres. Evergreens have an advantage over other trees as they can begin to grow as soon as temperatures rise above 6°C. This gives them a head start over other species which have to use precious time growing new leaves every spring.

Most coniferous forests grow on a type of soil known as **podsol**, shown in diagram **B**. The ground surface is covered with a thick layer of pine needles as very few bacteria, which help to rot dead plants, can live in these cold conditions. Under the needles there is a pale grey sandy layer. This is caused by the rain moving down through the soil and washing out all the coloured minerals. Lower down there are dark layers where these minerals are deposited. The soil is very acid and few worms or micro-organisms can live in it.

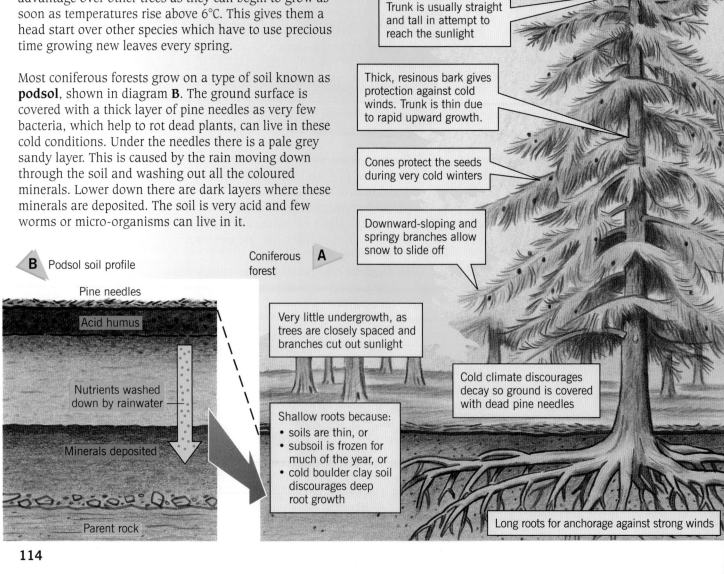

Spruce can grow 30 metres high

Whirls

Distance between whirls indicates 1 year's growth

Evergreen – no need to renew leaves for the short growing season

Compact conical shape gives stability in the wind

Needles to reduce moisture loss

Trunk is usually straight and tall in attempt to reach the sunlight

Thick, resinous bark gives protection against cold winds. Trunk is thin due to rapid upward growth.

Cones protect the seeds during very cold winters

Downward-sloping and springy branches allow snow to slide off

B Podsol soil profile

A Coniferous forest

Pine needles

Acid humus

Nutrients washed down by rainwater

Minerals deposited

Parent rock

Very little undergrowth, as trees are closely spaced and branches cut out sunlight

Cold climate discourages decay so ground is covered with dead pine needles

Shallow roots because:
• soils are thin, or
• subsoil is frozen for much of the year, or
• cold boulder clay soil discourages deep root growth

Long roots for anchorage against strong winds

114

C Coniferous forest in the Rocky Mountains, Canada

Few plants or animals have been able to adapt to the harsh conditions found here. Beneath the canopy of forest trees there is a layer of low-growing plants such as bilberry, cowberry and coarse grass, while lichens and mosses carpet the forest floor. Cold conditions and a shortage of food make life difficult for animals. There are just a few birds in the woods and, in some areas, wolves, deer, bears and beavers can be found.

The coniferous forests are the world's main source of softwood timber. In Canada and Scandinavia, exports of timber products are high and forests are largely managed on a **sustainable** basis (pages 120–121). Once felled and the main branches cut off, some of the logs are floated downriver to sawmills. Here the logs are cut into different shapes and sizes for export. Most of the timber goes to pulp mills where it is made into paper.

D

Activities

1 Match the following beginnings to the correct endings to give six ways that coniferous trees are adapted to the climate.

Characteristics
- Conical shape:
- Downward sloping branches:
- Needle leaves:
- Thick bark:
- Evergreen:
- Fir cones:

Adaptations
- reduce loss of moisture.
- protect seeds from winter cold.
- flexible and bend in strong winds.
- enables summer growth to begin quickly.
- snow slides off more quickly.
- protects tree from great winter cold.

2 a) Make a larger copy of drawing **D**.
  b) Label the drawing with your completed statements from Activity **1**. Give your drawing a title.

3 Why do coniferous forests have relatively few species of plants and animals?

4 Few people live in coniferous forests and the ecosystem is little affected by human activity. Suggest reasons for this.

Summary The natural vegetation of the coniferous forests has had to adapt to the cold, dry conditions. It is relatively little affected by human activity.

What is acid rain?

Acid rain is a term used to describe rainfall that has a higher than normal acid level. All rainfall is naturally slightly acidic, but as map **A** shows, many places around the world are now recording pH values (the scale on which acidity is measured) that show acid levels to be well above average.

Acid rain is a type of air pollution and is mainly caused by power stations and industries burning **fossil fuels** which give off sulphur dioxide and nitrogen oxide. Nitrous oxide from car exhausts also adds to the problem. Diagram **B** shows how these chemicals are carried up into the atmosphere and spread over wide areas by prevailing winds. Some are deposited directly onto the Earth's surface as dry deposits. The majority are converted into acids which fall as acid rain. Acid rain can be very damaging to the environment and can cause serious harm to forests, soils, lakes, rivers, and even buildings (photo **C** page 52).

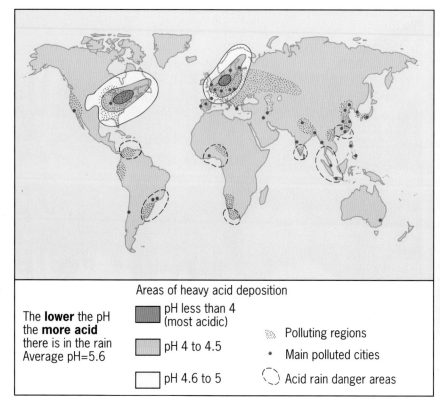

The **lower** the pH the **more acid** there is in the rain Average pH=5.6

Areas of heavy acid deposition

▨ pH less than 4 (most acidic)

▨ pH 4 to 4.5

☐ pH 4.6 to 5

▨ Polluting regions

• Main polluted cities

⌒ Acid rain danger areas

Global variations in acid rain **A**

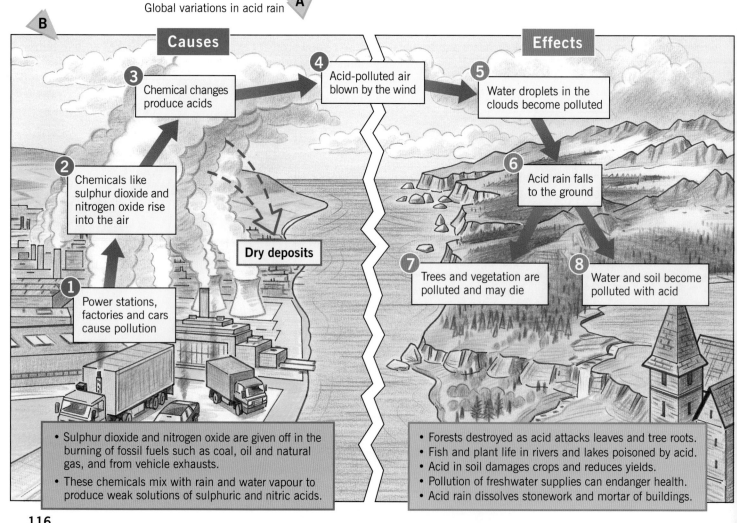

B

Causes

3 Chemical changes produce acids

4 Acid-polluted air blown by the wind

2 Chemicals like sulphur dioxide and nitrogen oxide rise into the air

Dry deposits

1 Power stations, factories and cars cause pollution

Effects

5 Water droplets in the clouds become polluted

6 Acid rain falls to the ground

7 Trees and vegetation are polluted and may die

8 Water and soil become polluted with acid

• Sulphur dioxide and nitrogen oxide are given off in the burning of fossil fuels such as coal, oil and natural gas, and from vehicle exhausts.
• These chemicals mix with rain and water vapour to produce weak solutions of sulphuric and nitric acids.

• Forests destroyed as acid attacks leaves and tree roots.
• Fish and plant life in rivers and lakes poisoned by acid.
• Acid in soil damages crops and reduces yields.
• Pollution of freshwater supplies can endanger health.
• Acid rain dissolves stonework and mortar of buildings.

Acid rain is an international problem because it is blown across oceans and continents ignoring political boundaries. As map **A** shows, many countries produce acid rain. Some, like Britain, Germany and the USA, 'export' it while others, such as Norway, Sweden and Canada, 'import' it. Any solution to the problem requires international co-operation. This is not easy, as the countries most affected by acid rain are often not the same ones as are responsible for causing it.

There are many different ways of tackling the acid rain problem. One short-term answer is to spray affected trees and add powdered limestone to polluted lakes and rivers. In the long term, however, the only really effective solution is to reduce the emissions of the offending gases. This can be achieved in a variety of ways:

* burn less fossil fuels by conserving energy
* use non-fossil fuels such as nuclear energy, or power from the wind or sun
* remove sulphur from coal before burning
* use new, more efficient boilers in power stations
* remove sulphur from waste gases
* reduce car emissions by using unleaded petrol.

Unfortunately, all of these methods are expensive and have been difficult to implement. A European Union (EU) directive in 1988 called for a 71 per cent reduction in sulphur dioxide levels by 2005. Despite considerable effort, no country is yet near to achieving that target. So far, Britain's emissions are down by just 42 per cent.

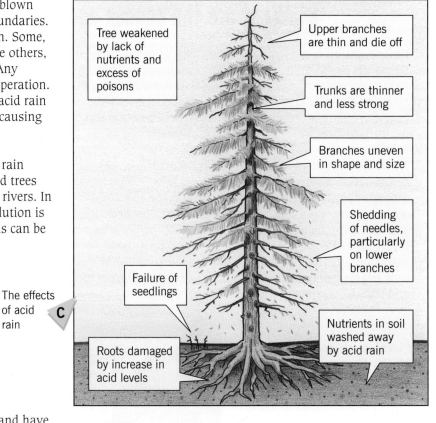

The effects of acid rain **C**

Upper branches are thin and die off

Tree weakened by lack of nutrients and excess of poisons

Trunks are thinner and less strong

Branches uneven in shape and size

Shedding of needles, particularly on lower branches

Failure of seedlings

Nutrients in soil washed away by acid rain

Roots damaged by increase in acid levels

We have the technology to reduce acid rain and to manage our environment.

It will all cost a lot of money which few people or countries have. We will have to pay much more for our energy and cars.

D

Activities

1 **a)** What is *acid rain*?
  b) With the help of a diagram, explain the causes of acid rain.

2 Describe the problems caused by acid rain. List them under the headings: **Vegetation**, **Water**, **Others**.

3 Look at map **A** showing areas affected by acid rain.
  a) Describe the location of the two regions most affected by acid rain.
  b) Why do countries like Norway, Sweden and Canada want a decrease in the emissions of sulphur dioxide and nitrogen oxide?
  c) Why are countries like Britain, Germany and the USA reluctant to reduce these emissions?

4 **a)** Explain how using energy more efficiently could reduce acid rain.
  b) Why is international co-operation necessary in dealing with acid rain?

5 Two views are expressed in drawing **D**. Which of these views is likely to be supported by:
  a) an industrialist
  b) a coalminer
  c) a conservationist?
Give reasons, in each case, for your answer.

Summary

Acid rain is a major form of atmospheric pollution and can cause great damage to the environment. International co-operation is needed to reduce the problem.

117

What are the causes and effects of deforestation?

Deforestation is the felling and clearance of forest land. Deforestation began in the Mediterranean lands and north-west Europe many centuries ago. Today it is mainly taking place in those less economically developed countries that have tropical rainforests as their natural vegetation (pages 92, 110–111).

The rapid clearances in recent years have made deforestation a key global environmental issue. Some estimates suggest that one-fifth of the Amazon rainforest – an area the size of Great Britain – was cleared between 1960 and 1990. The fastest clearances are still taking place in Brazil, and satellite photographs show that up to 5000 individual fires may be burning at the same time. These fires release carbon dioxide which pollutes the atmosphere and reduces visibility.

Deforestation can be the result of several types of activity, as shown in drawing **A**. These are mainly:
- **government policies** aimed at helping local people and developing the region's economic potential
- **transnational companies** that wish to make money from the natural resources of the region
- **local people** who need more land on which to grow crops if they are to be able to feed themselves.

To continue deforestation at the present rate can best be described as tragic. The rainforest is a **fragile environment**, which once destroyed is unlikely to be replaced. Almost a million people live in the rainforest and their way of life would end. The forest ecosystem would also change and the plants and animals living there could be lost forever. As drawing **C** shows, it could also damage our world in many other ways.

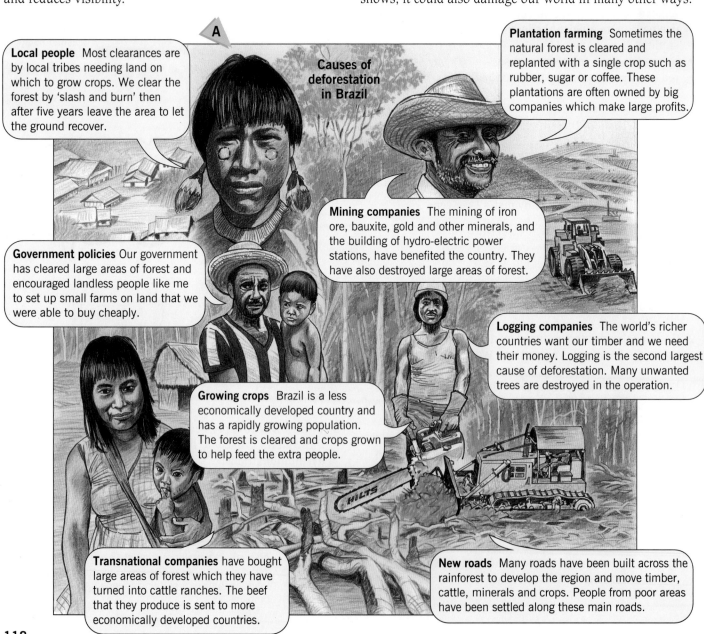

A

Causes of deforestation in Brazil

Local people Most clearances are by local tribes needing land on which to grow crops. We clear the forest by 'slash and burn' then after five years leave the area to let the ground recover.

Plantation farming Sometimes the natural forest is cleared and replanted with a single crop such as rubber, sugar or coffee. These plantations are often owned by big companies which make large profits.

Government policies Our government has cleared large areas of forest and encouraged landless people like me to set up small farms on land that we were able to buy cheaply.

Mining companies The mining of iron ore, bauxite, gold and other minerals, and the building of hydro-electric power stations, have benefited the country. They have also destroyed large areas of forest.

Logging companies The world's richer countries want our timber and we need their money. Logging is the second largest cause of deforestation. Many unwanted trees are destroyed in the operation.

Growing crops Brazil is a less economically developed country and has a rapidly growing population. The forest is cleared and crops grown to help feed the extra people.

Transnational companies have bought large areas of forest which they have turned into cattle ranches. The beef that they produce is sent to more economically developed countries.

New roads Many roads have been built across the rainforest to develop the region and move timber, cattle, minerals and crops. People from poor areas have been settled along these main roads.

Activities

1 Briefly describe eight reasons why the tropical rainforests are being cleared in countries like Brazil. List your reasons under the headings:
   • government policies
   • transnational companies
   • local people.

2 Diagram **B** suggests eight consequences of deforestation in Brazil. Describe how each one affects the environment and the local community.

3 Deforestation is taking place so that rainforests may be economically developed. Which of the following do you consider to be the most important?
   • Economic development
   • Rainforest conservation
   Give reasons for your answer.

B

Loss of homes and land belonging to the Amazon Indians

Decrease in soil fertility

Loss of wildlife habitat

Forests, once felled, are burnt. This reduces oxygen given out by the trees

Loss of plants for medicinal purposes

Hardwoods becoming endangered

Increase in soil erosion

Land spoilt by mining activities or flooded for hydro-electricity

Summary Despite its luxuriant vegetation, the rainforest is a fragile environment. There are fears that if deforestation for economic development continues at the same rate, the rainforest will soon be totally destroyed.

C

Effects of deforestation in Brazil

Ecosystem The loss of forest has changed the ecosystem. Large numbers of plants and animals have already been destroyed. Some of the plants once used to make medicines have disappeared.

Soil fertility With the increase of rain reaching the ground, many nutrients have been washed out of the soil. This has made the soil less fertile. Some farms have had to be abandoned.

Soil erosion The clearing of trees means that there is no canopy to protect the soil from heavy rain, or plant roots to bind it together. The result is that much of the soil has been washed away.

Amerindians There has been a huge reduction of people from local tribes living in the area. Their numbers are down from around 6 million to the present number of just 200 000.

Traditional culture Many Amerindians have been forced off their land to make way for new developments. In many cases their traditional culture and way of life have been destroyed.

World climate change The burning of forest releases carbon dioxide into the air. This has caused global warming and an increase in world temperatures.

Decrease in hardwood Only about one tree in twenty is of economic value to timber companies. Other trees are destroyed and some, like mahogany and rosewood, are becoming endangered.

Can rainforest development be sustainable?

Deforestation is of international concern. The rainforests are a vital natural resource which provide a wide variety of products needed by all countries. Their destruction is not only losing these resources, it is also changing world climates.

- The burning of the rainforests, and the subsequent release of carbon dioxide (a greenhouse gas), is a major cause of global warming (pages 104–105).
- A greatly reduced number of trees will mean a decrease in evapotranspiration. With less water vapour in the air, rainfall is expected to decrease (diagram **A**). Some scientists believe this could eventually turn places like the Amazon Basin into desert.
- Trees take in carbon dioxide (CO_2) and, through photosynthesis, give out oxygen (diagram **A**). It is estimated that nearly one-half of the world's supply of oxygen comes from trees in the Amazon Basin. It takes one large tree to provide enough oxygen for two people for one day, and 150 large trees to absorb the carbon dioxide produced by one small car.

A

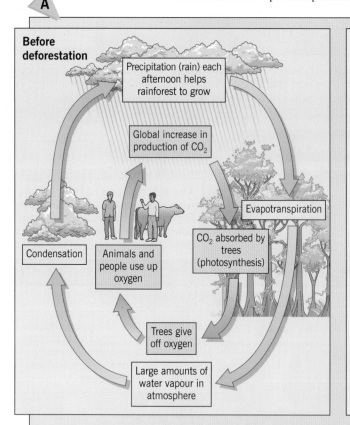

Before deforestation

Precipitation (rain) each afternoon helps rainforest to grow

Global increase in production of CO_2

Evapotranspiration

CO_2 absorbed by trees (photosynthesis)

Condensation

Animals and people use up oxygen

Trees give off oxygen

Large amounts of water vapour in atmosphere

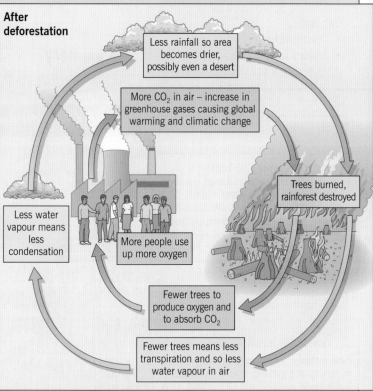

After deforestation

Less rainfall so area becomes drier, possibly even a desert

More CO_2 in air – increase in greenhouse gases causing global warming and climatic change

Trees burned, rainforest destroyed

Less water vapour means less condensation

More people use up more oxygen

Fewer trees to produce oxygen and to absorb CO_2

Fewer trees means less transpiration and so less water vapour in air

There are two extreme conflicts of interest in the rainforest. On one hand there are those groups of people who wish to use the forest to make a quick profit. On the other there are those who wish to protect the forest and leave it exactly as it is. Caught in the middle are the people who actually live there. To them the forest is their home. They need to preserve the forest as well as being able to use its resources if they are to find work and improve their standard of living. The solution is to manage the forest in a **sustainable** way, using the resources carefully. Trees are a renewable resource, but only if they are used and managed carefully. At present nobody is taking responsibility for the rainforest. The forest is no longer managed in the traditional way by small-scale shifting cultivators, nor by government agencies or commercial companies. This leaves the forest open to illegal logging, mining and conversion to other land uses. Such an approach benefits nobody.

Diagram **B** suggests several proposed methods of managing the rainforest. Extract **C** quotes an example of an attempt to manage the rainforest in a sustainable way.

Emilio Sanchoma lives in the Amazon region of Peru where rapid deforestation is taking place. He is involved in a scheme aimed at managing the forest sustainably. A similar sustainable logging project in Mindanao in the Philippines is shown in photo **D**.

B
- Create more National Parks and Forest Reserves similar to the Korup Park in the Cameroons (Africa).
- Only give logging grants to transnational companies on the condition that they replant an equal number of trees to those they fell.
- Reduce international trade in such endangered hardwoods as mahogany.
- Limit the mass burning of trees to reduce global warming and climate change.

C

I was born in the rainforest and have lived here all my life. Like all rainforest people, I know how to harvest the forest sustainably to feed myself, my family and my tribe. I grow crops like rice to sell and other crops and fruit to eat. I also keep poultry and other animals. We use the plants of the rainforest for medicines.

As rainforest people we are worried about our own way of life and about the future of the planet. We have been trying to find better ways of using the forest. In our area, the Palcazú Valley, the government has given us title to the land. We have set up the Yanesha Forestry Co-operative to log the forest on a sustainable basis.

We clear-cut a strip of rainforest 20–50 metres wide, the same size as would be cleared if a large tree were blown down. It is wide enough to enable sunlight to penetrate the canopy and narrow enough for the plants to reseed themselves from the surrounding forest.

D

The trees are felled with chainsaws and then taken out by oxen onto a main logging road where they are loaded onto the co-op truck. The oxen don't damage the soil and surrounding vegetation and are less expensive than heavy machinery. We use every stick in the forest. Logs greater than 30 cm in diameter are sawn into lumber. Smaller logs between 5 and 30 cm are used for construction in our area. Smaller trees and odd-shaped scraps are made into charcoal and sold locally. We get 250 cubic metres of wood per hectare under this system, compared with a typical logger's yield of 3–5 cubic metres.

Source: Global Environment, *BBC/Longman*

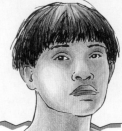

Activities

1 Deforestation is an international problem. How is it affecting:
   a) global temperatures
   b) rainfall in equatorial areas
   c) the balance of oxygen and carbon dioxide in the atmosphere?

2 a) What is meant by the term 'to manage the rainforests in a sustainable way'?
   b) Describe one scheme that is trying to 'manage the rainforest in a sustainable way'.
   c) Why is it difficult to get international agreements that would enable the forests 'to be managed in a sustainable way'?

3 Divide your class into four groups. Each group should represent the views of one of the following:
   • the Brazilian government
   • an international logging company
   • a group of forest Indians
   • a conservation group.
Each group should take their turn to put forward their case for or against the development of the rainforest.

Summary There is a need to balance the protection and the economic development of the rainforest. Sustainable development can only come about through international co-operation and management.

What causes soil erosion?

The first stage in the formation of soil is when physical and chemical weathering break down the underlying rock of a place into small particles. The second stage includes the addition of water, air (including oxygen), humus (material from decayed plants and animals) and living organisms (bacteria, worms). There are many different types of soil. They can vary in **depth**, **colour**, **texture** and **organic content** as well as in drainage, nutrients (humus) and acidity. Yet there is one thing that all soils have in common – the long period of time needed for them to form. Estimates suggest that, in Britain, it takes about 400 years for 1cm of soil to form, and between 3000 and 12 000 years for soil to become deep enough for farming. In contrast **soil erosion** can be very rapid. Several centimetres can disappear within minutes during a severe storm, while human mismanagement can cause soil to lose its fertility within a few years. Soil is a renewable source but, like water and trees, it needs careful management.

A

Himalayas (worst for water erosion)

Central Asia

Middle East

Mongolia

USA (Dust Bowl)

North Africa

Mexico

Sahel (worst area for wind erosion)

North China (most fragile soils)

East Africa

High Andes

North-east Brazil

Key Soil erosion
Mainly by wind
Mainly by water in mountainous areas

Botswana and Namibia

Australia

Soil erosion is a process by which soil is removed by the wind and running water. Soil erosion is not a major problem in places where there is permanent cover of grass or forest. It does become a problem when human activity removes this protective vegetation cover either to plough the land or through deforestation. If soil is left exposed to the weather it can be washed away during times of heavy rainfall or blown away when it dries out during times of drought. Map **A** shows those places most vulnerable to soil erosion.

- Where the land is mountainous with very steep slopes: one quarter of a million tonnes of topsoil are washed off the deforested mountain slopes of Nepal and northern India each year (extract **B**) only to be deposited into the Bay of Bengal.
- Where the climate includes a pronounced dry season and where annual rainfall totals are unreliable (page 96). When adverse human activity takes places in areas of unreliable rainfall, such as the Sahel countries in Africa, the resultant loss of soil and vegetation is known as **desertification** (page 126).

B

Much of the land has been deforested. The lower hillsides have been left almost bare of vegetation, and goats are busy eating what is left. It is not simply that the trees that once grew here cannot grow again, but that without its protective tree cover the soil itself is washed away. When the heavy monsoon rains come, both rain and earth are lost. Now, with the trees gone, the water runs quickly off the land, carving out new erosion channels and creating bare rock above and floods and landslides below. Great ravines are rapidly gouged out creating a man-made wilderness.

Adapted liberally from Nigel Nicolson, Himalayas, Time-Life Books

Diagram **C** shows the main causes of soil erosion. Most result from deforestation, overgrazing and overcultivation.

C

Steep slopes
e.g. mountainous areas in various parts of the world

Areas with unreliable rainfall
e.g. tropical continental/savanna grasslands

Tourism
Walkers enlarge footpaths. New runs for skiers.

Deforestation
a) Increases surface run-off and throughflow
b) Decreases interception and evapotranspiration

Silt (soil) blocks river. Increases flood risk and erosion of banks.

Overgrazing – Rearing too many animals in relation to amount of grass available

Heavy machinary
compacts ground, reducing infiltration

Burning grass
a) By people, to force new growth for grazing
b) By lightning strikes

Autumn ploughing
leaves soil unprotected during winter storms

Removing hedges or shelter belts to meet demand for fuelwood

Overcropping and monoculture
(Growing crops intensively, or a single crop year after year) – crop needed for export (cash) or to feed a growing population. Lack of manure – used as fuel instead of as fertiliser.

Ploughing up and down hill
creates channels down which rain-water can flow. Increases amount and speed of surface run-off.

- Washed downhill by water
- Moves slowly downwards under gravity

EXPOSED SOIL

- Washed away during wet season storms
- Blown away during dry season

Summary

It can take several centuries for soil to form but only a short time for it to be destroyed. Soil erosion, which is greatly accelerated by human activity, is most serious on steep-sided mountains and in places where the rainfall is unreliable.

Activities

1 a) How does soil form?
   b) How long does it take for soil to form?

2 a) Name two mountainous parts of the world where soil erosion is a serious problem.
   b) Give three possible causes of soil erosion in mountainous areas.
   c) Name two parts of the world with an unreliable rainfall.
   d) How can unreliable and seasonal rainfall be a cause of serious soil erosion?
   e) Give three possible causes of soil erosion in places with an unreliable rainfall.

What can be done to prevent or reduce soil erosion?

As the world's population continues to increase, farmers are going to have to produce more food in order to feed the extra numbers. This can only be done if the soil is protected and carefully managed. Evidence suggests that in the year 2000, 20 per cent of land that was arable in 1985 had been lost through erosion, desertification and conversion to non-agricultural uses.

By 2020 the same amount could again disappear. Although the loss is greatest in tropical less economically developed countries, it is by no means limited to those places. It is important that greater attempts are made internationally, similar to those described in case studies **A** to **D**, to reduce erosion and sustain productivity.

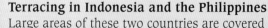

Terracing in Indonesia and the Philippines
Large areas of these two countries are covered in volcanic mountains which have steep slopes and fertile soil. Over 2000 years ago terraces, which resemble giant steps, were first built on many of the hillsides. Each terrace is flat and is fronted by a mud or stone wall known as a 'bund'. The bund traps both rainwater and soil. By allowing rainwater time to infiltrate into the ground, surface run-off and the removal of topsoil is prevented.

Animal welfare in Kenya
Large herds of cattle, goats, sheep and camels have long been considered a source of wealth and prestige in several African countries. Unfortunately quantity, rather than quality, has tended to result in overgrazing. The problem of overgrazing has increased partly because rainfall has become even less reliable and partly because of the rapidly growing population. Intermediate Technology Development Group, a British organisation, is working with local people in several parts of Kenya. In those areas it is helping to train one person from each a village to become a 'wasaidizi' or animal care worker. By recognising and being able to treat basic animal illnesses, the wasaidizi is improving the quality of local herds. As the quality improves there should be less need for so many animals so that, hopefully, overgrazing will be reduced.

Contour ploughing and strip cropping in the USA
Contour farming is ploughing around hillsides rather than up and down the slope. By ploughing parallel to the contours, the furrows trap rainwater and prevent the water from washing soil downhill. Strip cropping is when two or more crops are planted in the same field. Sometimes one crop may grow under the shelter of a taller crop. It is harvested at a different time of year and uses different nutrients from the soil. Often the crops are rotated from year to year.

Stone lines ('magic stones') in Burkina Faso

This project, begun by Oxfam in 1979, uses appropriate technology, local knowledge and local raw materials. It involves villagers, of all ages and both sexes, collecting some of the many stones lying around their village. The stones are laid across the land to stop surface run-off following the all too rare heavy rainstorms. Water and soil are trapped. The water now has the time to infiltrate instead of being lost immediately through surface run-off. The soil soon becomes deep enough for the planting of crops. Erosion is reduced and crop yields have increased by as much as 50 per cent. The only equipment needed is a simple level, developed by Oxfam, to help keep the lines parallel to the contours.

D

Activities

1 Diagram **E** shows several methods aimed at reducing soil erosion.
   a) Briefly describe each method.
   b) Which methods are appropriate to Britain?

E

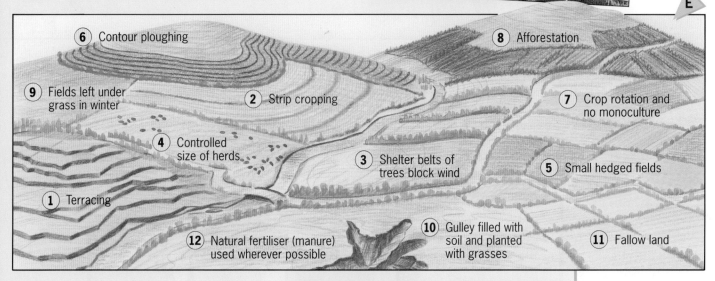

6 Contour ploughing
8 Afforestation
9 Fields left under grass in winter
2 Strip cropping
7 Crop rotation and no monoculture
4 Controlled size of herds
3 Shelter belts of trees block wind
5 Small hedged fields
1 Terracing
12 Natural fertiliser (manure) used wherever possible
10 Gulley filled with soil and planted with grasses
11 Fallow land

2 Figure **F** is a soil conservation poster used in Kenya. It was designed in cartoon form by a local artist who used local examples of afforestation, terracing and controlled grazing.
   a) What are the advantages of providing educational information in this way?
   b) Which methods of soil conservation are likely to be of most value to an economically developing country with unreliable rainfall and a growing population such as Kenya?
   c) Why is it often difficult for an economically developing country to introduce these methods?

F

Tueneze umuhimu wa kuhifadhi udongo wetu
Tell your friends about soil conservation

Summary

Soil is a renewable resource but only if it is managed carefully. All countries must make greater efforts if soil erosion is to be prevented and reduced so that soil productivity can be sustained.

What causes desertification?

'Turning land into desert' is the simplest of several definitions of desertification. Desertification occurs mainly in semi-arid lands which border the world's major deserts. Map **A** is often misinterpreted. It locates places that are **at risk** from desertification, **not** places where desertification has actually occurred. The area at greatest risk is the Sahel, a narrow belt of land extending across Africa and lying to the south of the Sahara Desert.

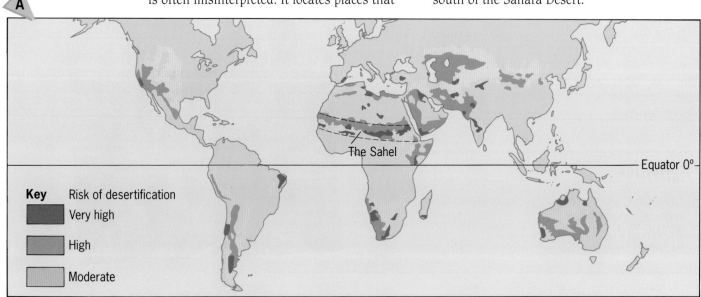

A

The Sahel

Equator 0°

Key   Risk of desertification

Very high

High

Moderate

The causes of desertification are complex. They appear to result from a combination of climatic changes (e.g. decreased rainfall and global warming), and increased human activity and pressure upon the land (e.g. overgrazing, overcultivation and deforestation). How these factors may have contributed to desertification are shown on diagram **B**. During the 1980s it was claimed, and accepted, that the Sahara was advancing southwards at a rate of between 6 and 10 km a year. However, many claims were little more than estimates based upon short-term observations and were made at the height of one of Africa's worst ever recorded droughts. The drought, which began in the early 1970s, followed two wet decades (graph **C**).

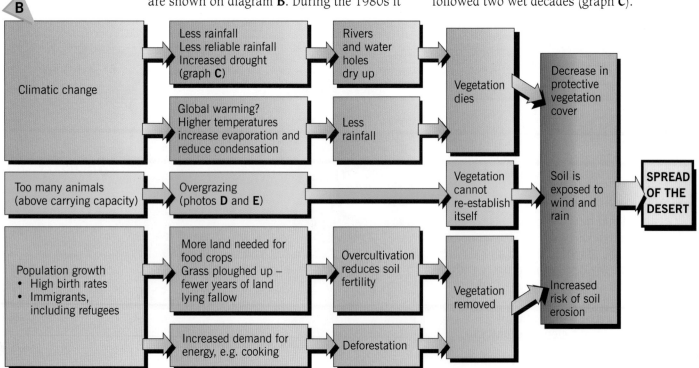

B

Climatic change → Less rainfall / Less reliable rainfall / Increased drought (graph **C**) → Rivers and water holes dry up → Vegetation dies → Decrease in protective vegetation cover

Climatic change → Global warming? / Higher temperatures increase evaporation and reduce condensation → Less rainfall → Vegetation dies

Too many animals (above carrying capacity) → Overgrazing (photos **D** and **E**) → Vegetation cannot re-establish itself → Soil is exposed to wind and rain → **SPREAD OF THE DESERT**

Population growth
• High birth rates
• Immigrants, including refugees
→ More land needed for food crops / Grass ploughed up – fewer years of land lying fallow → Overcultivation reduces soil fertility → Vegetation removed → Increased risk of soil erosion

→ Increased demand for energy, e.g. cooking → Deforestation

During this wetter period farmers began to grow crops on land that had not been cultivated for several centuries, and to crowd larger herds of livestock onto smaller areas of pasture. When this was replaced by a much drier period it looked as if the land was being overcultivated and overgrazed. Within a few years over 100 000 people and millions of animals died. The initial investigations for those deaths and, later, for many more in Ethiopia, Sudan and Somalia, put the blame on desertification.

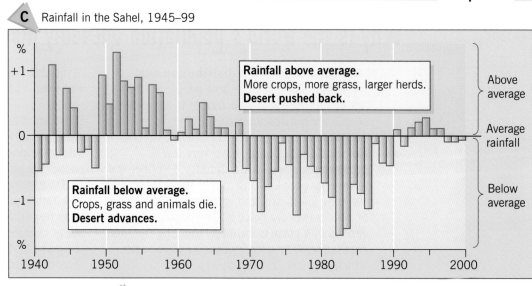

C Rainfall in the Sahel, 1945–99

> **Rainfall above average.**
> More crops, more grass, larger herds.
> **Desert pushed back.**

> **Rainfall below average.**
> Crops, grass and animals die.
> **Desert advances.**

Above average

Average rainfall

Below average

Recently the claim that the Sahara is advancing has been disputed. Evidence, based mainly on satellite images, does show annual changes resulting from variations in rainfall, but no permanent advance. This does not, of course, mean that the risk has disappeared. Hopefully the threat of desertification has increased people's awareness of the semi-arid lands as a fragile environment. Their boundaries are constantly changing as a result of variations in rainfall, and it is difficult to separate natural causes from human activity. (For example, is overgrazing a result of increased drought or increased human activity – photos **D** and **E**?)

Regardless of whether desertification is already a major hazard or whether it is a future risk, one thing is certain: increased desertification is likely, as with all environmental issues, to result from people's misuse of natural resources.

D Goats, sheep and camels drinking at a well, Sudan

E Overgrazed area, Burkina Faso

Activities

1 **a)** What is *desertification*?
   b) With the help of an atlas, name six Sahel countries.
   c) Desertification is blamed upon physical processes and human activity. Explain how the following might have combined to cause desertification:
     i) changes in rainfall and global warming
     ii) overgrazing, overcultivation and deforestation.

2 What recent evidence seems to contradict the claim that desertification is increasing?

Summary

Desertification in semi-arid lands may result from a combination of physical processes and human activity. Recent evidence suggests that, although the risks remain high, the increase in desertification may be less than was previously claimed.

Why is the world's population unevenly distributed?

Map **A** is a **population distribution** map of the world, and describes how people are spread out across the Earth's surface. It is obvious that people are not spread evenly. Some places, like Bangladesh, are very crowded, while others, like the Sahara Desert, have very few inhabitants. **Population density** describes the number of people living in a given area, usually a square kilometre, and is, therefore, a measure of how crowded a place is. The population density of a place is found by dividing the total population of an area by the size of the area in which they live. For example:

$$\frac{\text{Population of the UK}}{\text{Area of the UK (km}^2)} = \frac{58\ 306\ 000}{244\ 880} = 243 \text{ per km}^2$$

Places that are crowded are described as **densely populated** and have a high population density. Places that have few people living there are said to be **sparsely populated** and have a low population density.

There are many reasons for the population patterns shown on map **A**. Some reasons are said to be **positive factors** because they encourage people to live in an area (diagram **B**). Positive factors tend to create high population densities. Other reasons are referred to as **negative factors** and these often discourage people from living in a place (diagram **C**). Negative factors are usually responsible for low population densities. There are generally several reasons why an area is either densely or sparsely populated.

Positive and negative factors can each be divided into physical and human factors. **Physical factors**, which form the natural part of the environment (page 106), include relief, climate, vegetation, soils and natural resources. **Human factors**, which result from people's activities, may be economic, political or social.

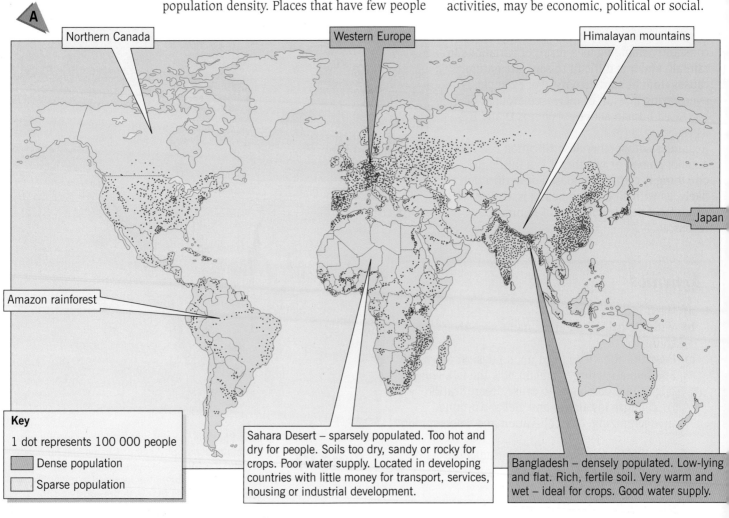

A

Northern Canada

Western Europe

Himalayan mountains

Japan

Amazon rainforest

Sahara Desert – sparsely populated. Too hot and dry for people. Soils too dry, sandy or rocky for crops. Poor water supply. Located in developing countries with little money for transport, services, housing or industrial development.

Bangladesh – densely populated. Low-lying and flat. Rich, fertile soil. Very warm and wet – ideal for crops. Good water supply.

Key

1 dot represents 100 000 people

Dense population

Sparse population

PHYSICAL FACTORS

No **climate extremes** – not too hot, too cold, too dry or too wet

Flat, low-lying **relief**

Natural vegetation is grass

Soils are either deep with humus, deposited by rivers, or found around volcanoes

Natural resources of minerals, energy supplies, water supplies

Positive factors giving areas with high population densities

HUMAN FACTORS

Economic – money and technology to create industry and jobs and to improve transport

Social – better housing and cultural opportunities, health and education services, retirement areas

Political – governments invest money, build new towns, add services, reclaim land

C

PHYSICAL FACTORS

Climate extremes – too hot, too dry, too cold, too wet. Short growing season

Relief may be high, steep and rugged mountains

Natural vegetation is desert or dense forest

Soils are either too thin, too dry, lacking in humus or badly eroded

Lack of **natural resources** – few minerals or energy supplies, lack of water

Negative factors giving areas with low population densities

HUMAN FACTORS

Economic – poor transport, lack of money for industrial development or to create new jobs

Social – poor housing, limited health and education services, few cultural opportunities

Political – lack of government investment, rural depopulation, loss of land

Hadrian's Wall, Northumberland

D

Hong Kong

E

Activities

1 What is the difference between:
   a) *population distribution* and *population density*
   b) *positive factors* and *negative factors*
   c) *physical factors* and *human factors*?

2 Map **A** shows seven parts of the world that have either a high population density or a low population density. Two of these areas have been described on the map. For each of the remaining five areas:
   a) state whether it has a high population density or a low population density
   b) give as many reasons as possible why its population density is high or low.

3 What negative factors and what positive factors have affected the density of population in your local area?

Summary

People are not spread evenly throughout the world. Positive factors encourage settlement and create high population densities. Negative factors discourage settlement and give low population densities.

What are the present and predicted trends in population growth?

Population growth

Graph **A** shows that the world's population increased slowly but steadily until the early nineteenth century. After that time it increased at a much faster rate, a process referred to as a **population explosion**. However, during the last two decades of the twentieth century there were signs that the rapid rate of increase was beginning to slow down. Even so, in 2001 the United Nations (UN) estimated that the world's population was still growing by 140 persons per minute. This means that there an extra 78 million people on Earth each year – more than the present populations of the UK and Canada added together.

In October 1999, the UN claimed that the world's population had reached 6000 million (that is, 6.0 billion). However, the accuracy of this claim needs to be treated with some caution. This is because, for example, census data for most countries is only collected every ten years; many people are unlikely to return census forms or complete them accurately; and groups of people like refugees (page 143) and shanty town dwellers (page 184) are likely to be excluded or missed.

According to the present rate of growth, the UN are now suggesting that the world's population will reach 8.9 billion by 2050. This figure is much lower than the 11.0 billion that the UN were predicting in the mid-1970s (graph **A**).

Population growth has not been even throughout the world. Graph **A** also shows:

* The fastest growth has been in the world's poorer, less economically developed countries. At present 86 per cent of the world's population now live in Asia, Africa and South America (two out of every five global citizens live in China and India).
* There is a very slow growth rate in the world's richer, more economically developed countries. Some countries in north-west Europe even have a zero growth rate, indicating that they no longer have an increase in population.

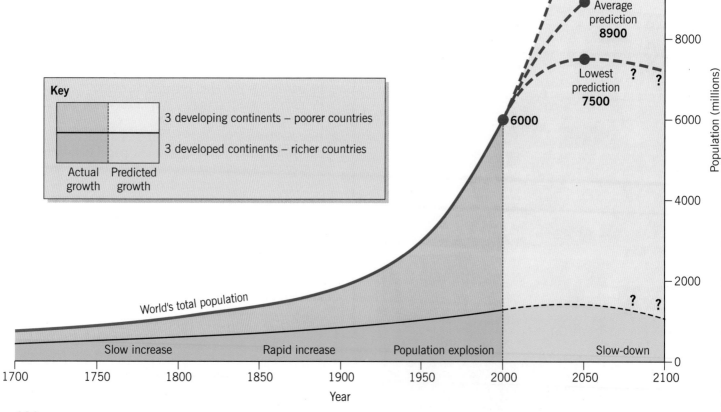

A

Key

3 developing continents – poorer countries

3 developed continents – richer countries

Actual growth | Predicted growth

World's total population

Slow increase | Rapid increase | Population explosion | Slow-down

2050 prediction

Highest prediction **10 700**

Average prediction **8900**

Lowest prediction **7500**

6000

Population (millions)

12 000
10 000
8000
6000
4000
2000
0

1700 1750 1800 1850 1900 1950 2000 2050 2100

Year

Population change

Population change, whether it is in a country or the whole world, depends mainly upon the balance between the birth rate and the death rate. It can also be affected by migration (page 142).

- The **birth rate** is the average number of live births in a year for every 1000 people in the total population.
- The **death rate** is the average number of deaths in a year for every 1000 people in the total population.
- The **natural increase** (or **decrease**) is the difference between the birth rate and the death rate.

If the birth rate is higher than the death rate then the total population will increase (diagram **B**). If the death rate is higher than the birth rate then the total population will decrease (diagram **C**).

Throughout history, birth rates have usually exceeded death rates. Exceptions have occurred:

- during major outbreaks of disease (bubonic plague in the Middle Ages, Aids in present-day southern Africa)
- as a result of wars (the former Yugoslavia, Afghanistan)
- at present in several Western European countries (due to improvements in family planning and female education)
- due to the one-child policy in China (page 138).

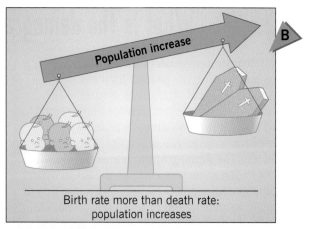

Birth rate more than death rate: population increases

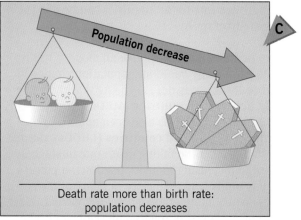

Death rate more than birth rate: population decreases

Activities

1 Refer to diagrams **A** and **D**.
   a) What was the world's population in:
   i) 1700   ii) 2000?
   b) Describe the rate of population growth at K, L, M and N.

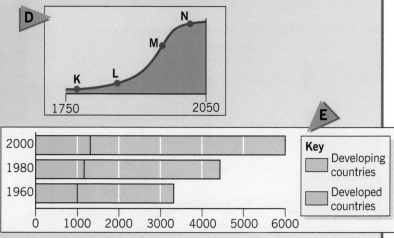

2 Look at graph **E**.
   a) i) name the three developed continents
   ii) name the three developing continents.
   b) Why is the population of developing continents growing more quickly than the population of developed continents?

3 Graph **F** shows birth rates and death rates in England and Wales since 1900.
   a) What is meant by the terms:
   • *birth rate* • *death rate* • *natural increase*?
   b) Which of the letters W, X, Y and Z refers to the natural increase?
   c) According to the graph:
   i) which year had the most rapid population growth
   ii) what was the natural increase in 2000?

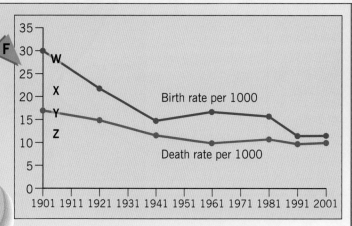

Summary   Population growth, which is more rapid in developing continents, depends mainly upon changes in birth rates and death rates.

What is the demographic transition model?

Birth rates and death rates for all countries change over a period of time, so affecting the natural increase in population. A study of these changes in several industrialised countries of Western Europe and in North America led to the suggestion that each country had experienced a common sequence of change. This resulted in a model, known as the **demographic transition model**, being produced to show the sequence. (A **model** is when difficult or complex real-world situations are simplified to make them easier to understand or to explain.) The demographic transition model shows that the population (or demographic) growth rate for all countries can be divided into four stages (diagram **A**).

The demographic transition model, like all models, has its limitations. As it was only based on population change in industrialised countries, it is not surprising that the **more economically developed countries (MEDCs)** have reached stage 4. However, the model has also been applied to the **less economically developed countries (LEDCs)**, despite the model assuming that the falling death rate in stage 2 was a response to increasing industrialisation – a process now accepted as unlikely to take place in many of the poorest LEDCs. While many of the more developed LEDCs have reached stage 3, most of the least developed LEDCs remain at stage 2.

Meanwhile, several MEDCs appear to be entering a fifth stage, a stage not predicted by the model. This is where the death rate is exceeding the birth rate which will give, in time, a natural decrease in population growth.

B

Crowded city scene, India

A

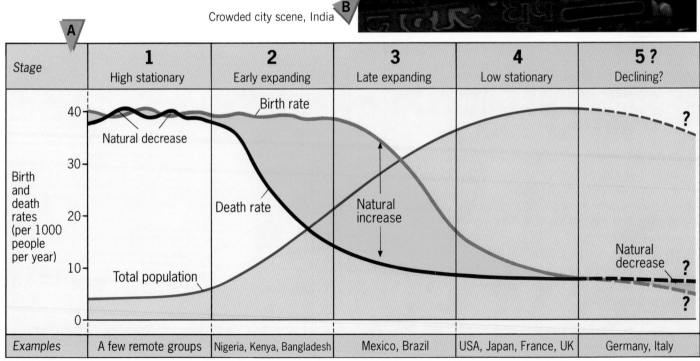

Stage	**1** High stationary	**2** Early expanding	**3** Late expanding	**4** Low stationary	**5 ?** Declining?
Examples	A few remote groups	Nigeria, Kenya, Bangladesh	Mexico, Brazil	USA, Japan, France, UK	Germany, Italy

Birth and death rates (per 1000 people per year)

Natural decrease

Birth rate

Death rate

Natural increase

Natural decrease

Total population

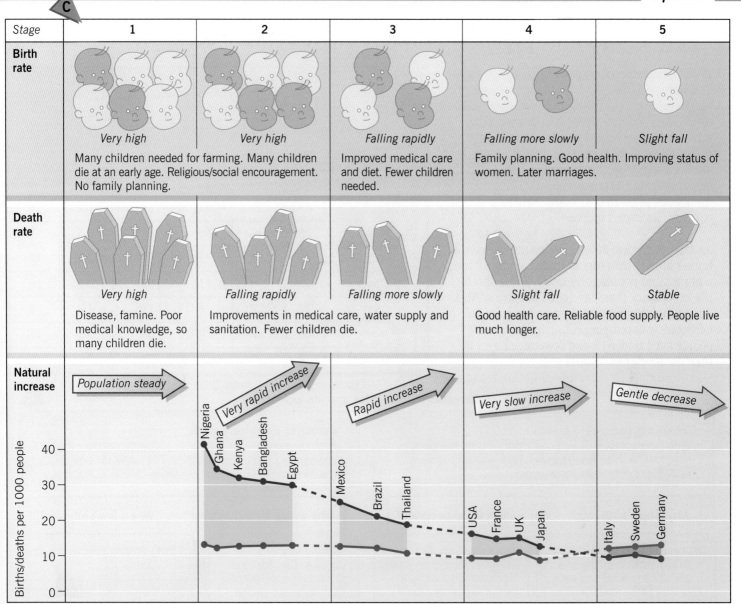

C

Stage	1	2	3	4	5
Birth rate	*Very high*	*Very high*	*Falling rapidly*	*Falling more slowly*	*Slight fall*
	Many children needed for farming. Many children die at an early age. Religious/social encouragement. No family planning.		Improved medical care and diet. Fewer children needed.	Family planning. Good health. Improving status of women. Later marriages.	
Death rate	*Very high*	*Falling rapidly*	*Falling more slowly*	*Slight fall*	*Stable*
	Disease, famine. Poor medical knowledge, so many children die.	Improvements in medical care, water supply and sanitation. Fewer children die.		Good health care. Reliable food supply. People live much longer.	

Natural increase:
- Population steady
- Very rapid increase
- Rapid increase
- Very slow increase
- Gentle decrease

Births/deaths per 1000 people (y-axis: 0, 10, 20, 30, 40)

Countries: Nigeria, Ghana, Kenya, Bangladesh, Egypt, Mexico, Brazil, Thailand, USA, France, UK, Japan, Italy, Sweden, Germany

Activities

1 Diagram **D** shows, in a simplified way, the demographic transition model. Make a copy of the diagram and the table beneath it.

a) On your diagram label the birth rate, death rate and natural increase.

b) Complete the table to describe the changes in birth rate, death rate and population change (diagram **C** should give help you).

c) What causes the change from:
i) stage 1 to stage 2   ii) stage 2 to stage 3 and
iii) stage 3 to stage 4?

2 How does the demographic transition model help explain the five changes in population growth as shown on diagram **C**?

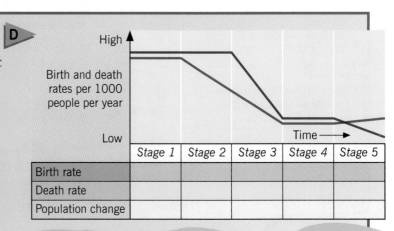

D

High

Birth and death rates per 1000 people per year

Low

Time →

	Stage 1	Stage 2	Stage 3	Stage 4	Stage 5
Birth rate					
Death rate					
Population change					

Summary
Population growth depends upon changes in birth and death rates. Changes in a country's population seem to pass through several stages known as the demographic transition.

How do population changes differ between countries?

It has already been suggested that birth rates and natural increases differ between developed and developing countries. Likewise there are differences between them in infant mortality rates and life expectancy. **Infant mortality** is the number of children out of every 1000 that are born alive but who die before they reach the age of one year. **Life expectancy** is the average number of years that a person born in a country can expect to live.

As table **A** shows, developing countries have a higher birth rate, a higher infant mortality rate, a more rapid population growth and a shorter life expectancy than developed countries. Also, due to their higher birth rate and shorter life expectancy, developing countries have more people aged under 15 and fewer aged over 60 than do developed countries.

A

	Country	Birth rate	Natural increase	Infant mortality rate	Life expectancy	% under 15 years	% over 60 years
D E V E L O P I N G	Nigeria	43	31	70	56	47	4
	Kenya	32	21	55	55	50	4
	Bangladesh	30	19	100	56	44	5
	Egypt	28	19	71	64	39	6
	India	25	19	69	61	37	7
D E V E L O P E D	USA	15	6	7	79	21	17
	France	13	4	6	81	20	19
	UK	13	2	6	79	19	21
	Japan	10	2	4	83	18	17
	Germany	9	–2	6	79	16	6

Differences in birth rates and life expectancy

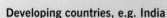

B

Developing countries, e.g. India
Birth rates are high often because people want and need large families. The relatively short life expectancy is more likely to result from a lack of wealth.

C

Developed countries, e.g. UK
Birth rates are low often because people do not need many children and prefer small families. The longer life expectancy results from the greater amount of wealth that is available.

Why a high birth rate?   **Why a short life expectancy?**

We need many children:
• to help us work on the land and to carry wood and water,
• to care for us when we are ill or old and cannot work,
• because so many die from disease. Four of my eight children died before their first birthday.

Both my parents died when they were quite young. My mother died during a famine. My father caught cholera from dirty water. There was no hospital near and we could not afford medical care.

One child might get a job in the city and send us money

My religion forbids birth control

Having a big family increases my importance in the village.

Why a long life expectancy?   **Why a low birth rate?**

Both my parents are still alive. They live near to a doctor and a hospital. Their home has central heating. They are very comfortable.

Family planning controls the size of our family

We only wanted two children and we are sure they will live a long life, free from disease.

We con afford to spend more money on our car, holidays and entertainment.

We have pensions for when we are old.

I wanted to return to my career and not stay at home.

Why are some places overpopulated?

Many people, especially in developing countries, have to rely upon natural resources in the environment in order to live. Natural resources include soils, plants, animals, minerals, water and sources of energy. However, like people, natural resources are not evenly spread across the Earth's surface. Places where the number of people living there outweigh the availability of resources are said to be **overpopulated**. Overpopulation can result from either:

- an increase in population, perhaps due to a high birth rate (fact file **D**) or to large numbers of people moving into an area (fact file **E**)
- a decrease in resources, perhaps resulting from soil erosion or the exhaustion of a mineral.

Overpopulation makes it difficult for people to improve their standard of living and quality of life.

D

Fact file: Sudan, Somalia and Ethiopia – an area with a low population density

- Birth rates high – more people to feed and care for.
- Soils overused for food production and now eroded by rain and wind (page 122).
- Deforestation (for fuelwood and more farmland) and overgrazing have led to desertification (page 126).
- Frequent droughts have led to disease, malnutrition and famine.
- No minerals, and limited fuel supplies.
- Lack of money and technology to develop or buy resources.
- Civil wars have destroyed resources and increased the number of refugees (page 143).

As birth rate and total population increase, the few natural resources are rapidly being used up.

E

Fact file: Cities in developing countries – places with a high population density

The rapid increase in urban population, mainly due to high birth rates and movement from rural areas (page 144) creates numerous problems in cities (pages 180 and 184):

- Decrease in food supplies.
- Shortage of building materials, houses and jobs.
- Inadequate supplies of fresh water and electricity.
- Inadequate methods of disposing of sewage and rubbish.
- The increase in population outpaces the building of hospitals and schools.
- Congested and inadequate transport.

There is neither money nor the resources to keep pace with the increase in population.

Activities

1 What is meant by the terms *infant mortality rate* and *life expectancy*?

2 Give four reasons why many families in:
   a) developing countries have four or more children
   b) developed countries have two or fewer children.

3 a) What is meant by the term *overpopulation*?
   b) Choose **two** areas of the world that are overpopulated. One should have a low population density, the other a high population density. Draw a star diagram to show why each is overpopulated.

Summary

Developing countries have higher birth and infant mortality rates and a shorter life expectancy than developed countries. Many become overpopulated with too many people for the resources available.

How do population structures differ?

The **population structure** shows the number of males and females within different age groups in the population. Information, ideally collected through a census, is plotted as a graph. Because of the traditional shape of the graph, and the information shown, it is commonly referred to as either an **age–sex** or **population pyramid**.

Population pyramids show:
- the total population divided into five-year age groups
- the percentage of people in each of those age groups
- the percentage of males and females in each age group

- trends in the birth rate, death rate, infant mortality rate and life expectancy
- the proportion of elderly and young people who are dependant upon those of working age – the economically active
- the results of people migrating into or out of the region or country.

The pyramid is useful when trying to predict both short and long-term population changes. Graph **A** shows how to interpret the population pyramid for the United Kingdom.

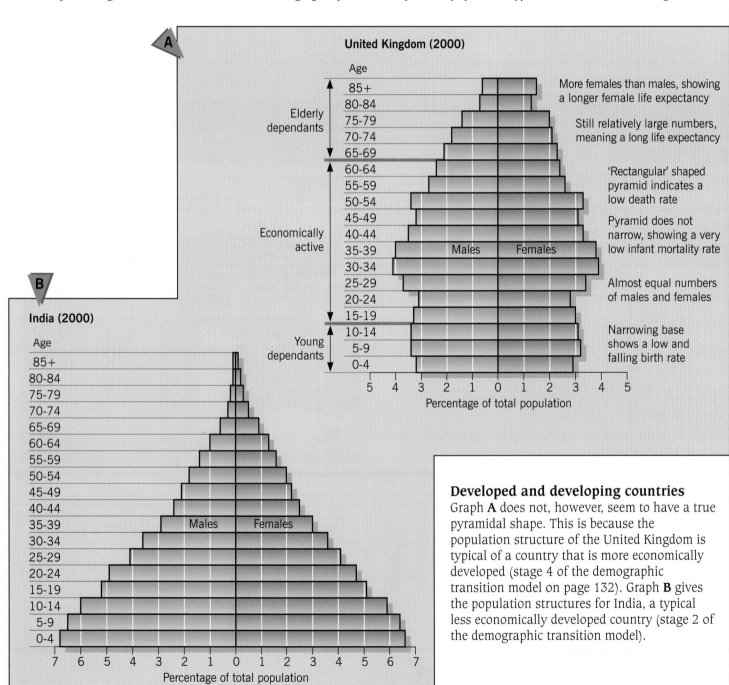

A United Kingdom (2000)

Age

Elderly dependants: 85+, 80-84, 75-79, 70-74, 65-69

Economically active: 60-64, 55-59, 50-54, 45-49, 40-44, 35-39 (Males / Females), 30-34, 25-29, 20-24, 15-19

Young dependants: 10-14, 5-9, 0-4

5 4 3 2 1 0 1 2 3 4 5
Percentage of total population

More females than males, showing a longer female life expectancy

Still relatively large numbers, meaning a long life expectancy

'Rectangular' shaped pyramid indicates a low death rate

Pyramid does not narrow, showing a very low infant mortality rate

Almost equal numbers of males and females

Narrowing base shows a low and falling birth rate

B India (2000)

Age

85+, 80-84, 75-79, 70-74, 65-69, 60-64, 55-59, 50-54, 45-49, 40-44, 35-39 (Males / Females), 30-34, 25-29, 20-24, 15-19, 10-14, 5-9, 0-4

7 6 5 4 3 2 1 0 1 2 3 4 5 6 7
Percentage of total population

Developed and developing countries

Graph **A** does not, however, seem to have a true pyramidal shape. This is because the population structure of the United Kingdom is typical of a country that is more economically developed (stage 4 of the demographic transition model on page 132). Graph **B** gives the population structures for India, a typical less economically developed country (stage 2 of the demographic transition model).

Most Indian families are large. The wide base to the graph indicates a high birth rate and a rapidly growing population. The pyramid narrows rapidly, initially as a result of a high infant mortality rate and later due to a relatively short life expectancy. The shape of this graph is much closer to that of a true pyramid. India's population structure is typical of a developing country as it is 'bottom heavy' in contrast to the 'top heavy' population structure of a more developed country such as the UK.

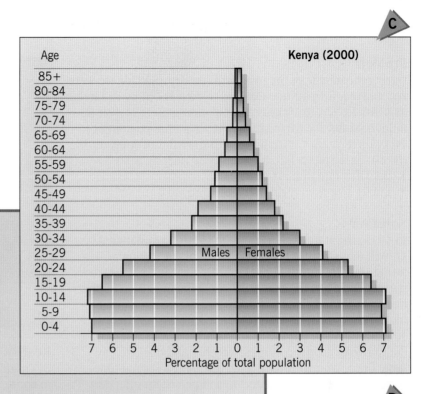

Activities

1 Copy out table **E** and complete it by using information from graphs **C** and **D**.
   a) The answers in part **X** all need a percentage figure.
   b) For part **Y** put a tick in the correct column.

2 What would happen to the shape of Kenya's pyramid if there was a rapid decline in its:
   a) birth rate
   b) death rate?

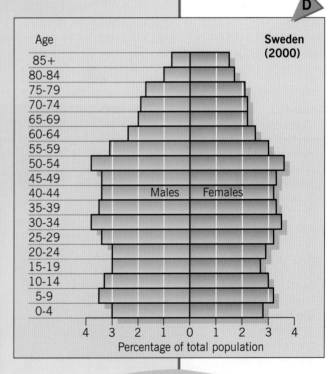

	Sweden	Kenya
Part X		
% males aged 0–4	7.3	3.0
% females aged 0–4		
Total % aged 0–4		
Total % aged 0–14		
Total % aged 15–64		
Total % aged 65 and over		
Part Y		
Highest birth rate		
Fastest-growing population (natural increase)		
Highest infant mortality rate		
Highest % living to middle age		
Lowest life expectancy		
Most people living over 65 years		
More economically developed country		

Summary

Population structures divide males and females into different age groups. Population pyramids show considerable differences in population structures between developed and developing countries.

How has China tried to control population growth?

Problem

During the middle of the twentieth century, the Chinese were encouraged, for patriotic reasons, to have as many children as possible. The result was a population growth of over 55 million (about the same size as the UK's total population) every three years. Concern over this growth rate grew during the 1970s despite a fall, between 1950 and 1975, in:

- the birth rate from 44 to 31
- family size from 5 children per family to 3.

Solution

In 1979, the state decided to try to control population growth and stabilise numbers. To do this, it forcibly introduced its 'one child per family' policy and set the marriageable age for men at 22 and women at 20. Couples had to apply to be married and again before attempting to have a child. Couples who conformed were given free education, priority housing and family benefits. Those who did not were deprived of these benefits and made to pay large fines. Women who became pregnant a second time were forced to have an abortion, and persistent offenders were 'offered' sterilisation. According to human rights activists, this resulted in thousands of forced abortions together with evidence of female infanticide by parents who wanted their 'one child' to be a boy.

Reality

Of China's **ethnic** population, 92 per cent belong to the Han majority group and 8 per cent to the remaining 56 minority groups. The 'one-child' policy was aimed mainly at the Han. However, there were notable exceptions to the policy as applied to both the Han (photo **B**) and the minority groups (photo **C**).

The Han

- They could have a second child if the first was mentally or physically handicapped, or died.
- In most rural areas, farmers could have a second child if the first-born was a girl. Although girls do much of the farmwork, they are considered less useful than boys for working in the fields. (If the second child was also a girl, then hard luck!)
- In some rural areas, local officials allowed a second child on payment of a fine (as for the younger brother here) – or a bribe.

These two brothers and their cousin live in central China.

Ethnic minority groups, most of whom live in remote provinces like this family in south-west China, were allowed at least two children. Some groups, whose numbers were small and where there were fewer officials to keep a check, were permitted up to four.

Success

By the late 1990s, the state felt that the policy had been more than successful if judged by:

- the birth rate – which had fallen from 31 to 19 in 20 years
- the size of the overall population – estimated to be 230 million less than it would have been had the one-child policy not been introduced (graph **D**).

Indeed, the authorities are now concerned that, with the birth rate being so low, there are insufficient children being born to maintain the population.

Relaxation

In 1999, several restrictions in birth control were either lifted or relaxed. These included:

- as the first single-family children had now reached marriageable age, then if two married they could have two children
- allowing all families in rural areas to have two children
- abolishing quotas for child births in 300 trial districts and replacing them with voluntary family planning (permission to have children is still needed)
- allowing women, for the first time, an informed choice between different kinds of contraception.

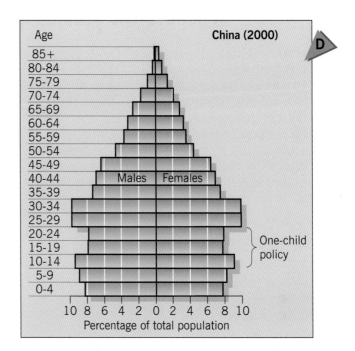

D China (2000) population pyramid. Percentage of total population. One-child policy marked at ages 15-19 down to 0-4.

Activities

1 Describe China's one-child policy by answering the questions in drawing **E**.

2 **a)** What effect has the one-child policy had on China's population structure as shown in graph **D**?

  b) Should the Chinese government tell people how many children they can have? Give reasons for your answer.

  c) How would you feel if the British government adopted a similar policy? Explain your answer.

3 Extract **F** suggests a different approach to the one adopted in China.

  a) How does the UN believe birth rates should be controlled?

  b) Why is it difficult for developing countries to reduce the birth rate?

E

a) Why did China want to control its population?

b) What was the *one-child policy*?

c) What were the main exceptions to the policy?

d) In what ways was the policy successful?

e) What were the main criticisms of the policy?

f) Why has the policy been relaxed?

g) How has it been relaxed?

F

The UN states two basic needs that must be met if birth rates are to be controlled:

1 To improve the status of women and to recognise their right (not accepted in many countries) to make the decision between having more children, or birth control.

2 To provide further education, especially for women, on birth control.

It used to be believed that high birth rates were a result of poverty. However, where the status of women has been improved, birth rates have declined even though there has been no obvious reduction in poverty.

Summary China is one of the few developing countries that has managed to reduce its birth rate although it had to resort to extreme measures to achieve this.

What are the problems of an ageing population?

Life expectancy is increasing in almost every country in the world (page 134 and table **A**). Initially this increase occurred in the more economically developed countries – the MEDCs. More recently, however, it has also begun to affect many of the less economically developed countries – the LEDCs. This increase, together with a falling birth rate, means an increasingly higher proportion of the population is living beyond the age of 65, and even beyond 80. This process is called **ageism**. Diagram **B** gives some reasons for the increase in life expectancy.

Life expectancy, males/females			
	1970	*2000*	*2025 (est.)*
Japan	71/76	77/83	82/85
Italy	69/75	75/82	79/84
UK	69/75	74/79	76/82
USA	68/75	73/79	76/82
China	63/64	69/72	74/78
India	51/49	60/61	66/68
Kenya	49/53	54/58	61/63

A

B

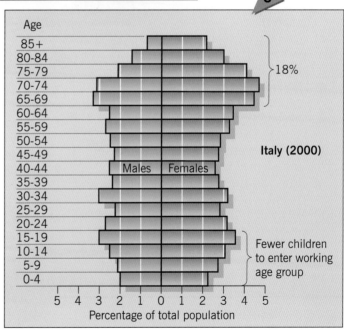

Life expectancy has increased due to:

- improved primary health care
- development of new drugs and vaccines
- advances in medical knowledge and techniques – some diseases and conditions that were considered untreatable a few years ago are now routine
- improved standards of hygiene
- better diet.

There are three main effects of an increase in life expectancy:

- An increase in the **old age dependency ratio**: this is a measure of the number of retired people or pensioners per 100 people of working age. In the UK this ratio was 15 in 1950, 20 in 2000 and is predicted to be 40 by 2040. This means that as the proportion of people aged over 65 increases, the number of working people in the 16–65 age group decreases.
- Changes in the population structure, especially in the MEDCs. Italy is the extreme example (diagram **C**). Already 18 per cent of its population are aged over 65, and 19 per cent of its GNP (page 266) is spent on them. Estimates suggest that by 2030 over 25 per cent of Italians will be over 65 and the proportion of the country's GNP will be an unsustainable 33 per cent (20 per cent on pensions alone).
- The UN predict that by 2025 there will be more elderly people in the world than there will be children aged under 15.

C

Italy (2000)

18%

Fewer children to enter working age group

140

Consequences of an ageing population

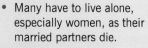

Problems created by the increase

- An increasing amount of money is needed for extra
  - residential homes and sheltered housing
  - health care, including home visits and free prescriptions
  - social services, including home helps and meals on wheels
  - subsidies, e.g. TV licences and bus passes.
- Elderly people take up an increasing amount of the family doctor's financial budget and time.
- There is an increase in long-term and house-bound illnesses such as Parkinson's and Alzheimer's.
- There is an increasing dependence upon people of working age to provide money, by taxes.
- Less money is available for younger age groups – this can affect education and the provision of leisure/social amenities.

Problems facing the elderly

- Many have to live alone, especially women, as their married partners die.
- Most cannot afford the cost of residential homes which, in some parts of the country, can cost over £2000 per month.
- Some people who cannot afford residential homes have to sell their own homes and use up their lifetime savings.
- There are long waits for hospital operations such as hip replacements.
- Many people are fit enough and willing to work but face prejudice against their employment on account of their age.
- Those living in urban areas are fearful of crime and traffic.
- Those in rural areas who cannot drive have problems getting to the doctor, hospital and shops due to a lack of public transport.

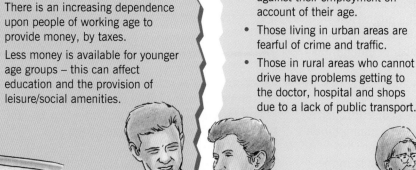

Activities

1. a) Why has there been an increase in life expectancy in the MEDCs?
   b) What is the *old age dependency ratio*?

2. How does an increase in life expectancy affect:
   a) the old age dependency ratio in a country such as the UK
   b) the population structure of a country such as Italy (diagram **C**)?

3. What are some of the problems created by an increase in life expectancy for:
   a) people in the working age group
   b) people in the over-65 age group
   c) the government of the country?

Summary   As the life expectancy of a country increases, so too do problems associated with providing adequate services and amenities.

What is migration?

Migration is a movement of people from one place to another to live or to work and usually involves a change of home. Table **A** gives a classification of migration. These movements are for different reasons:

- Movement to a better home in a nearby street or town, or to a better job elsewhere in the same country. This is known as **internal migration**. Internal migration affects the distribution of people within a country (page 128).

- Movement to a different country. This is referred to as **external** or **international migration**. It affects both the total population and the population structure of a region or country (page 136).

- Shorter-term movements that can be **temporary**, **seasonal** or **daily**.

International migration

People who leave a country are called **emigrants**. Those who arrive in a country are known as **immigrants**. The **migration balance** is the difference between the number of emigrants and the number of immigrants.

Assuming that birth rates and death rates are evenly balanced:

- countries that lose more people through emigration than they gain by immigration will have a declining population, for example in the former Yugoslavia in the early 1990s

	Type	Example
Permanent	External – international (between countries)	
	1 Voluntary	• Jews into Israel (page 148) • West Indians into the UK
	2 Forced	• Refugees (page 143) • Slaves to America
	Internal (within a country)	
	1 Rural depopulation	Most developing countries (Brazil, page 144)
	2 Counter-urbanisation	Away from British conurbations (page 162)
	3 Regional	North-west of Britain to the south-east of Britain (page 276)
Semi-permanent	For several years	Migrant workers – North Africans into France (page 146)
Seasonal	For several months or weeks	Holidaymakers to Greece (page 236), university students
Daily		Commuters in south-east England (page 176)

- countries that receive more people through immigration than they lose by emigration will see an increase to their population – two examples are France, page 146 and Israel, page 148.

The reasons for international migration can be divided into two types – **voluntary** and **forced** (diagram **B**).

B

Voluntary migration is when people choose to move. This is usually because of the **'pull'** or attraction of a better standard of living and quality of life elsewhere. Reasons may include:

- to improve their standard of living, e.g. more and better-paid jobs

- to improve their quality of life, e.g. retiring to live in a warmer climate or working in a more pleasant environment

- to have good, available services and amenities, e.g. schools, hospitals, shops and entertainment

- for increased personal freedom, e.g. greater religious or political tolerance

- to be with friends and relatives or with people of a similar culture.

Forced migration is when people have no choice and either have to, or are made to, move. These people are **'pushed'** out of their homes and country. Reasons may include:

- natural disasters, e.g. earthquakes, volcanic eruptions, drought and floods

- man-made disasters, e.g. war and ethnic cleansing (creating large numbers of refugees)

- overpopulation or a lack of resources, e.g crop failure causing famine

- racial discrimination or religious and political persecution

- government schemes, e.g. inner city redevelopment, the building of a motorway or dam.

Refugees

Refugees are people who have been forced to leave their homes:

- for fear of persecution (race, religion or politics)
- due to wars, many of which have been internal civil wars
- due to environmental disasters, especially earthquakes and famine.

The UN estimated that, in the late 1990s, there were over 15 million refugees worldwide. Of these, over half had been forced to leave their own country (diagram **C**). Most of this latter group had to leave all their personal possessions behind, now live in extreme poverty, have no rights or prospects, and are unlikely to return to their homeland. Most are forced to live in large refugee camps, relying upon their host country or world charity organisations for food, water, clothing, medical supplies and shelter (photo **D**). Over half of the world's refugees are children and most of the adults are women. Over 80 per cent of refugees live in developing countries – countries which, due to a lack of resources and money, can least afford to help.

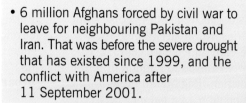

Some recent refugee movements

- 6 million Afghans forced by civil war to leave for neighbouring Pakistan and Iran. That was before the severe drought that has existed since 1999, and the conflict with America after 11 September 2001.

- 1.5 million Ethiopians, Sudanese and Somalis driven from their homes in Africa by drought, famine and civil war.

- The Rwandan civil war in central Africa in the early 1990s created 1.5 million refugees. An estimated 300 000 died en route or later in refugee camps due to starvation and cholera.

- In the early 1990s, 600 000 people from the former Yugoslavia fled to other European countries, with another 2 million displaced within the country due to ethnic cleansing.

Extract from a letter written by a UNICEF helper at a refugee camp in western Tanzania (March 2001).

"The conditions here are appalling. Every day heavy rains flood the camp and most nights are very cold. The children living here are suffering terribly. Along with the rain and cold comes malaria and respiratory infections like pneumonia. Many children are sick, hungry and alone. Separated from their families by civil wars in Burundi, Rwanda and the Congo, they have been weakened by war, drought and disease. The camp is short of medicines, food, clean water and blankets. Without these, many of the children might die. What is worse, there seems to be no end to their plight."

Activities

1 **a)** What is the difference between:
   - an *emigrant* and an *immigrant*
   - *voluntary* and *forced* migration?
   b) Give two examples each of *voluntary* and *forced migration*.

2 **a)** What is a *refugee*?
   b) Give three reasons why refugees are forced to leave their home country.
   c) Describe the likely conditions in a refugee camp.

Summary

Migration is the movement of people within or between countries, either voluntary or forced. Refugees are a major international problem.

What is rural–urban migration?

Rural–urban migration is the movement of people from rural areas to towns and cities. Rural–urban migration began in the UK during the early nineteenth century. This was a time when many people moved to London or to the 'new' industrial cities of northern England in order to find work. Today there tends to be more of a balance between people moving into and out of British cities.

Rural–urban migration became increasingly important in developing countries, especially those in Asia and South America, during the twentieth century. This type of migration is due to two factors: **rural push factors**, and **urban pull factors**.

A

RURAL 'PUSH' – negative reasons for migration

Governments spend very little money on us (lack of investment).

Rural areas become overpopulated due to high birth rates. The average family size is eight people. There are too many of us to find jobs on the farms.

Villages and crops can be destroyed by such natural disasters as droughts, floods, earthquakes and insect pests.

Drought and poor soils mean we cannot grow enough food to feed ourselves and so we have to move to the city.

Our village does not have a school, a doctor or a hospital. Our houses do not have electricity, running water or sewage disposal facilities.

We are farmers but not many of us own any land, and if we do the plots are very small.

We have learned some skills at school but we cannot use them in our local village.

As farmers, we have poor seed, few tools and little fertiliser. Where machinery has been introduced it means there are fewer jobs on farms.

Overgrazing by animals and overcultivation of crops has increased soil erosion.

Many villages are in remote areas and have little or no transport.

Activities

1 What is the meaning of the terms:
   a) *rural–urban migration*
   b) *rural push factors*
   c) *urban pull factors*?

2 Make a copy of diagram **C** and complete it by:
   a) giving five reasons why people who live in a developing country may decide to move away from the countryside (rural push factors)

   b) giving five reasons why people who live in a village in a developing country may wish to move into a nearby large city (urban pull factors).

3 Do you think that conditions found by new migrants to the city will meet their expectations? Give two reasons for your answer.

1 **Rural push factors** are those that encourage – and sometimes force – people to leave the countryside (diagram **A**). Most people who move do so because of extreme poverty in rural areas. They migrate to the cities in the hope of improving their living conditions and quality of life.

2 **Urban pull factors** are those that attract people to the cities (diagram **B**). Most of these migrants are attracted by their **perception** of what they think, or were led to believe, the city is like. The reality is often very different (pages 180 and 184–185). On arrival, they are likely to find a considerable gap between themselves and the rich, a gap that is far greater than in developed countries.

URBAN 'PULL' – positive reasons for migration

There are plenty of 'bright lights' – shops and entertainment.

Cities have a more reliable food supply.

Business people and politicians live here, so there is plenty of investment.

Life expectancy is, on average, ten years longer in cities.

Transport is much better.

There are plenty of schools, hospitals and doctors.

There are more jobs in the city. They are also better paid.

Housing is much better, with electricity and a water supply.

PUSH FACTORS

1 ------------
2 ------------
3 ------------
4 ------------
5 ------------

PULL FACTORS

Hope for:
1 ------------
2 ------------
3 ------------
4 ------------
5 ------------

Summary

The movement from rural to urban areas is the main type of migration in most developing countries. This may be due to poverty in rural areas or to the perception that cities offer a better way of life.

How does migration affect different countries?

North Africans into France

Like several other Western European countries in the late 1940s, France had many more job vacancies than it had workers. This was partly due to the higher than average death rate experienced during the Second World War, and partly due to the need to rebuild the country after that war. Later, as France became increasingly more prosperous, it attracted workers from the poorer parts of Europe (notably Portugal) and from North Africa (especially its former colonies of Algeria and Morocco).

Initially many migrants went into farming or found unskilled jobs in Marseille – the port of entry into France for most North Africans. As these migrant workers became more settled, they turned to relatively better-paid jobs in factories and the construction industry. Even so, the majority of them were forced to take jobs that the French themselves did not want – work that was dirty, unskilled, poorly paid or demanded long and unsociable hours (photo **A**). In time, many of these workers decided to stay permanently in France, usually having to live in poor accommodation (photo **B**). Since then, some of them have become French citizens, and others have been joined by their families.

The population pyramid in diagram **C** shows the characteristics of a country such as Algeria, from where many migrant workers come. The population pyramid in diagram **D**, in contrast, shows the characteristics of a country such as France that receives a large number of migrant workers.

These characteristics include:
- more males than females – this is because males are usually the first to travel abroad to seek work, with the intention later of either returning home or being joined by their families
- more aged between 20 and 34 years, i.e. the younger levels of the economically active age group

- a relatively large number of children, because the economically active groups coincide with the reproductively active age groups
- few elderly people, as they are least likely to migrate.

The movement of migrant workers from the 'losing' country (e.g. Algeria or Turkey) to the 'receiving' country (e.g. France or Germany) presents benefits and creates problems for both countries. Some of these advantages and disadvantages are shown in diagrams **C** and **D**.

C Population pyramid of a typical 'losing' country, e.g. Algeria, Turkey

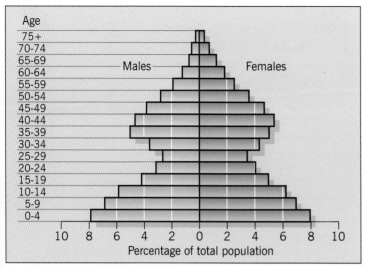

D Population pyramid of a typical 'gaining' country, e.g. France, Germany

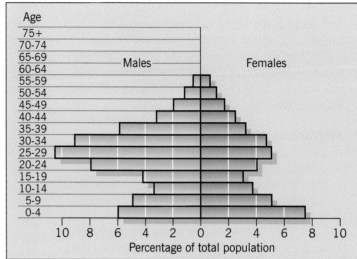

Advantages

• Reduces pressure on jobs and resources, e.g. food.
• Loses people of child-bearing age, causing a decline in the often high birth rate.
• Migrants develop new skills which they may bring back to their home country.
• Money earned may be sent back to their home country.

Disadvantages

• Loses people in the working-age group.
• Loses people mostly likely to have some education and skills.
• Left with an elderly population and so a high death rate.
• Increasing dependency on money sent home by the workers.

Advantages

• Overcomes labour shortage.
• Prepared to do dirty, unskilled jobs.
• Prepared to work long hours for a low salary.
• Cultural advantages and links.
• Some highly skilled migrants.
• In a developing country these migrants could increase the number of skilled workers.

Disadvantages

• Immigrants are most likely to be the first unemployed in a recession.
• Low-quality, overcrowded housing lacking in basic amenities (inner city slums – *bidonvilles* in France)
• Ethnic groups tend not to integrate – racial tension
• Limited skills/education, language difficulties.
• Lack of opportunities to practise own religion, culture, etc.

Activities

1 a) Why did France originally need foreign workers?
   b) Name, in rank order, four countries from where most migrant workers came.
   c) What types of job were given to the migrant workers?

2 a) If you were from North Africa, what might be some of the:
      i) reasons why you wanted to find work in France
      ii) problems you might leave behind in North Africa?
   b) If you were French, what would be some of the:
      i) advantages of having North Africans coming to work in your country
      ii) problems which you think the migrant workers may create?

E Source of immigrants to France

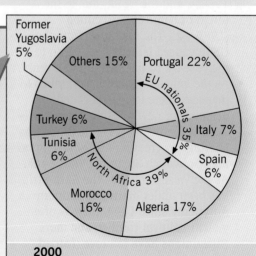

2000
Total population 58 800 000
93% French
4% non-EU born
3% EU born

Summary The movement of workers from poorer countries to more wealthy countries can create problems as well as having benefits to both.

147

Israel – an example of international migration

Few parts of the world have experienced more migrations, either **forced** or **voluntary** (page 142), than the region that is the present-day state of Israel. Diagram **A** gives a very brief history of the area, which was often referred to as 'Palestine' before Israel came into being in 1948.

Migrations since 1948

- The state of Israel was proclaimed in May 1948. Immediately thousands of Palestinians, fearing religious and political persecution, fled the country (**forced migration**). They were forced to settle in refugee camps in the Gaza Strip, the West Bank and in nearby countries such as Jordan (map **B**). This was the first time, certainly in modern times, that the refugee problem was to become both permanent and insoluble (page 143). The surrounding Arab states declared war on Israel, but failed to capture any land.

- In 1950, the Law of Return provided free citizenship for all immigrant Jews. During the next two decades, many Jews decided to come to Israel (**voluntary migration**) from Europe, North Africa and what was then the USSR.

- The number of 'returning' Jews increased rapidly during the 1980s and early 1990s. This was partly due to large numbers of Jews having to leave (**forced migration**) Ethiopia, due to famine and civil war (early 1980s), and the USSR, due to the threat of political persecution (1989–92). The migration from the USSR caused Israel's population to increase by 10 per cent in three years.

- As the Israeli population continues to increase, there is an associated demand for new housing. The Israelis have attempted to solve their housing shortage by creating new settlements, many of which are in the West Bank. As the West Bank, like the Gaza Strip, has been under limited Palestinian Autonomy since 1994, there is often violent political conflict, and more Palestinians have to move (**forced migration**).

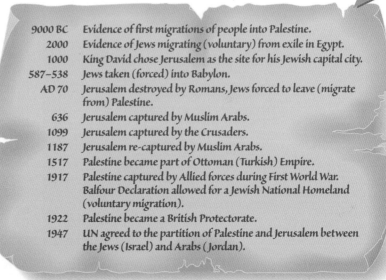

9000 BC	Evidence of first migrations of people into Palestine.
2000	Evidence of Jews migrating (voluntary) from exile in Egypt.
1000	King David chose Jerusalem as the site for his Jewish capital city.
587–538	Jews taken (forced) into Babylon.
AD 70	Jerusalem destroyed by Romans, Jews forced to leave (migrate from) Palestine.
636	Jerusalem captured by Muslim Arabs.
1099	Jerusalem captured by the Crusaders.
1187	Jerusalem re-captured by Muslim Arabs.
1517	Palestine became part of Ottoman (Turkish) Empire.
1917	Palestine captured by Allied forces during First World War. Balfour Declaration allowed for a Jewish National Homeland (voluntary migration).
1922	Palestine became a British Protectorate.
1947	UN agreed to the partition of Palestine and Jerusalem between the Jews (Israel) and Arabs (Jordan).

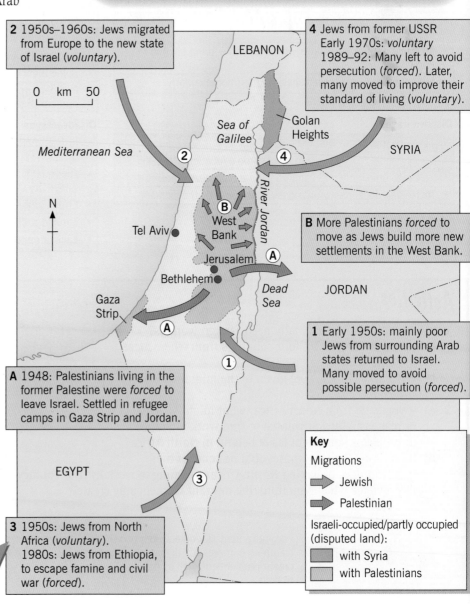

2 1950s–1960s: Jews migrated from Europe to the new state of Israel (*voluntary*).

4 Jews from former USSR Early 1970s: *voluntary* 1989–92: Many left to avoid persecution (*forced*). Later, many moved to improve their standard of living (*voluntary*).

B More Palestinians *forced* to move as Jews build more new settlements in the West Bank.

1 Early 1950s: mainly poor Jews from surrounding Arab states returned to Israel. Many moved to avoid possible persecution (*forced*).

A 1948: Palestinians living in the former Palestine were *forced* to leave Israel. Settled in refugee camps in Gaza Strip and Jordan.

3 1950s: Jews from North Africa (*voluntary*). 1980s: Jews from Ethiopia, to escape famine and civil war (*forced*).

Migrations since 1948 **B**

Key
Migrations
➡ Jewish
➡ Palestinian
Israeli-occupied/partly occupied (disputed land):
with Syria
with Palestinians

How has this migration affected the Jews?

We have got our own country but:

- We have been attacked several times by our Arab neighbours. We have to spend a lot of money defending ourselves.
- The large number of immigrants means a big demand for new settlements. We have had to build some of these in the occupied territories (map **B**).
- As our population increases so too does the need for schools, hospitals and other services. This increases our cost of living and causes high inflation.
- Many recent Jewish immigrants are from Ethiopia. They have large families, do not speak much Hebrew and have limited skills.
- With so many migrants it is becoming harder for people to find jobs.

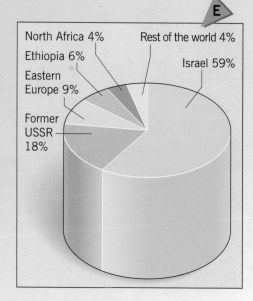

C

How has this migration affected the Palestinian Arabs?

Views of the Palestinian Arabs:

- We have lived in this area as long as the Jews. It is our home.
- Over 1.75 million of us lost our homes when the state of Israel was created. We have become refugees in places like the Gaza Strip and the West Bank. Many of us were born in and have lived all our lives in camps.
- The Israelis are building new settlements in areas where we were meant to live.
- The land they were given was the best for farming. Where we live there are very few job opportunities or natural resources. Many of us have to seek work within Israel.
- Our settlements do not have many services, and jobs are hard to find.

D

Activities

1 a) Using diagram **E**, name three areas from which most of the Jewish migrants to Israel have come.
  b) Using map **B**, give three reasons why many Jews have migrated back to Israel. State, for each reason, if the movement was voluntary or forced.
  c) Using map **B**, give two reasons why many Palestinians have had to move. State, for each reason, if the movement was voluntary or forced.

2 a) Using diagram **C**, describe four effects of these migrations upon the Jews.
  b) Using diagram **D**, describe four effects of these migrations upon the Palestinians.
  c) Why is it so difficult to maintain peace between the Jews and the Palestinians?

E

North Africa 4%   Rest of the world 4%
Ethiopia 6%
Eastern
Europe 9%   Israel 59%

Former
USSR
18%

Summary

One recent large-scale international migration has been the voluntary return of Jews to, and the forced movement of Palestinians out of, Israel.

How were sites for early settlements chosen?

The location and growth of an individual settlement depended upon its **site** and **situation**. The site was the actual place where people decided to locate their settlement. The growth of that settlement then depended upon its situation in relation to natural resources (physical features) and other settlements (human features).

What did early settlers consider to be important when choosing a site? They needed to:
- be near a reliable supply of water
- be away from marshy areas or places which flooded
- be able to defend themselves in case of attack

- be near to materials for building their homes
- be able to feed themselves
- have access to other places
- have shelter from bad weather
- have a supply of fuel for cooking and heating.

Although the ideal site needed more than one of these **location factors**, it was unlikely to have them all. For example, while a hilltop site may have been excellent for defence, it probably would not have had a supply of water nor would it have been sheltered from strong winds. Diagram **A** shows several factors that influence the choice of site for a settlement.

A

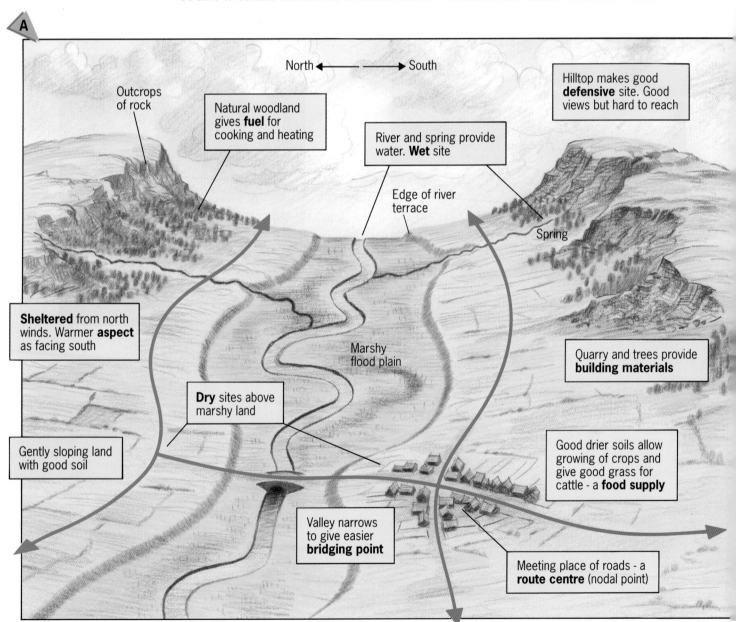

North ← → South

Outcrops of rock

Natural woodland gives **fuel** for cooking and heating

River and spring provide water. **Wet** site

Hilltop makes good **defensive** site. Good views but hard to reach

Edge of river terrace

Spring

Sheltered from north winds. Warmer **aspect** as facing south

Marshy flood plain

Quarry and trees provide **building materials**

Dry sites above marshy land

Gently sloping land with good soil

Valley narrows to give easier **bridging point**

Good drier soils allow growing of crops and give good grass for cattle - a **food supply**

Meeting place of roads - a **route centre** (nodal point)

In the years since the settlement was founded it is likely to have grown in size and developed into a different shape. Several of the early factors are less important today, for example water is brought to our homes by pipe and marshy land can be drained. Even so, several of the original location factors may still be visible not only when visiting a place but also when studying an OS map. Diagram **B** is an **annotated** or labelled **sketch map** showing how the site and location of a town is related to physical and human features (page 288). The sketch map, which is for Carlisle, is taken from the 1:50 000 OS map on pages 278–279.

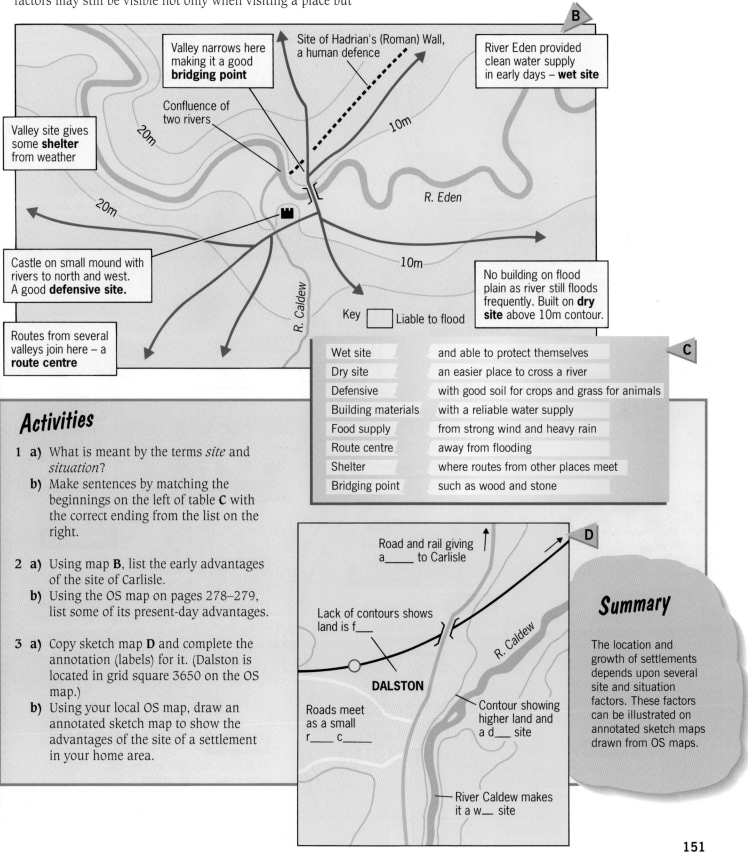

B

Valley narrows here making it a good **bridging point**

Site of Hadrian's (Roman) Wall, a human defence

River Eden provided clean water supply in early days – **wet site**

Confluence of two rivers

20m

10m

Valley site gives some **shelter** from weather

20m

R. Eden

Castle on small mound with rivers to north and west. A good **defensive site.**

10m

R. Caldew

Routes from several valleys join here – a **route centre**

Key ☐ Liable to flood

No building on flood plain as river still floods frequently. Built on **dry site** above 10m contour.

C

Wet site	and able to protect themselves
Dry site	an easier place to cross a river
Defensive	with good soil for crops and grass for animals
Building materials	with a reliable water supply
Food supply	from strong wind and heavy rain
Route centre	away from flooding
Shelter	where routes from other places meet
Bridging point	such as wood and stone

Activities

1 **a)** What is meant by the terms *site* and *situation*?

  b) Make sentences by matching the beginnings on the left of table **C** with the correct ending from the list on the right.

2 **a)** Using map **B**, list the early advantages of the site of Carlisle.

  b) Using the OS map on pages 278–279, list some of its present-day advantages.

3 **a)** Copy sketch map **D** and complete the annotation (labels) for it. (Dalston is located in grid square 3650 on the OS map.)

  b) Using your local OS map, draw an annotated sketch map to show the advantages of the site of a settlement in your home area.

D

Road and rail giving a_____ to Carlisle

Lack of contours shows land is f___

R. Caldew

DALSTON

Roads meet as a small r____ c_____

Contour showing higher land and a d___ site

River Caldew makes it a w__ site

Summary

The location and growth of settlements depends upon several site and situation factors. These factors can be illustrated on annotated sketch maps drawn from OS maps.

151

What are the different functions of settlements?

The term **function** describes what a settlement did, or still does. The function, or purpose, of early settlements may have been, for example, to defend people, to act as a religious centre, or to collect farm produce from the surrounding area. The term also relates to the subsequent development of a settlement and refers to its main activity or activities (diagram **A**).

A

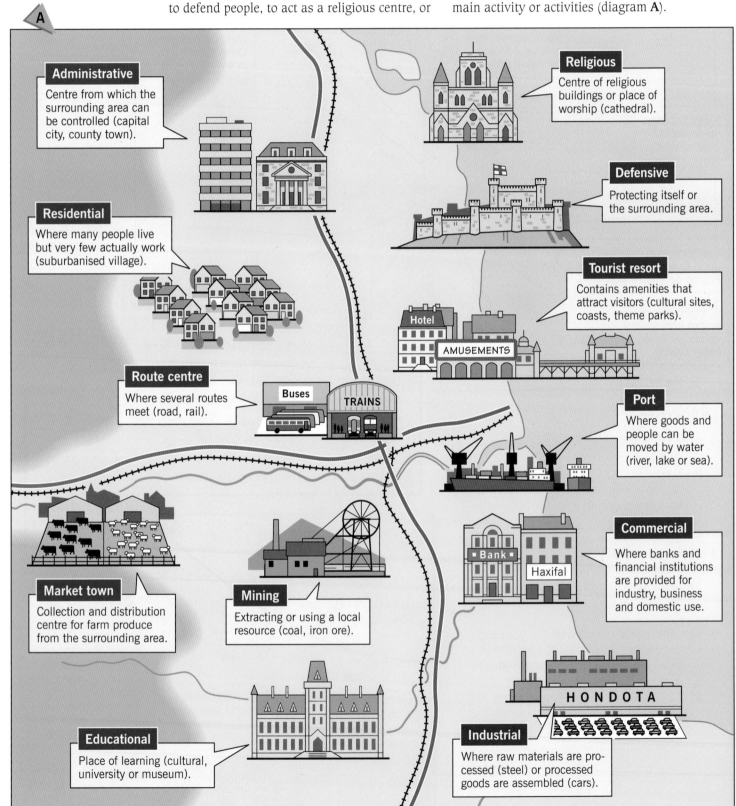

Administrative
Centre from which the surrounding area can be controlled (capital city, county town).

Religious
Centre of religious buildings or place of worship (cathedral).

Defensive
Protecting itself or the surrounding area.

Residential
Where many people live but very few actually work (suburbanised village).

Tourist resort
Contains amenities that attract visitors (cultural sites, coasts, theme parks).

Hotel

AMUSEMENTS

Route centre
Where several routes meet (road, rail).

Buses

TRAINS

Port
Where goods and people can be moved by water (river, lake or sea).

Commercial
Where banks and financial institutions are provided for industry, business and domestic use.

Bank

Haxifal

Market town
Collection and distribution centre for farm produce from the surrounding area.

Mining
Extracting or using a local resource (coal, iron ore).

Educational
Place of learning (cultural, university or museum).

Industrial
Where raw materials are processed (steel) or processed goods are assembled (cars).

H O N D O T A

In some cases, the original function of British settlements may:
- no longer be applicable, e.g. they no longer have a defensive function (diagram **B**)
- have changed over a period of time, e.g. a fishing village has become a tourist resort, or a mining town has used up its resource and developed high-tech industries.

Most settlements, especially those that are larger, tend to be multi-functional. This means that they have several functions even if one or two tend to predominate (diagram **B**).

B

Original function	Defensive *(no longer applies)*
Later function (still applies)	Farming village
Present-day functions	Residential
	Tourist centre

C

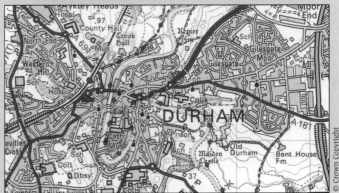

Original function	Defensive *(no longer applies)*
Later functions (still applies)	Religious (cathedral)
	Educational (university)
	Market town
	Administrative (county town)
Present-day functions	Residential
	Route centre
	Tourist centre

Activities

1 What is meant by the term *function of a settlement*?

2 With reference to Corfe (map **B**) and Durham (map **C**):
   a) What was the original function of both settlements?
   b) Why is this no longer a function of either settlement?
   c) List three functions of present-day Corfe.
   d) List six functions of present-day Durham.

3 a) With reference to diagram **A**, describe the main functions of the nearest large settlement to where you live.
   b) Have these functions changed over a period of time?

Summary

The term 'function' describes the main purpose and activity, or activities, of a settlement. The function can change over a period of time.

What is a settlement hierarchy?

A **settlement hierarchy** is when settlements are put into order based upon their size or the services which they provide for people. A settlement hierarchy can be produced using three different methods:

1 **Population size** – the larger the settlement the fewer there will be of them.

2 **Distance apart** – the larger the settlement the further it will be from other large settlements.

3 **Range and number of services** – the larger the settlement the more services it will provide.

These methods, and the settlement hierarchy, are summarised on diagram **A**. Note that the figures quoted are given for comparisons to be made and are not actual figures.

SETTLEMENT IN THE UK

Hierarchy	Distance apart (km)	Population size	Services (non-shopping)
One capital	—	Over 5 000 000	Government offices; several universities; main line railway stations; international airport; large, specialist hospitals; national events
Two or three conurbations		Up to 1 000 000	
Several cities	100	Up to 500 000	County hall; cathedral; luxury hotel; university; many cinemas; theatres; hospitals; main railway station; several football teams
Many large towns	50	Up to 100 000	Small hospital; large restaurants; hotels; cinema and small theatre; several secondary schools; large bus and railway stations; large football team
Hundreds of small towns	20	Up to 20 000	Town hall; doctor; several churches/chapels; several public houses; cafés and restaurants; small secondary school; railway station; bus station, football team
Thousands of villages	7	Up to 1 000	Village hall; church; public house; small primary school
Several thousand hamlets	2	Up to 20	Public telephone

N.B. All places in the hierarchy have all the services of the settlements below them

Activities

1 Find the seven places named in table **B** on the OS map on pages 278 and 279. Using map evidence only, list the services found in each settlement. Complete a copy of table **B** (the first settlement has been done for you).

2 What other services may the settlements have which are not shown by symbols on the map?

B

Settlement	List of services provided	Total number	Rank	Type of settlement
Cotehill	Post office, public house, chapel	3		Village
Dalston				
Scotby				
Rockcliffe				
Wetheral				
Beaumont				
Carlisle				

Is there a hierarchy of shopping centres of different sizes?

Shops can also be placed into a hierarchy based upon the services they provide. At the bottom of the hierarchy are small shops selling low order, convenience goods which are needed daily, such as food and newspapers. At the top are shops selling high order, specialist goods bought less frequently, such as furniture and video recorders. The same three methods used to produce the settlement hierarchy can also be applied here.

1 **Population size** – the larger the settlement the greater the number of high order shops.

2 **Distance apart** – the larger the shopping centre the further it will be to other large centres.

3 **Range of services** – the larger the shopping centre the more services it will provide.

These methods, and the shopping hierarchy, are summarised on diagram **C**.

SHOPPING IN THE UK				
Hierarchy	Distance apart (km)	Types of shop	Goods sold	Frequency of visits
Major national shopping centre Several central covered areas Several suburban and edge-of-city centres		All major national and some international chain and department stores	Highest order, specialist, luxury and comparison goods (top furniture and fashions)	Yearly
One covered area in city centre Many shopping streets Several edge-of-city regional centres	100	Large national chain stores, department stores, several hypermarkets	High order, increasingly specialist (fashion, jewellery)	Monthly
Several shopping streets One or two edge-of-city centres	50	Large chain stores, hypermarkets	Middle order, some specialist and comparison shops (clothes, books)	Weekly
One main shopping street and market	20	Small chain stores, superstores	Mainly low order – more volume and a bigger range	Two or three times per week
One village shop	7	Village shop, post office	Low order, convenience (bread, newspapers)	Daily
None	2	(Mobile shop)	(Meat, fish, groceries)	Weekly
N.B. All shopping centres in hierarchy have all the services of the centres below them				

3 **a)** What are the differences between *low order* and *high order shopping centres*?

b) Why are there many more lower order shopping centres than higher order centres?

c) Name the nearest low order and the nearest high order shopping centre to your school. For each, name the types of shop found there, and the goods they sell.

Summary

Settlements can be placed in order in a hierarchy based on their size and the services which they provide. The same concept can be applied to account for the distribution of shopping centres of different sizes.

What is a typical urban land use model?

It has been suggested that towns and cities do not grow in a haphazard way but tend to develop recognisable shapes and patterns. Each town will have developed:
- its own distinctive pattern, making it different from other towns
- some characteristics shared by other urban settlements.

Several geographers have offered theories as to how these characteristic patterns and shapes develop. These theories are illustrated as **urban land use models**. Remember that a model is used to simplify complex real-world situations, making them easier to explain and to understand. The earliest, and still the easiest to understand and apply, is the Burgess model.

The Burgess model

Burgess claimed that the focal point of a town was the **central business district or CBD** (page 158). He then suggested that as towns developed they grew outwards from the CBD. This means that buildings become increasingly more recent towards the urban boundary (assuming no changes occur in, or near, the CBD). Burgess showed this outward growth on his model by four circular **zones** (diagram **A**). The first, and adjacent to the CBD, was a transition zone where Burgess suggested that industry had replaced, or developed along with, the oldest of houses. The other three zones, represented as concentric circles, were based upon the age of the houses and the wealth of their occupants.

Functional zones in a city

As towns grew, each of the zones shown on diagram **A** developed its own special type of land use or **function** (page 152). The three major types of land use in a town are shops and offices, industry, and housing. Other significant types of land use include open space, transport, and services (e.g. schools, hospitals and shops). As each city develops its own pattern of land use, that pattern is likely to be more complex than the one shown in the Burgess model. Diagram **B** is, therefore, a more realistic map showing land use and functional zones in a city.

Key

	Central Business District (CBD)
	Wholesale light manufacturing (transitional)
	Low-class residential (old inner city areas)
	Medium-class residential (inter-war areas)
	High-class residential (modern suburbia)

A The Burgess concentric model

B Typical land use and functional zones in a city

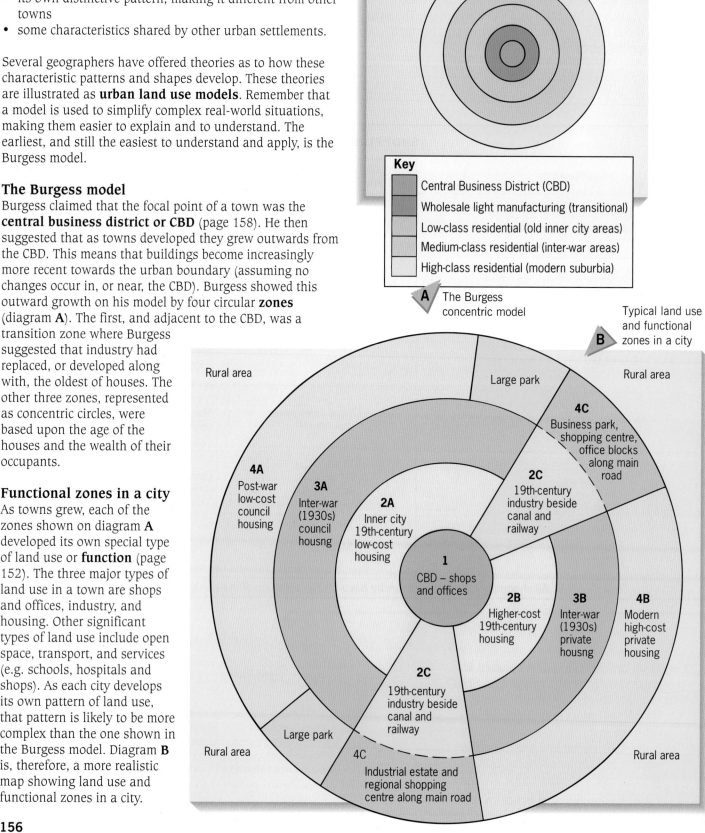

Rural area

Large park

Rural area

4A Post-war low-cost council housing

3A Inter-war (1930s) council housng

2A Inner city 19th-century low-cost housing

2C 19th-century industry beside canal and railway

4C Business park, shopping centre, office blocks along main road

1 CBD – shops and offices

2B Higher-cost 19th-century housing

3B Inter-war (1930s) private housng

4B Modern high-cost private housing

Large park

2C 19th-century industry beside canal and railway

Rural area

4C Industrial estate and regional shopping centre along main road

Rural area

You must remember that, by making generalisations, the Burgess model is a simplification of reality. If you make a study of your local town or city, you must avoid the temptation to say that it fits the model. At best, it is only likely to show one or more *characteristics* of the model. This should be made clearer to you if, for example, you study photo **C** and land use map **D** below, and the OS map of Carlisle on pages 278–279.

Key
A = Castle
B = Civic centre
C = Cathedral
D = Railway station

C Carlisle from the air, looking north

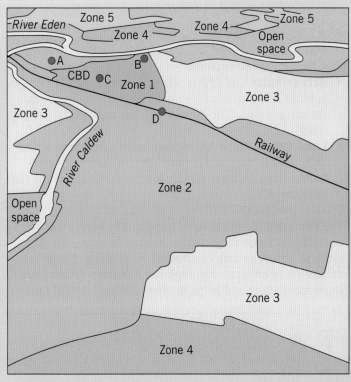

D Carlisle's land use zones

Activities

1 Look at diagram **E**.
   a) What do the letters *CBD* mean?
   b) Give two main functions of the CBD.
   c) In which zone is industry the main function?
   d) In which zone is the most expensive housing?
   e) In which zone is the cheapest housing?

2 Compare the land use and functions of Carlisle in photo **C** and map **D** with those of the Burgess model in diagram **A**.
   a) Give three similarities between Carlisle and the model.
   b) Give three differences between Carlisle and the model.

3 Draw a simple land use model of your nearest town or city. Does it have any obvious similarities to or differences from the Burgess model?

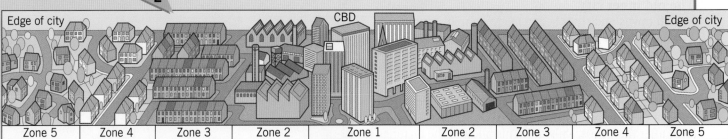

E

| Zone 5 | Zone 4 | Zone 3 | Zone 2 | Zone 1 | Zone 2 | Zone 3 | Zone 4 | Zone 5 |

Summary It is possible to recognise patterns of land use and functional zones within a city. These zones may be shown more simply as an urban land use model.

What were the CBD and inner city zones like?

The location of each functional zone and the pattern of land use in a city are related to such factors as accessibility, land values, competition for land, age of buildings, wealth of the residents and changes in demand.

The CBD

The CBD is the commercial and business centre of a town or city. This is for two reasons:

- **Accessibility** The CBD is where the main roads from the suburbs and surrounding towns meet. It has been, until the congestion of the present day, the easiest place to reach.
- **Land values** These are highest in the CBD where space to build is limited and competition for land is greatest. Land values decrease rapidly towards the edge of the city (diagram **A**).

The two main functions have traditionally been shopping and offices. Shops in particular need to be accessible in order to attract as many customers as possible. Shops and offices are often located in high-rise buildings which gives them more space and helps to offset the high cost of land

(photo **B**). Even so it is often only the large department stores and specialist shops, together with banks, building societies and other commercial companies, that make a large profit and so can afford to locate here.

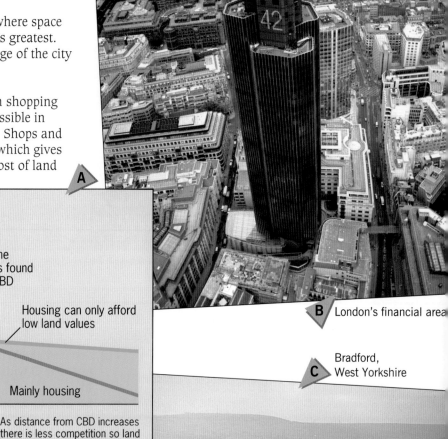

A

High → Price of land (land values) → Low

Shops and offices can afford high land values of CBD

Industry cannot afford the highest land values so is found further away from the CBD

Housing can only afford low land values

Shops, offices | Mainly industry | Mainly housing

CBD ——— Distance ——→

As distance from CBD increases there is less competition so land values fall

B London's financial area

C Bradford, West Yorkshire

The old inner city

British towns and cities began to grow rapidly at the beginning of the Industrial Revolution in the early nineteenth century. Until this time, few settlements extended beyond the limits of their present-day CBD. As industry developed, it located on land that happened to be next to the city centre but which was, at that time, on the edge of the urban area. This land was also next to canals and, after the 1840s, railways. This allowed the factories to receive heavy and bulky raw materials and to transport manufactured goods. As urban areas continued to grow outwards, these early industrial sites found themselves in old inner city areas (photo **C**).

The early industries needed large numbers of workers, and the workers needed somewhere to live. The answer was to pack as many workers and their families into as small a space as possible. This was achieved by creating streets in a grid-iron pattern, and building houses in long, tightly packed rows (photo **D** and the eastern part of grid square 4055 on the OS map on pages 278–279). The small terraced houses, built before the days of planning, had few amenities either inside (no indoor toilet, bathroom, running water, sewerage or electricity) or outside (no gardens or open space).

The houses were built as close as possible to the factories. This was to allow the workers, who in those days had no other form of transport, to walk to work. It was common for a factory to be found at one end of the street and either a corner shop or a public house at the other. No land was wasted, and there was no open space (diagram **E**). During the 1960s and 1970s many of old inner cities were bulldozed and cleared. Unfortunately, although the high-rise tower blocks that replaced them had more amenities, they proved to be a social failure.

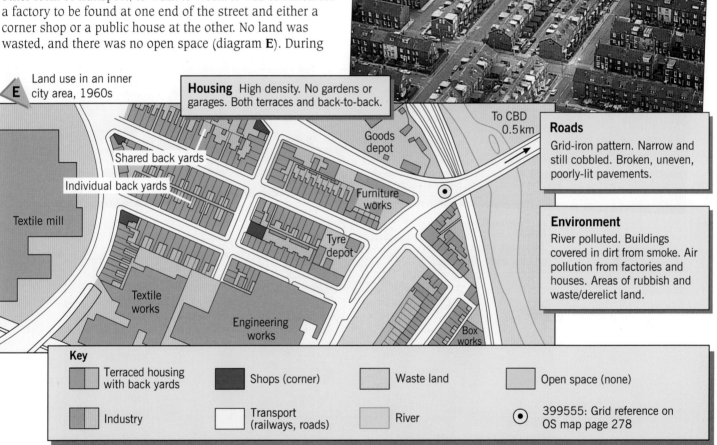

E Land use in an inner city area, 1960s

Housing High density. No gardens or garages. Both terraces and back-to-back.

Roads Grid-iron pattern. Narrow and still cobbled. Broken, uneven, poorly-lit pavements.

Environment River polluted. Buildings covered in dirt from smoke. Air pollution from factories and houses. Areas of rubbish and waste/derelict land.

Key

Terraced housing with back yards	Shops (corner)
Industry	Transport (railways, roads)
Waste land	River
Open space (none)	⊙ 399555: Grid reference on OS map page 278

Activities

1 How has the development of shops and offices in a city centre been affected by:
   a) accessibility b) land values?

2 With reference to old inner city areas:
   a) What were the three main types of land use?
   b) Which important type of land use was often absent?
   c) Give two reasons why industry located here.
   d) Describe the likely living conditions.
   e) Give two advantages of living here.
   f) Give two disadvantages of living here.
   e) What attempt was made in the 1960 and 1970s to try to improve living conditions? How successful was it?

Summary

The main characteristic of the CBD is the presence of shops and offices. In the inner city the main feature was the development of industry, housing and transport.

What were the suburbs and the rural–urban fringe zones like?

The suburbs

The outward growth of cities began with the introduction of public transport. It accelerated in the 1930s with the increased popularity of the private car and, in London, the extension of the Underground. This outward growth, known as **urban sprawl**, led to the construction of numerous private housing estates in car-based suburbs. These inter-war estates corresponded with Burgess's zone of medium-cost housing (diagram **A** page 156).

Inter-war housing was mainly semi-detached (photo **A**). The houses were characterised by their bay windows, front and back gardens and, often, a garage. The roads were wider, quieter and safer than those in the inner city, and crescent shapes and culs-de-sac replaced the grid-iron pattern (grid square 3957 on the OS map on page 278).

Roads
Tree-lined, wide. Crescent shapes and culs-de-sac.

Land use in the suburbs in the 1970s **C**

Housing
Low-density semi-detached with gardens and some garages.

Key
- Semi-detached houses with gardens
- Garages
- Transport
- Open space
- Tree-lined pavements
- 402577: grid reference on OS map page 278

Lansdowne Crescent
Beech Grove
Knowefield

Environment
Relatively clean, as less traffic and no industry. More open space. Quiet.

As well as individual gardens, the suburbs have larger amounts of public open space. This is mainly because land becomes cheaper away from the city centre, and partly to satisfy the growing demand for recreation and a more pleasant environment in which to live. The inter-war suburbs were characterised by their lack of industry, and residents had to travel increasing distances to their place of work (map **C**).

As cities expanded, each new housing estate was located further from the shops of the CBD. Consequently small parades of shops were built within each housing estate (photo **B**). Each parade was likely to include a sub-post office and a small chain store (providing mainly convenience goods), several specialist shops such as a butcher and chemist, and one or two non-essential services such as a hairdresser. Most parades had limited car parking space.

The rural–urban fringe

The rural–urban fringe is located at the edge of a town or city. It forms a transition zone where there is competition for land between the built-up area and the countryside. Urban sprawl continued during the 1960s, and still continues. Initially the land was used mainly for either private or council-built housing (photos **D** and **E**).

Private estates **D**

Low-density, high-quality housing. Most houses are large and detached with spacious gardens and often a double garage. They are likely to have modern amenities that include central heating, double-glazing and deluxe bathrooms and kitchens. The winding roads and numerous cul-de-sacs are usually wide, tree-lined and relatively traffic free (grid square 3958 on the OS map on page 278).

Outer city council estates **E**

High-density, average-quality housing. Created during the 1950s, 1960s and 1970s as local councils cleared the worst inner city areas and re-located residents on large edge-of-city estates (grid squares 4254 and 3854 on page 278). Housing was often in high-rise tower blocks, low-rise flats or single-storey link-houses. Most homes were small and lacked gardens and garages, but had modern amenities such as a bathroom and kitchen.

The rural–urban fringe often included large parks and woodland (grid square 3858 on page 278). More recently the range in land use has increased with the development of industrial estates and business parks (grid square 3959 on page 278), out-of town shopping centres, hotels and office development.

Activities

1 Make a larger copy of diagram **F**. The diagram, based on the OS map on page 278, has six of its grid squares numbered. Add the six labels from the following list to the appropriate grid square (pages 159–161 should help with your answer).
   - Small area of terraced housing
   - Open space with farmland
   - Inter-war suburb with open space
   - Inter-war suburb
   - Modern estate with private housing
   - Industrial estate, shopping centre and hotel

2 How do **a)** housing and **b)** other types of land use differ between the inter-war suburbs and the rural–urban fringe? Give reasons for your answers.

F
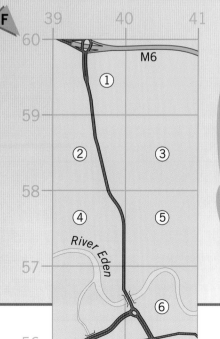

Summary

Whereas housing is the main characteristic of the inter-war suburbs, competition for land at the rural–urban fringe has led to range of land uses including modern housing, open space, business parks and shopping centres.

What are the main urban problems in London?

Most cities in the developed world grew rapidly during the nineteenth and twentieth centuries. Their main attraction was the large number of jobs available, together with people's perception that housing, transport, food supplies and services were all better in urban areas (page 145). The trend for these cities to grow continued until the late twentieth century when the increasing problems of living there began to outweigh the earlier advantages.

London has long attracted people from other parts of the British Isles and the wider world. It was the world's first city to have one million inhabitants, and reached a peak of 8 million just after the Second World War. At the beginning of the twenty-first century, that figure has fallen to 7 million as the movement out of the city, called **counter-urbanisation**, now exceeds the number of new arrivals.

Figure **A** describes some of the problems facing residents of London and their local authorities. By concentrating only on problems, rather than advantages, the diagram gives a rather a negative picture. In studying it you should remember that:
* many people still want to live in London and most do not want to move elsewhere
* the problems are often accentuated by London's age and size
* the extent and severity of the problems are not evenly spread across London and are greatest in the CBD and inner city areas
* these problems are typical of any large city in the developed world.

Urban deprivation and the cycle of poverty
Deprivation is a measure of how either individuals or groups of people are at a disadvantage compared with those living elsewhere. Deprivation can be measured by using four indicators: economic, social, housing and environmental (diagram **B**). The Department of the Environment suggested that there is an ongoing **cycle of poverty**. In this cycle, poverty is transmitted from one generation to the next, making escape from deprivation very difficult (diagram **C**).

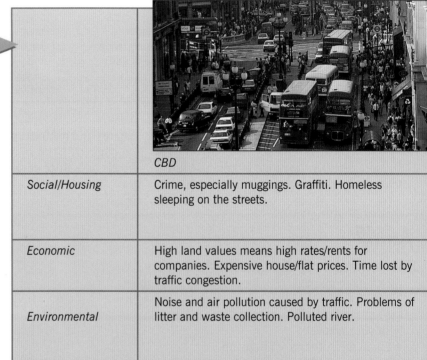

A		
		CBD
Social/Housing		Crime, especially muggings. Graffiti. Homeless sleeping on the streets.
Economic		High land values means high rates/rents for companies. Expensive house/flat prices. Time lost by traffic congestion.
Environmental		Noise and air pollution caused by traffic. Problems of litter and waste collection. Polluted river.

 B Deprivation factors

Economic stress
* unemployment, limited skills
* low-income families

Social stress
* single parents, large families
* crime, racial tension
* no family earner, many pensioners
* families without a car

Housing stress
* lacking one or more basic amenities (WC, running water, bathroom)
* overcrowding – more than 1 person per room

Environmental stress
* noise, pollution
* derelict land, empty buildings

C

THE CYCLE OF POVERTY

START

Low incomes → Poor living conditions, low standard of living → Environmental problems can create stress and cause poor health → Low educational prospects. Lack of educational resources, few opportunities for young people → Many young people leave school early with few job prospects → Poor living conditions and few prospects can lead to a rise in crime → Area becomes run down. Negative image created which discourages help and investment → Conditions get worse → Low incomes

INNER CITY

SUBURBS | *RURAL–URBAN FRINGE*

INNER CITY	SUBURBS	RURAL–URBAN FRINGE
Racial tension. Large families. Residents mainly immigrants, ethnic groups, pensioners, disabled and single-parent families. Few car owners. Poor-quality housing/high-density/lacking amenities. Health problems. Crime, vandalism, graffiti.	Crime (burglary). Increasingly expensive housing.	Crime (burglary). Very expensive housing.
Lack of jobs, high unemployment, lower-paid jobs. Mainly semi-skilled/unskilled jobs. Expensive to maintain services (health, education) and to pay/keep nurses and teachers.	Time–distance–cost of commuting to work if in city centre. Underground expensive, unreliable, dirty.	
Noise and air pollution from traffic. Litter. Problems of waste collection. Graffiti. Derelict buildings (houses/factories) and waste land.	Noise and air pollution from traffic on main roads leading to city centre.	Noise and air pollution from traffic on main roads leading to city centre. Loss of greenfield sites due to urban sprawl.

Activities

D

London's problems
- Problems of growth
- Social problems
- Economic problems
- Environmental problems
- The cycle of poverty

1 **a)** What is meant by the terms *urban deprivation* and *cycle of poverty*?
  b) Briefly describe the cycle of poverty.

2 Write a report on the urban problems of London, using the headings in diagram **D**.

Summary

The growth of large cities such as London has created social, economic and environmental problems.

Who makes the decisions in urban planning?

A Who are the decision makers?

Until fairly recently towns and cities grew in a haphazard way. Nowadays their growth is planned and controlled. No change is meant to take place unless **planning permission** has been given. Getting planning permission involves several groups of people. Some of these groups are likely to be in favour of the plan, and some will be against it. As each group has a role to play before the final decision is made, it can often take a long time before the plan is finally accepted or rejected.

B How is a planning decision made?

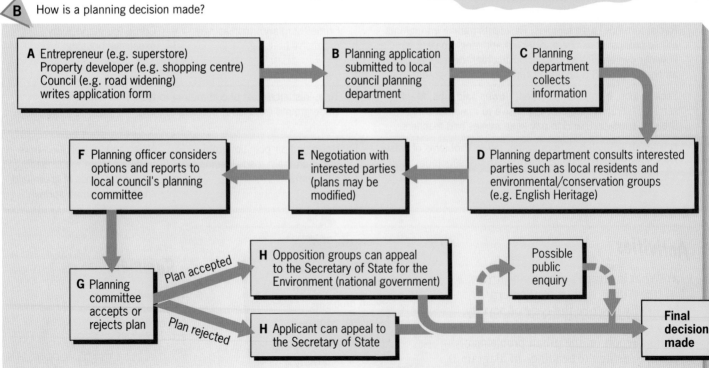

We all are likely to be involved at some time in the planning process. For many, participation may be little more than submitting an application for a house extension, or passing comment by means of a questionnaire on the design of a new shopping centre. For others, it means becoming more actively involved in, for example, the siting or improvement of a sports stadium, the construction of a road by-pass, or building on greenfield sites (page 175).

After the football disaster at Hillsborough in Sheffield in 1989, the government decreed that all major stadiums had to become all-seater venues. Each football club had, initially, to make the decision:
- either to reconstruct its existing stadium, or
- to move to a different site and develop a new one.

The club then had to go through the planning procedures illustrated in diagram **B**.

C

Photo **C** shows what one Premier club ground looked like in 1992. The club decided to improve its existing stadium, which was near to the city centre, rather than move to a new one. The main objections to the proposed development came from residents and others wishing to preserve the character of the Victorian houses seen in the background (behind what was then an all-standing area). The plan was accepted and completed by 1993.

Several years later, the club made a further planning application to increase the ground's capacity from 36 000 to over 50 000. Their preferred option was to build a new stadium 1 km away (to bottom left of photo **D**). When this led to opposition from local conservation groups, the club applied to enlarge the existing ground. This application was accepted, although the stand in front of the houses has had to remain at its 1992 height (on far side of photo).

D

Activities

1 **a)** Why does any change in a town or city have to get planning permission?
   b) What is the role of these groups of people in the planning procedure?
   - *entrepreneurs* • *property developers*
   - *local community groups* • *local government* • *central government*

2 Choose a recent plan put forward in your nearest town or city. The plan might be for a shopping centre on farmland on the edge of the town or a housing development on land at present used as a play area.
   a) Which groups of people are likely to support the plan and which groups are likely to be against it?
   b) What would be *your* decision?

Summary

Various groups of people will have different opinions and will play different roles when making decisions that affect urban development and planning.

What changes have taken place in the CBD?

The CBD has already been defined as the commercial and business centre of a town or city (page 158). As such, it is the centre for:

- shopping, with large department stores and numerous specialist shops (e.g. clothes, shoes and opticians)
- most offices, many of which are located in high-rise buildings or above shops in the main streets
- services such as banks and building societies, insurance companies and estate agents
- public buildings, including entertainment (theatres and cinemas), culture (museums and art galleries) and civic amenities (town or city hall).

The CBD is usually a place of change with new shops and shopping centres, taller buildings and altered traffic schemes.

Shopping in the CBD – before the 1980s

- Commercial and shopping centre. Accessible, as most main roads meet here.

- Area of largest number of shops, biggest shops, and most shoppers.

- Large department stores and superstores which can afford the high land values.

- Comparison shops (e.g. clothes, shoes) where style and prices can be compared.

- Specialist shops (e.g. jewellery, furniture, electrical goods).

- Small food shops (e.g. bakers, grocers, butchers and fishmongers).

- Some out-of-door pedestrian precincts, but most streets shared between cars, buses, delivery lorries and shoppers.

Oxford Street, London

Why have CBDs changed?

- As the CBD became busier, there was an increase in traffic congestion and air pollution. This led to several shops moving to out-of-town locations, and to the introduction of traffic-free areas.
- The development of hypermarkets and out-of-town shopping centres (page 172) led to a decline in the number of shoppers visiting the city centre. Some former shops have had to change their function to non-retailing activities.
- The decline in the number of shoppers has continued with the increase of internet shopping.
- People visiting the city centre wanted a safer and a more pleasant environment. This included protection from traffic and the weather (photo **B**), the planting of trees and shrubs, and the provision of seats (photo **C**).
- There was an increasing demand for leisure amenities, including places for eating, drinking and entertainment.
- Many taller buildings were built in order to offset the costly rates and rent resulting from the high land values (see diagram **A** and photo **B** on page 158).

How have CBDs changed?

- There has been an increase in both pedestrianised precincts and covered shopping malls. The undercover malls protect shoppers from:
    - traffic, by reducing accidents and air pollution
    - the weather, especially rain, cold and the wind.

  They also reduce the distance shoppers have to walk between shops.

- An increase in the number of food supermarkets has caused the closure of many specialist shops such as butchers, fishmongers and small grocers.

- Many furniture and carpet shops have moved out of the city centre. They have relocated either on vacant space in inner city areas or on cheaper land at the rural/urban fringe (page 172).

- There has been an increase in nationwide chain stores at the expense of family-run businesses.

- The space created by those shops being forced to close or opting to move to edge-of-city sites has been filled by building societies, estate agents, restaurants and open-air cafés, theme bars and even more clothes shops.

B The Whitgift Centre, Croydon

Sheffield ◄ **C**

D

- a delivery van driver
- the head of a small family-owned shop
- an elderly shopper
- an assistant in a shopping mall shop
- a teenager

Activities

1. **a)** What types of shop are most likely to be found in the city centre?
   b) What types of land use, other than shops, can be found in the city centre?
   c) Give three reasons why shopping malls have been created in city centres.
   d) What are the main types of shops likely to be found in a city centre shopping mall?
   e) Describe three changes, other than the creation of shopping malls, that have taken place recently in city centres.

2. How may recent changes in the CBD have affected the groups of people named in diagram **D**?

Summary

The CBD is constantly undergoing change as solutions are sought to overcome its social, economic and environmental problems.

What changes have taken place in London's Docklands?

During the nineteenth century the port of London was the busiest in the world. Surrounding the dock were:

- numerous industries using imported goods
- high-density, poor-quality housing typical of old inner city areas.

During the 1950s the size of many ships increased so much that they were unable to reach London's docks. By the 1970s, the area had become virtually derelict, with few jobs, few amenities and poor living conditions. Many families were forced to leave the area to look for work and a better quality of life elsewhere. This was because:

- traditional jobs in the docks, which had been manual, unskilled, unreliable and poorly paid, had almost disappeared
- much of the housing was substandard, lacking basic amenities and located in a poor-quality environment.

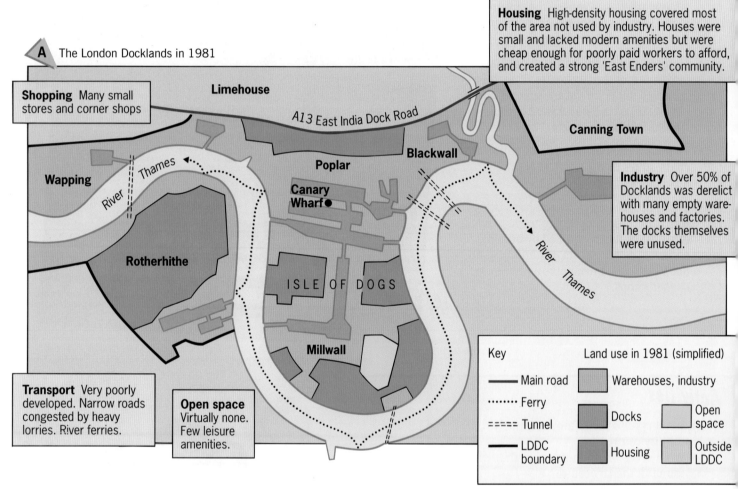

A The London Docklands in 1981

Shopping Many small stores and corner shops

Housing High-density housing covered most of the area not used by industry. Houses were small and lacked modern amenities but were cheap enough for poorly paid workers to afford, and created a strong 'East Enders' community.

Limehouse

A13 East India Dock Road

Canning Town

Blackwall

Poplar

Wapping

River Thames

Canary Wharf●

Industry Over 50% of Docklands was derelict with many empty warehouses and factories. The docks themselves were unused.

Rotherhithe

River Thames

ISLE OF DOGS

Transport Very poorly developed. Narrow roads congested by heavy lorries. River ferries.

Open space Virtually none. Few leisure amenities.

Millwall

Key	Land use in 1981 (simplified)
—— Main road	Warehouses, industry
······· Ferry	
===== Tunnel	Docks / Open space
—— LDDC boundary	Housing / Outside LDDC

The London Docklands Development Corporation

In 1981 the London Docklands Development Corporation (LDDC) was set up to try to improve the social, economic and environmental conditions of the area (map **A**). The LDDC was given three main tasks:

1 To improve **social conditions** by creating new housing and recreational amenities and improving shopping facilities.

2 To improve **economic conditions** by creating new jobs and improving the transport system both to and within the area.

3 To improve **environmental conditions** by reclaiming derelict land, cleaning up the docks, planting trees and creating areas of open space.

SOCIAL IMPROVEMENTS	LONDON DOCKLANDS 2001	ECONOMIC IMPROVEMENTS

SOCIAL IMPROVEMENTS

1 Housing
- 22 000 new homes created. Many are luxury flats in former warehouses.
- 10 000 refurbished local authority former terraced houses.

Population rose from 40 000 in 1981 to 85 000 in 2000.

2 Services
- Several huge new shopping malls.
- Post-16 college and campus for the new University of East London.
- Leisure facilities including watersports marina and a national indoor sports centre.

LONDON DOCKLANDS 2001

ENVIRONMENTAL IMPROVEMENTS
- 750 hectares of derelict land reclaimed.
- 200 000 trees planted and 130 hectares of open space created.

ECONOMIC IMPROVEMENTS

1 Employment
- Number of jobs rose from 27 000 in 1981 to 90 000 in 2000.
- Many new firms and financial institutions, e.g. Stock Exchange, ITV studios, newspaper offices.
- Many high-rise office blocks, especially at Canary Wharf.

2 Transport
- Docklands Light Railway links the area with central London.
- Jubilee Line Underground extension.
- City Airport.
- Many new roads, including M11 link.

B

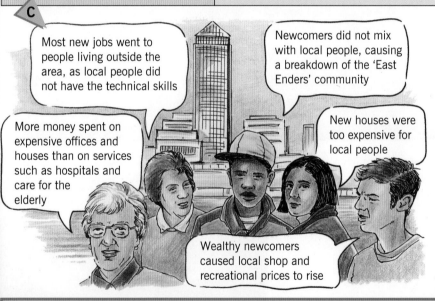

Most new jobs went to people living outside the area, as local people did not have the technical skills

Newcomers did not mix with local people, causing a breakdown of the 'East Enders' community

More money spent on expensive offices and houses than on services such as hospitals and care for the elderly

New houses were too expensive for local people

Wealthy newcomers caused local shop and recreational prices to rise

C

The London Docklands in 2001

Although the LDDC was wound down in 1998, changes continue at a rapid rate. Diagram **B** shows some of the social, economic and environmental improvements that had been made in the former Docklands by 2001. However, not everyone has benefited from the changes. Diagram **C** expresses the views of many local people.

D

Social	Housing	1
		2
	Services	1
		2
Economic	Transport	1
		2
	Jobs	1
		2
Environmental		1
		2

Activities

1 a) Using a star diagram, describe five problems facing people who lived and worked in the London Docklands in about 1980.

b) Describe the three main reasons why the London Docklands Development Corporation was set up in 1981.

2 a) On a larger copy of table **D**, describe some of the changes that had taken place by 2001.

b) What have been the advantages and disadvantages of the changes to the following people?
- *A school leaver*
- *A local shopkeeper*
- *A former docker*
- *A local retired couple*
- *A worker in a high-tech office*

Summary

The London Docklands is an example of how land use, jobs and the way of life of people who live there have changed.

How has Trafford Park in Manchester been regenerated?

It was the opening of the Manchester Ship Canal in 1894 that led to the development of Trafford Park. The park, surrounded by inner city terraced housing, became the world's first planned industrial estate. By the early 1980s, however, there was increasing concern over its decline. There was an urgent need for the area to be **regenerated**, meaning 'brought back to life'. In order to tackle the underlying problems of an outdated infrastructure and a lack of investment, the Trafford Park Development Corporation was created in 1987. Photo **A** shows how the park had developed by 1997.

The Corporation's first task was to transform an out-dated, semi-derelict industrial estate into a modern and thriving business location where leading companies would want to invest. The plan involved:

*   identifying four major development areas:
    1   **Wharfeside** – commercial development alongside the Manchester Ship Canal (photo **Ba**)
    2   **Village** – a former housing area in the centre of the park to become a thriving focal point (photo **Bb**)
    3   **Hadfield Street** – over 100 small to medium-sized firms in an improved industrial zone (photo **Bc**)
    4   **Northbank Industrial Park** – a modern industrial park on the site of a former steelworks 4 km to south-west of Trafford Park
*   improving transport – 50 km of new/upgraded roads; extending the Metrolink within the park; improving communications to the city centre and to adjacent motorways
*   improving the environment – creating the Ecology Park and canal walkways; improving frontages to buildings
*   improving services and recreation facilities – White City and the Trafford Centre retail parks (photo **Bd**); restaurants, pubs and a hotel; a regional sports complex and, for 2002, the Commonwealth Games site.

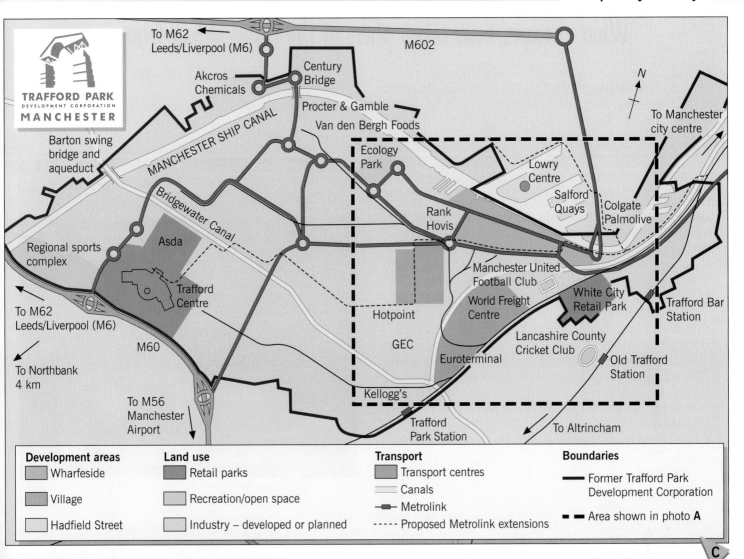

Map showing Trafford Park, Manchester.

Development areas
- Wharfside
- Village
- Hadfield Street

Land use
- Retail parks
- Recreation/open space
- Industry – developed or planned

Transport
- Transport centres
- Canals
- Metrolink
- Proposed Metrolink extensions

Boundaries
- Former Trafford Park Development Corporation
- Area shown in photo **A**

By the time the Development Corporation was wound up in 1998, Trafford Park had 46 000 new jobs and over 1800 new companies, 38 from overseas.

Map **C** shows the major features of Trafford Park in 2000.

Activities

1 Photo **A** covers part of the area shown in map **C**. With the help of the map, name:
   a) the transport features numbered 1 and 2
   b) the open space/recreation areas numbered 3 and 4
   c) the firms with factories at 5 and 6
   d) the types of land use at 7, 8 and 9
   e) the major development area at 10.

2 Describe the improvements made so far at Trafford Park under the headings:
   - *Transport*
   - *Employment*
   - *The environment*
   - *Retailing*
   - *Recreation*.

Summary

Trafford Park is an example of an old inner city area that has undergone regeneration and changes in land use.

What changes have taken place at the rural—urban fringe?

Recently there has been increasing competition for the use of land at the rural—urban fringe. The main reasons are that at the fringe there is:
- cheaper land (diagram **A** page 158)
- less traffic congestion and pollution
- easier access and a better road infrastructure
- a pleasant environment with more open space.

Places at the rural—urban fringe that have not been built upon are known as **greenfield sites** (photo **A**). These sites, still used for farming and recreation, are under constant threat for:
- housing developments as urban sprawl continues and nearby villages become suburbanised (page 174)
- science and business parks with their mainly high-tech firms (page 218)
- hypermarkets, superstores, retail parks and regional shopping centres – diagram **B** gives several location factors explaining why large shopping centres such as Meadowhall (photo **D**) are attracted to these sites
- office development now that modern technology and easier access to information allows firms a freer choice of location
- hotels and conference centres which, like science parks and office blocks, often stand in several hectares of landscaped grounds

- road development schemes, including motorways and urban by-passes
- sewage works and landfill sites for urban waste
- recreational areas such as country parks, playing fields and new sports stadiums.

A Settlement on the rural—urban fringe

B

Out-of-town location where rates and rents are cheaper due to lower land values. Allows for a larger floor area and cheaper prices.

Near to several large urban areas for customers.

Plenty of space for possible future expansion.

Near to suburban estates which provide the workforce – many employees are women working part-time.

Hypermarkets can stock a large volume and a wide range of goods.

Out-of-town site to avoid opposition from CBD shop-owners.

Large free car parks. Hypermarkets are aimed at motorists. MetroCentre and Meadowhall each have over 12 000 parking spaces.

Large single-storey buildings, over 6000 m².

Near a main road or, ideally, a motorway intersection for easy delivery of goods and access for shoppers from several urban areas. No traffic congestion as in CBD.

Meadowhall – a regional shopping centre

Meadowhall is located on the site of a former steelworks on the urban fringe of Sheffield (map **C**). The site covers 56 hectares, has direct access to the M1 motorway, and was a vital part of the Don Valley regeneration scheme.

Nine million people live within one hour's driving time of the centre. For those arriving by car, there are 12 000 free parking spaces. The purpose-built bus station can handle 120 buses an hour, while Sheffield's new Supertram links the site with the city centre (page 179). The majority of shoppers are relatively young and are car owners. They are prepared to travel considerable distances to find free, easy parking and to buy in bulk.

Other regional shopping centres include Gateshead MetroCentre, Manchester Trafford Park (pages 170–171), Thurrock Lakeside, Dartford Bluewater, and Dudley Merry Hill. Like Meadowhall, all these centres provide a wide range of associated amenities all under cover in a pleasant environment (photo **D**). Apart from the huge range of shops, the centres provide cinemas, restaurants, children's play areas and petrol stations. The emphasis is to provide families with a day out.

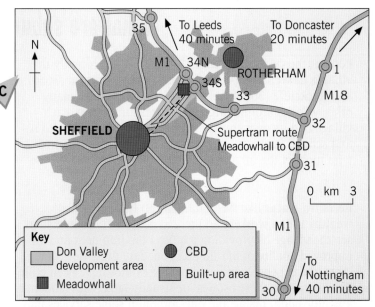

C

Inside Meadowhall, Sheffield

D

Activities

1 **a)** What is a *greenfield site*?
   b) Why are greenfield sites attractive for urban development?
   c) Name six likely competitors for greenfield sites.

2 A large national supermarket chain wishes to build a hypermarket in the area shown on map **E**. Five available sites have been shown as J to N. Some sites will not be acceptable to the supermarket chain, and other sites will not be acceptable to local people and the planners.
   a) Rank the five sites in the order which you think that the supermarket chain will prefer. Give reasons for your answer.
   b) Which of the sites do you think will:
   i) cause most opposition from local people
   ii) be rejected by the planners?
   Give reasons for your answer.
   c) On which site do you think that the hypermarket will eventually be built? Suggest reasons.

3 With the help of the information on diagram **B** and map **C**:
   a) Give five advantages of building a regional shopping centre at Meadowhall on the edge of Sheffield.
   b) Why are regional shopping centres like the one at Meadowhall a good place for:
   i) family shopping
   ii) a day out for the whole family?

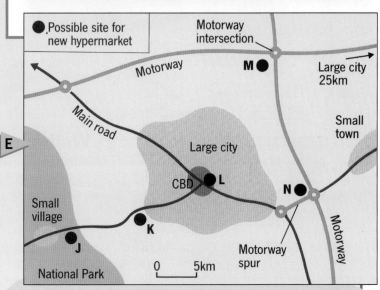

E

Summary   There is continuous competition for the use of greenfield sites at the rural–urban fringe, including that from regional shopping centres.

What are suburbanised villages?

Look at the OS map on pages 278–279. Until recently the villages of Wetheral (4654), Scotby (4454), Dalston (3650) and Rockcliffe (3561) were all quiet villages. Like many other villages around large urban areas in Britain, each village had its own church, shop-cum-post office (photo **A**) and school. Such settlements were largely self-contained with most of the inhabitants finding jobs in or near to the village.

Recently, places like Wetheral (map **B**) have attracted wealthy urban workers and retired people. These in-comers see the village as providing a quieter environment and an improved quality of life. To accommodate these newcomers, large and expensive private housing estates have been built. The houses on these estates are usually large with modern amenities both inside and out. Although some services may have improved, the rural appearance of the villages has changed. They increasingly begin to look like an extension to the suburbs of adjacent towns – hence the term **suburbanised village**. Many settlements are also **commuter villages** from where many of the inhabitants travel to work in nearby towns. The local community may be swamped by newcomers and quite often is divided into two groups. The narrow roads are rarely suited to the increase in traffic.

A Wetheral

Photo **A** was taken near the northern end of the village green. It shows, from left to right: an advert for a restaurant; a signpost pointing to both a conference centre in a hotel and the railway station; a noticeboard for a doctors' branch surgery; and the village shop/post office.

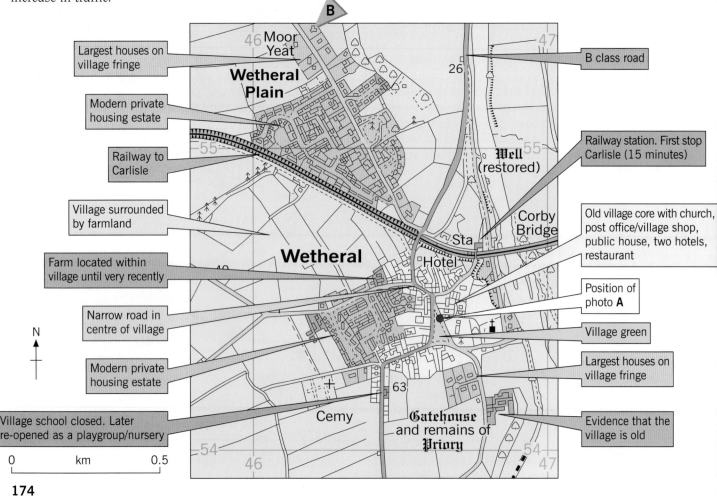

B

- Largest houses on village fringe
- Modern private housing estate
- Railway to Carlisle
- Village surrounded by farmland
- Farm located within village until very recently
- Narrow road in centre of village
- Modern private housing estate
- Village school closed. Later re-opened as a playgroup/nursery
- B class road
- Railway station. First stop Carlisle (15 minutes)
- Old village core with church, post office/village shop, public house, two hotels, restaurant
- Position of photo **A**
- Village green
- Largest houses on village fringe
- Evidence that the village is old

Moor Yeat · Wetheral Plain · Well (restored) · Corby Bridge · Sta · Hotel · Wetheral · Cemy · Gatehouse and remains of Priory

N

0   km   0.5

Should planners favour greenfield or brownfield development?

Britain is short of houses. Estimates suggest that over 4 million new homes could be needed by 2016. Assuming this prediction to be correct, the question then is, 'Where will all these new houses be built?' In 1998 the government announced that:

- 50 per cent would be on **brownfield sites**, i.e. disused land within existing urban areas
- 50 per cent would be on **greenfield sites**, i.e. in the countryside, including possible sites in **green belts**.

A green belt is land surrounding a large urban area that is protected from urban development.

After a major public outcry to 'save the countryside', the government revised the figures to 60 per cent brownfield and 40 per cent greenfield sites. However, the National Data Base shows a mismatch between the South East of England, where most houses are needed but where brownfield sites are limited, and the Midlands and the North, where most brownfield sites are available but where demand for new homes is less. Figure **C** shows some of the arguments as to why some groups of people would prefer most of the new developments to be on brownfield sites and why others favour greenfield sites.

C

Why build on brownfield sites?	**Why build on greenfield sites?**

Groups such as the Council for the Protection of Rural England (CPRE) and Friends of the Earth argue that:
- there are already three-quarters of a million unoccupied houses in cities that could be upgraded
- a further 1.3 million could be created by subdividing large houses or using empty space above shops/offices
- 1.6 million could be built on derelict land or re-using old industrial/commercial premises
- urban living reduces reliance upon the car and maintains city centre services, e.g. retailing.

(The government assumes that 80 per cent of the demand for new houses will come from single-parent families who prefer, or need, to live in cities.)

Developers claim that:
- most British people prefer to own their own home set in a rural/semi-rural location
- people are healthier and often have a better quality of life in rural areas
- at present, for every three people moving into cities, five move out into the countryside – a net loss to urban areas of 90 000 a year
- greenfield sites are cheaper to build on as they have lower land values and are less likely than urban land to need cleaning up
- only 11 per cent of Britain is urbanised – and still only 12.5 per cent if all new homes are built on greenfield sites.

Activities

1 **a)** What is a *suburbanised village*?
   b) Make a list of differences in i) a village and ii) its inhabitants **before** and **after** it became suburbanised.

2 **a)** What is the difference between a *brownfield site* and a *greenfield site*?
   b) Imagine that 500 new homes are needed in your local area. Why might some groups favour these houses being built in your nearest town? Why might other groups want the houses built on land surrounding a small nearby village?

Summary As more British people want to move away from cities and as the demand for new homes increases, there is growing pressure to build in villages and on greenfield sites.

Why is traffic a problem in urban areas?

The volume of traffic on British roads in 2001 was nearly ten times greater than it was in 1950. During that period the number of cars increased from 2.5 million to over 30 million and the number of heavy goods vehicles (HGVs) and buses rose from under 1 million to over 5 million. By 2030, the number of cars could double and the amount of freight carried could be three times greater than in 2001.

Traffic in general, and the car in particular, is the cause of many problems within urban areas. The volume of traffic has increased due to:
- greater affluence and increased car ownership – including more two-car families
- a reduction in public transport at the expense of private cars
- an increase in the volume of goods being moved by road, e.g. more delivery lorries
- more people commuting to work or travelling to city centres or new edge-of-city developments for shopping and entertainment (diagram **A**).

In many cities, in both Britain and across the world, the widespread use of the car has brought traffic to a near stand-still and is a major cause of environmental, economic and social problems (diagram **C**).

Commuting
Daily movement in urban areas shows a distinctive pattern with two peaks. The first is when most people travel to work (the morning rush hour), the second when they return home (the evening rush hour).

A **commuter** is a person who lives:
- either in the suburbs of a large city, or
- in a village or small town that surrounds a larger town or city.

In each case, that person travels to their place of work in, or near to, the city centre.

The increase in car ownership and the process of **counter-urbanisation**, which sees people moving away from large cities, means that more people are living further from their place of work. A recent trend has seen a partial reversal in the direction of movement of some commuters as firms close down in inner city areas and re-locate on the edges of cities (pages 172 and 218).

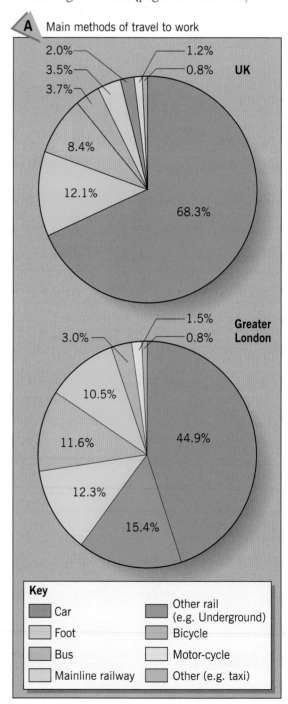

A Main methods of travel to work

UK
- 68.3%
- 12.1%
- 8.4%
- 3.7%
- 3.5%
- 2.0%
- 1.2%
- 0.8%

Greater London
- 44.9%
- 15.4%
- 12.3%
- 11.6%
- 10.5%
- 3.0%
- 1.5%
- 0.8%

Key
- Car
- Foot
- Bus
- Mainline railway
- Other rail (e.g. Underground)
- Bicycle
- Motor-cycle
- Other (e.g. taxi)

B

Air pollution – total percentage from vehicles:
- smoke   46%
- nitrogen oxide   51%
- carbon monoxide 90%
- carbon dioxide   21%
- sulphur dioxide   2%
- ground-level ozone   9%

| ECONOMIC | WHAT ARE THE DAMAGING EFFECTS OF TRAFFIC? | ENVIRONMENTAL |

ECONOMIC

- Congestion, especially at peak times. In central London, even though many people travel by public transport, the average speed of traffic in 2000 was the same as in 1900 – 18 km/hr.

- Time wasted when sitting in traffic jams and gridlock conditions, looking for parking places, or avoiding parked cars and unloading lorries.

- Cost of building and maintaining roads. Delays resulting from repairs to roads or to underground utilities.

- Cost of petrol or diesel and the using-up of a non-renewable resource (oil).

- Loss of property (domestic and industrial) for road widening and urban by-passes/ring roads.

WHAT ARE THE DAMAGING EFFECTS OF TRAFFIC?

C

ENVIRONMENTAL

- Air pollution from vehicle exhausts (diagram **B**) worst in large cities.

- Noise from cars, lorries and buses.

- Visual pollution of urban motorways and car parks.

- Loss of land for road widening and car parks.

SOCIAL

- Danger of accidents.

- Health problems:
  - stress to both drivers and pedestrians
  - respiratory illnesses caused by car fumes
  - asthma from low-level ozone.

- Property damaged by vibrations caused by traffic.

Activities

1 a) Draw a simple bar graph to show the increase in cars, lorries and buses between 1950 and 2000.

   b) Why has the number of cars in urban areas increased during this period?

2 a) What is meant by the terms *commuter* and *rush hour*?

   b) How do commuters add to traffic problems in urban areas?

3 a) How can traffic problems in urban areas affect:
   - the economy
   - people
   - buildings
   - the environment?

   b) Diagram **D** was taken from a London newspaper. Why are daily reports such as these thought to be necessary?

D

	Yesterday's reading	Forecast
Nitrogen dioxide	**40 ppb**	**Very good**
(Less than 50 = Very good, 50–99 = Good, 100-299 = Poor, 300+ = Very poor)		
Ozone	**28 ppb**	**Very good**
(Less than 50 = Very good, 51–89 = Good, 90–179 = Poor, 180+ = Very poor)		
Sulphur	**5 ppb**	**Very good**
(Less than 60 = Very good, 60–124 = Good, 125–399 = Poor, 400+ = Very poor)		
Air quality is calculated in ppb (parts per billion)		

London Evening Standard, 9 June 1997

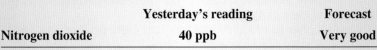

Summary   The increase in traffic in urban areas can create economic, environmental and social problems.

How may urban traffic problems be reduced?

Environmental considerations are likely to be given a high priority in all future urban transport plans. People have become increasingly concerned about the effects of transport upon:

- human health, such as the increased incidence of asthma and breathing problems linked to vehicle emissions
- the quality of their environment, including noise, congestion, fumes
- climatic change and global warming
- the economic cost of time wasted in traffic jams and in using a non-renewable source of energy.

Yet transport is not just an environmental issue. Future plans must allow all groups in the community to move around as safely, quickly and cheaply as possible. These plans must, however, be sustainable – they should improve people's mobility and safety without damaging the environment in which they live. Four such schemes at present in favour in the UK are described in fact files **A**, **B**, **C** and **D**.

A

Fact file A: Cycle tracks

- Encourage people to cycle instead of using their cars, which:
  - reduces car pollution
  - increases fitness of cyclists.
- Provide cycle tracks alongside existing main roads.
- Create new cycle paths that exclude the car.

B

Fact file B: Park and ride schemes

- Allow private car owners living in rural areas to travel where public transport systems are unavailable or inadequate.
- Provide free car parking, mainly at specially provided bus terminals, but sometimes at rail and metro stations on the edge of urban areas.
- Provide an efficient public transport system within urban areas.
- Reduce the number of vehicles entering the city centre.
- Reduce traffic fumes (air pollution), noise, congestion and land needed for car parking in the city centre.
- Are mainly bus-based, catering for commuters, shoppers and tourists working in or visiting the city centre.
- Are usually bus services that operate non-stop to the city centre charging a fare comparable with the cost of city centre parking.

Park and ride

C

Fact file C: Traffic in residential areas

Problems
- Increased volumes and speed of traffic:
  - creates safety problems for local residents
  - increases local air and noise pollution.
- Large vehicles use residential areas to avoid congestion on main roads caused by the volumes of traffic or by temporary roadworks.
- Parked cars limit the visibility of pedestrians.
- Joy riding.

Solutions
- Set up traffic calming schemes such as one-way streets, speed ramps, rumble strips and road narrowing.
- Limit parking to local residents by issuing residents-only permits.

Fact file D: Supertrams

The first supertrams ran in Manchester and Sheffield in the early 1990s. The one in Manchester was Britain's first purpose-built on-street light rail system.

- It operates on converted railway lines in the suburbs and on newly constructed routes in the city centre.
- Each supertram can carry 206 passengers (86 seated, 120 standing), greatly reducing the number of cars.
- The supertrams run at 6-minute intervals during weekdays.
- Being electrically driven, noise and air pollution is reduced although pylons and overhead power cables cause visual pollution.
- The supertrams and their stations are specifically designed to help the disabled/semi-disabled and pushchair users (each of these groups makes up 20 per cent of the total users).
- Cars have to give way to the trams in the city centre.

D

Activities

1 a) Describe four environmental factors that should be considered in new urban traffic plans.

b) Describe the advantages and disadvantages of three of the traffic schemes described in fact files **A**, **B**, **C** and **D**.

c) Describe two recent traffic schemes in your local area. How successful do you think each scheme has been?

2 a) Diagram **E** describes a scheme aimed at reducing congestion in a town centre.
- Where is the scheme?   • Why was it needed?
- Why do many people oppose the scheme?
- Why do many others support the scheme?
- Who are the 'winners'?   • Who are the 'losers'?

b) Do you think that, on balance, the scheme is good or bad? Give reasons for your answer.

A650 Bingley relief road

E

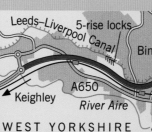

The problem
Bingley is 8 km north of Bradford in Yorkshire, on a bend in the narrow, high-sided valley of the River Aire. It has old, attractive stone buildings but its centre is blighted by congestion. Bingley is a bottleneck, with 38 000 vehicles passing through each day, 10–16 per cent of them HGVs. Long peak-time queues and increased pollution develop on the existing single carriageway.

Timetable
Work on the 5 km dual carriageway begins in July 2001 with a completion date in late 2003.

The conflict
Not a bypass, say opponents, but a 'duplicate trunk route' through the town. They claim it will blight other urban areas, generate extra traffic, spoil the canal conservation area, only shift the bottlenecks and ruin Bingley South Bog, a Site of Special Scientific Interest (SSSI).

The benefits
- Estimated to remove 60 per cent of traffic from the town centre.
- Will mean 730 fewer accidents over 30 years.
- Will reduce noise and air pollution.
- Will assist redevelopment and public transport facilities.
- New ponds created and better landscaping.

Winners and losers
- Less noise for 461 houses, more noise for 140–515.
- Improved air quality for 257 properties, reduced quality for 55.

Summary When attempting to reduce urban traffic problems, planners should aim to improve people's mobility and safety without further damaging the environment.

What are the problems of urbanisation?

The term **urbanisation** means the increase in the proportion of the world's population that live in cities. Urbanisation increased rapidly in developed countries between the mid-nineteenth and mid-twentieth centuries, and in developing countries from the mid-twentieth century (table **A**). Whereas between 1950 and 2000 the urban population in developing countries more than doubled, in developed countries the increase was less than half. Apart from urbanisation there have been two other rapid changes (map **B**).

A Proportion of the pollution living in cities

	1950	*2000*
World	30%	49%
Developed countries	53%	75%
Developing countries	17%	40%

1 The increase in **million cities**, i.e. places with over one million inhabitants. In 1850 the only two million cities in the world were London and Paris. By 2000 there were 324.

2 The increase in the number of million cities located in developing countries, especially those located within the tropics.

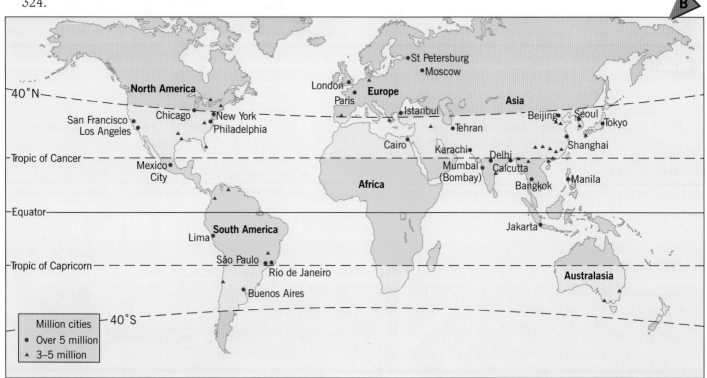

B

Million cities
- Over 5 million
▲ 3–5 million

Calcutta

Calcutta is an example of how problems are created when cities grow too quickly. The city is built on flat, swampy land alongside the River Hooghly which is part of the Ganges delta. Its population is believed to have grown from 7 million in 1970 to between 14 and 16 million in 2000. The increase is due to Calcutta's high birth rate and migration from the surrounding rural areas. The city authorities have little hope of providing enough new homes, jobs or services for the increasing population.

Housing Many families have no home and have to live on the streets (photo **C**). Nearly half a million people are reported to sleep in the open, covered only by bamboo matting, sacking, polythene or newspaper. Many more live in shanty settlements, which in India are called **bustees**. Bustee houses have mud floors, wattle or wooden walls and tiled or corrugated iron roofs – materials that are not the best for giving protection against the heavy monsoon rains. The houses are packed closely together and are separated by narrow alleys (photo **D**). Inside each house there is probably one small room in which the whole family, perhaps up to ten in number, live, eat and sleep. Bustee houses often lack electricity, running water and any means of sewage disposal.

Services There are relatively few schools and a lack of doctors and hospitals. Public transport is overcrowded and there is severe traffic congestion and pollution.

C Pavement dwellers, Calcutta

D A lane in the bustees

Water supply, sanitation and health
Although three-quarters of Calcutta's population has access to piped water, it is not uncommon in the bustees for a single tap to serve up to 40 families. Sewage often flows down the alleys. Where this contaminates drinking water it causes cholera, typhoid and dysentery. Rubbish, dumped in the alleys, provides an ideal breeding ground for disease. Many children have worms and suffer from malnutrition.

Employment Those with jobs tend to work in the informal sector (page 228), often using their homes as a place of work. The front of bustee houses can be opened up to allow the occupants to sell food, wood, clothes and household utensils (photo **E**). Few people in the bustees are totally unemployed, but most jobs only occupy a few hours a week.

Other problems include crime and the segregation of peoples of different caste,

language and religion. The Calcutta Metropolitan Development Agency, set up in 1970, has tried to make the bustees more habitable by paving alleys, digging drains and providing more taps and toilets. Prefabricated houses have been built and a better community atmosphere created. Even so, the lack of money has meant only relatively small areas have been improved.

E Spices stall in a Calcutta bustee

Activities

1 a) What is meant by *urbanisation*?
   b) Give three points to describe the distribution of million cities.

2 a) Describe the scenes in photos **C**, **D** and **E**. Mention the houses, alleys and jobs.
   b) What do you consider to be the worst problems of living in the bustees?

Summary

Urbanisation is the increase in the proportion of people living in cities. It is most rapid in cities in developing countries where it causes considerable problems.

What is a typical land use model?

We have already seen on page 156 that:
- towns and cities do not grow in a haphazard way but tend to develop recognisable shapes and patterns
- while each town or city will develop its own distinctive pattern, it is likely to share some characteristics with other settlements.

However, cities in developing countries tend to develop their own distinctive pattern which differs in several ways from a land use model in developed countries (diagram **A**). Three main differences are:

1 The gap between the relatively few rich and the numerous poor is much greater than in developed countries. This means that the contrast between well-off areas and the poorest areas, in terms of quality and the provision of amenities, is also greater.
2 Most of the better-off areas are located near to the city centre with increasingly poorer areas found towards the city boundary.
3 A large number of people, many of whom are migrants from surrounding rural areas, are forced to live as squatters in **shanty settlements** or, using the UN term, **informal settlements**.

As cities in developing countries grow, they are likely to develop a land use pattern that is more complex than the one shown in diagram **A**. Diagram **B** is therefore a more realistic map showing land use and functional zones.

Functional zones

The CBD Land use here is similar to that of a developed city with shops and office blocks, except that congestion and competition for space is even greater (photo **C**).

Industry Apart from the informal sector, which can be located in the CBD and the outer two residential areas, large factories tend to develop along main roads leading out of the city.

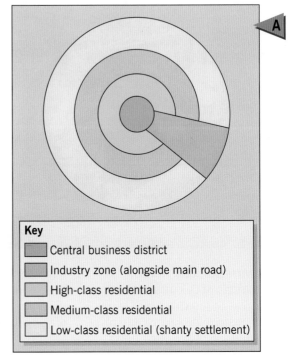

Key
- Central business district
- Industry zone (alongside main road)
- High-class residential
- Medium-class residential
- Low-class residential (shanty settlement)

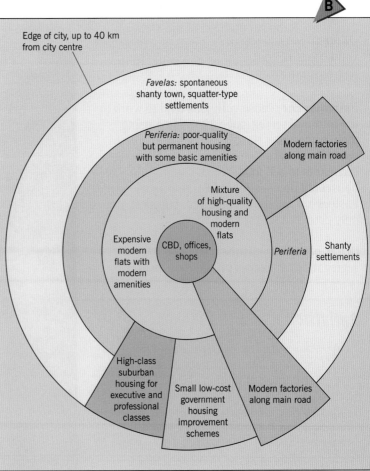

Edge of city, up to 40 km from city centre

Favelas: spontaneous shanty town, squatter-type settlements

Periferia: poor-quality but permanent housing with some basic amenities

Modern factories along main road

Mixture of high-quality housing and modern flats

Expensive modern flats with modern amenities

CBD, offices, shops

Periferia

Shanty settlements

High-class suburban housing for executive and professional classes

Small low-cost government housing improvement schemes

Modern factories along main road

Inner zone (high-class residential)

Many developing countries were former European colonies and it was in this inner zone that the wealthy landowners, merchants and administrators built their large and often luxurious homes. Although these houses have deteriorated over time, the well-off have continued to live here. Many now live in modern, high-rise, high-security apartments or in well-guarded detached houses (photo **D**).

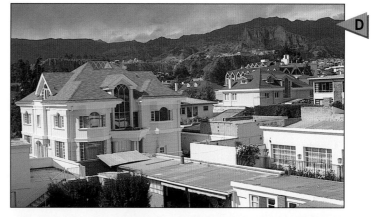

Middle zone (medium-class residential)

This zone, as in a city in a developed country, provides the 'in-between' housing, except that here it is of much poorer quality. In many cases it consists of self-constructed homes to which the local authorities may have added some basic amenities such as running water, sewerage and electricity (photo **E**). In some cases the authorities have encouraged self-supporting 'site and service' schemes (page 186).

Outer zone (low-class residential)

Unlike that in a developed city, the location of low-income housing is reversed as the quality of dwellings decreases rapidly with distance from the CBD. This is the zone where most of the recent arrivals from rural areas are forced to live, usually in shanty settlements. The inhabitants are mainly squatters who have no rights to the land upon which they build their flimsy, often temporary homes (photo **F**). These informal settlements are known as **favelas** in Brazil and **bustees** in India (pages 180, 184–185).

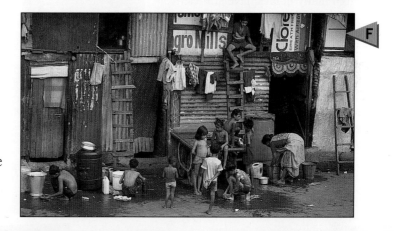

Activities

1 Make two copies of transect **G**.
  a) On the first, label the different types of residential land use that you might expect to find in a **developed city** (page 156).
  b) On the second, label the different types of residential land use that you might expect to find in a **developing city** (page 182).
  c) Describe the differences between the two transects.

2 Describe and suggest reasons for the location of:
  • poor-quality housing
  • high-quality housing
  in a city in a developing country.

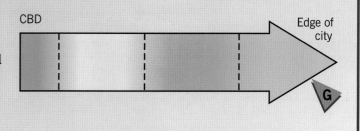

CBD   Edge of city

G

Summary

Cities in less economically developed countries have a different pattern of land use to those in more economically developed countries.

What is life like in shanty settlements?

Shanty settlements grow up well away from the CBD on land that previously had been considered unsuitable for building, for example:

- steep hillsides as in the case of Rio de Janeiro in Brazil
- swampy floodplains of rivers as in Nairobi in Kenya (map **A**).

Nairobi

The two largest shanty settlements are those that extend for several kilometres along the Mathare Valley (photo **B**) and in Kibera (photos **C** and **D**). Estimates suggest that over 100 000 people live in each area.

Housing Houses are built close together on any land that still remains vacant (photo **B**). Often it is difficult to squeeze between them. The walls are usually made from mud or wood and the roofs from corrugated iron (photo **C**). Inside there is often only one room. The earth floor is likely to be uncovered and furniture is limited. Although many homes have sufficient electricity to give a single light, few have running water or sewerage. Water comes from a tap that is likely to be shared by up to 40 families. Sewage is often allowed to run down the tracks between the houses (photos **C** and **D**). In the wet season, rain mixes with the sewage making the tracks muddy and, for young children, unusable. In the dry season, the tracks become very dusty.

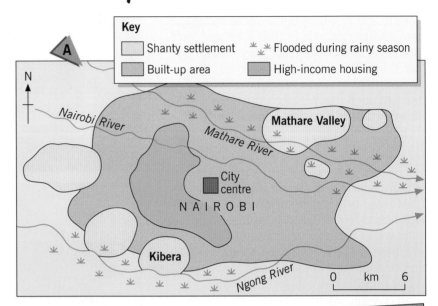

Key
- Shanty settlement
- Built-up area
- Flooded during rainy season
- High-income housing

Nairobi River · Mathare River · **Mathare Valley** · City centre · N A I R O B I · **Kibera** · Ngong River

0 km 6

Education and health Kenya has one of the highest birth and infant mortality rates in the world. These rates are even higher in the shanty settlements where it is not uncommon for families to have up to ten children. Very few children are likely to learn how to read or write as schools are few and far between or are too expensive to attend. Many children suffer from a range of illnesses: malaria due to the swampy land; dysentery and possibly cholera and typhoid from contaminated water; and various diet-deficiency diseases (page 275). There are no hospitals or qualified doctors in these settlements, and even if there were few residents would be able to pay for health care.

Shops Numerous stalls line the main roads. These sell everything from fresh fruit and vegetables, obtained from relatives still living in the surrounding rural areas, to clothes and household goods (photo **E**). Other residents sell goods from outside their homes.

Employment Virtually all jobs in shanty settlements are in the informal sector. This is where people have to find their own way to earn money (page 228). Some have small stalls from which they may sell fresh produce or cheaply made goods. Others, who have developed a skill, may collect waste material and recycle it in small workshops (photo **F**).

Transport Apart from those with bicycles, most people have to walk. Some, if they can afford to, may travel on dilapidated, overcrowded buses.

Community spirit Although there is crime because people are forced to extreme measures to survive, many shanty settlements have a good community spirit as survival can also depend upon living and working together.

Activities

1 **a)** With reference to map **A**, describe the location of shanty settlements in Nairobi.
   b) What is a shanty settlement and who lives in them?

2 Photos **B–F** show some aspects of life in a shanty settlement. With the help of diagram **G**, describe the living conditions and the problems that face people who live there.

a) Why are the houses built so close together?

b) What are the houses made from?

c) What household amenities are missing?

d) What are the main health problems?

e) Why are most children unable to read or write?

f) What are the shops like?

g) What types of jobs are people likely to do?

h) Why is transport a problem?

Summary The poorest people in a city in a developing country have to live under considerable hardship in shanty settlements.

185

How successful have self-help schemes been?

It is probable that local authorities in most developing countries would prefer, if given the choice, to remove shanty settlements altogether from their cities. However, few of them would be likely to find the necessary resources needed to provide alternative accommodation for so many people. The result is that shanty settlements become permanent. Low-cost housing and self-improvement schemes seem, therefore, the only hope for the squatters to improve their homes and quality of life. The most immediate needs of the poor are often simple: a small plot of land on which to build, a small loan to improve or extend the house, the availability of cheap building materials, and the provision of basic amenities such as electricity, running water and sewage disposal.

São Paulo, Brazil
Several schemes in São Paulo's *periferia* (diagram **B** page 182) have resulted, over a lengthy period of time, initially in the upgrading of living conditions (diagrams **A** and **B**) and, later, the introduction of shops and small-scale industries. Even so, most of the people remain poor.

A

Low-cost improvements
Existing homes may be improved by rebuilding the houses with quick, cheap and easy-to-use breeze-blocks. A water tank on the roof collects rainwater and is connected to the water supply and, in turn, to an outside wash basin and an indoor bathroom/toilet. Electricity and mains sewerage are added. Most inhabitants of this type of housing, which is found in the peripheral parts of São Paulo, will have some type of employment enabling them to pay a low rent.

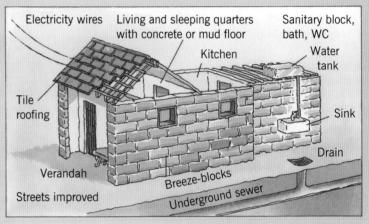

B

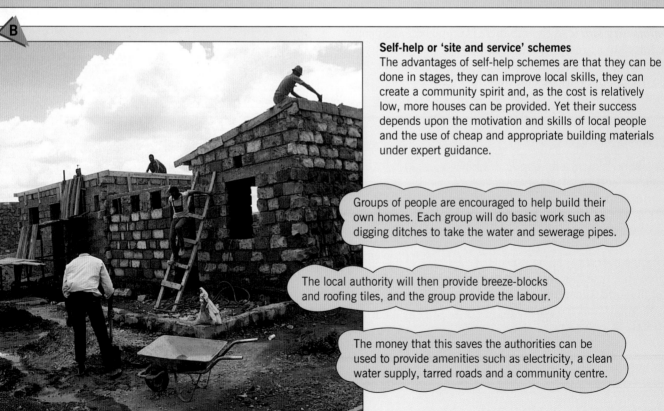

Self-help or 'site and service' schemes
The advantages of self-help schemes are that they can be done in stages, they can improve local skills, they can create a community spirit and, as the cost is relatively low, more houses can be provided. Yet their success depends upon the motivation and skills of local people and the use of cheap and appropriate building materials under expert guidance.

Groups of people are encouraged to help build their own homes. Each group will do basic work such as digging ditches to take the water and sewerage pipes.

The local authority will then provide breeze-blocks and roofing tiles, and the group provide the labour.

The money that this saves the authorities can be used to provide amenities such as electricity, a clean water supply, tarred roads and a community centre.

Intermediate technology in Nairobi, Kenya

The Intermediate Technology Development Group (ITDG) is a British charitable organisation that works with people in developing countries. It helps people in Africa, Asia and South and Central America to develop and use tools, skills and methods that are appropriate to their local environment and communities. It uses and adds to local knowledge by providing advice, training, basic equipment and financial support so that people can become more self-sufficient and independent. This allows local people more control over their lives and helps them to contribute to long-term sustainable developments.

In Kenya, several of the ITDG's projects involve the development and production of various building materials that are:
- suitable for the local environment (local raw materials and climate)
- affordable and usable in self-help housing schemes (diagrams **C** and **D**).

C

One scheme is the production of low-cost roofing tiles from locally obtained clay. This has reduced the cost of roofing and developed the skills of local people. In another scheme, lime and natural fibres are added to soil to produce 'soil blocks'. Soil is important because it can be obtained locally, can easily be compressed and, once heated, retains its warmth. Soil blocks are now replacing the more expensive concrete blocks and industrially produced bricks.

D

Other schemes have helped to improve ventilation and lighting in existing houses. Most city-dwelling Kenyans have, traditionally, cooked on wooden stoves in houses that had no chimneys and few windows. The result was a smoky and unhealthy atmosphere. To reduce the reliance upon wood and to improve living conditions, ITDG has helped train potters to produce improved cooking stoves. The stove, based on a traditional design, is made from scrap metal (page 229) to which potters add a ceramic lining. The new stove reduces smoke, improves women's health and pays for itself in a month.

Activities

1 a) What is a *self-help housing scheme*?
  b) Which of the features in diagram **A** will be provided by:
     • the shanty-town dweller • the local authority?
  c) From diagram **A** and photo **B**, list four of the most important features of self-help, low-cost building schemes. Give reasons for your answer.
  d) Why are some types of low-cost, self-help housing not much help to people who live in shanty settlements?

2 a) What are the main aims of the Intermediate Technology Development Group?
  b) Describe how two of the ITDG's projects in Nairobi may help people living in shanty settlements.

Summary Low-cost, self-help building schemes are often the only way by which people living in shanty settlements can improve their homes and quality of life.

What are employment structures?

Classifying employment

Most people have to work to provide the things they need in life. The various jobs or **activities** that people do are called **employment**. There are so many different types of work that it is helpful to put them into groups. Traditionally there were three main groups: **primary**, **secondary** and **tertiary**. Since the 1980s a fourth group has been added: this is called **quaternary** activities. The four groups, or sectors, of employment are shown in drawing **A**.

Employment structure

The proportion of people working in each of the primary, secondary and tertiary sectors is called the **employment structure**. Employment structures change over a period of time (graphs **B**) and vary from one place to another (map **C**). They also give an indication of how rich or how poor a place is. Poorer places tend to have most people working in primary industries such as farming. Richer places have their highest percentage in the tertiary sector.

Change over time in the UK (graphs **B**)

Up to about 1800, most people in Britain made a living from the land. The majority were farmers while many others made things either for use in farming, such as ploughs, or from items produced by farmers, such as bread. Most people at this time were on low wages and worked in rural areas.

During the nineteenth century the main types of jobs changed dramatically, mainly as a result of the Industrial Revolution. Fewer people worked on the land and many moved to towns to find work. Most mined coal or worked in heavy industries making things like steel, ships and machinery.

Further changes occurred in the twentieth century and up to the present day. Farming and industry have become more mechanised and need fewer workers. Coal and other natural resources are running out, while industry is facing increasing competition from other countries. However, there are now many more hospitals, schools and shops. Transport has also provided many jobs.

All industrialised, developed countries have experienced these changes which have brought increased wealth and and improved quality of life to these nations.

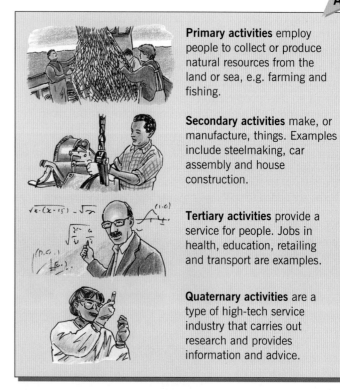

A

Primary activities employ people to collect or produce natural resources from the land or sea, e.g. farming and fishing.

Secondary activities make, or manufacture, things. Examples include steelmaking, car assembly and house construction.

Tertiary activities provide a service for people. Jobs in health, education, retailing and transport are examples.

Quaternary activities are a type of high-tech service industry that carries out research and provides information and advice.

B Employment structures in the UK, 1800–2000

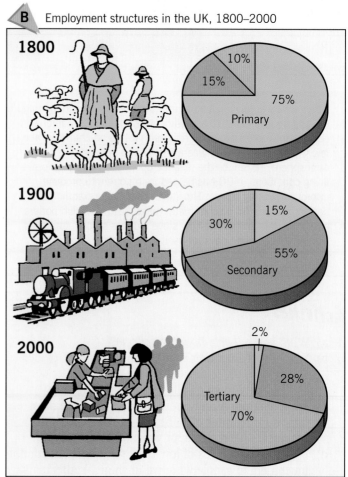

1800

10%
15%
75%
Primary

1900

15%
30%
55%
Secondary

2000

2%
28%
Tertiary
70%

Changes between places within the UK

Look carefully at map **C** which compares regional employment structures in the UK at the turn of the century. The map may look a little complicated but it contains some useful information. See if you can identify the following features.

1. The map shows two things for each region: the employment structure, and the number of people employed. The key has a scale to measure actual workforce numbers.
2. It is easy to see that the South-east has the largest workforce but a little more difficult to identify the region with fewest workers. It is Northern Ireland.
3. Every region has fewest workers in the primary sector and most in the tertiary sector. This pattern is typical of a richer, more developed country.
4. Most regions have an employment structure similar to the UK average. The South-east is probably most different, particularly in the secondary and tertiary sectors.

It must also be noted that within each region there may be big differences between types of town. For example, one may be a market town, one an industrial centre and one a holiday resort. Each will have a very different employment structure.

C Employment structures in the UK, 2000

UK average

Primary 3%
Secondary 28%
Tertiary 69%

Figures give percentage of total employed population in each division

8 million
5m
1m
0.5m

Area of circle proportional to total employed population in each region

Scotland
Northern Ireland
North
Yorkshire and Humberside
North-west
East Midlands
West Midlands
East Anglia
Wales
South-east
South-west

0   200 km
N

Activities

1. In your own words, give the definitions for *primary*, *secondary* and *tertiary* activities and *employment structures*.

2. a) Make a larger copy of table **D** below.
   b) Complete the table by putting the following jobs into the correct columns.
   - *teacher*
   - *bus driver*
   - *steelworker*
   - *pop singer*
   - *plumber*
   - *forestry worker*
   - *doctor*
   - *farmer*
   - *nurse*
   - *ambulance driver*
   - *quarry worker*
   - *shop assistant*.
   - *police officer*
   - *bricklayer*
   - *shopkeeper*
   - *bank manager*

3. Describe and explain the changes in the UK's employment structure between 1800 and 2000.

4. Look carefully at map **C**.
   a) Which three regions have the smallest workforce?
   b) How many people are employed in the South-east region?
   c) Which region has the highest proportion of people employed in the tertiary sector? Give the percentage.
   d) Name the region in which you live. Estimate the percentages for primary, secondary and tertiary industries. How many people are employed there?

D

Primary	Secondary	Tertiary
Lumberjack	Sawmill operator	Furniture shop assistant
Dairy farmer	Bricklayer	Car sales person
North Sea oilrig worker		

Summary

The proportion of people employed in primary, secondary and tertiary activities changes over time and differs from place to place.

How do employment structures vary between countries?

Map **A** shows the employment structures for 16 countries which are at different **stages of development**. It also divides the world into two economic parts. Countries to the north and east of the dividing line are the 'rich' countries and those to the south and west are the 'poor' countries (page 266).

- In most of the rich, or more economically developed, countries there are relatively few people employed in the primary sector, a higher proportion in the secondary sector and most in the tertiary sector.
- By contrast, in the poorer, or less economically developed, countries most people find jobs in the primary sector and very few are employed in the secondary and tertiary sectors. Usually it is the primary jobs that are worst paid.

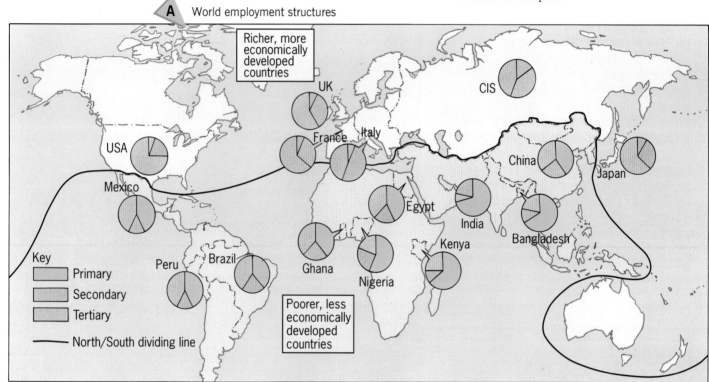

A World employment structures

As we have seen on page 188, the proportion of people working in primary, secondary and tertiary industries changes over time. This is especially true as individual countries develop economically.

India, Brazil and the USA are three countries at different stages of economic development. Their employment structures are given in graph **B**. India, the least economically developed, has by far the greatest proportion of its workforce employed in primary activities. By contrast, the USA, the most economically developed, has most of its workforce employed in the tertiary sector.

Two hundred years ago, countries like the USA and the UK had employment structures similar to that of present-day India. Diagram **C** explains how employment structures alter as a country reaches different levels of economic development.

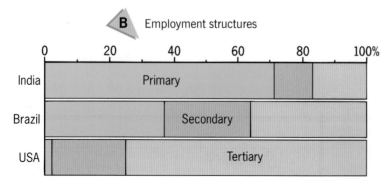

B Employment structures

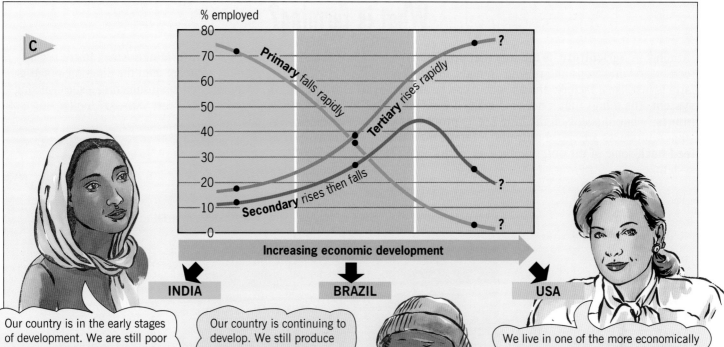

C

% employed

Primary falls rapidly

Tertiary rises rapidly

Secondary rises then falls

Increasing economic development

INDIA

BRAZIL

USA

Our country is in the early stages of development. We are still poor and have a low standard of living. Most of our labour force work in **primary** activities like farming, forestry and mining. Farming is very important because we have to try to produce enough food for our own people. We sell primary products like iron ore, timber and tea to rich countries. However, we rarely earn enough money from them to be able to buy many goods or much machinery in return.

Our country is continuing to develop. We still produce much of our own food but the use of machines has reduced the number of people needed to work on farms. Many of these people have now moved to towns and cities where they find jobs in **secondary** activities working in new factories. We are richer than before and our transport systems, health care and education have all improved. This is leading to an increase in tertiary jobs.

We live in one of the more economically developed countries. Many people are employed in **tertiary** activities. They work in hospitals, schools, offices, banks and in the leisure industry. We still have many secondary industries but these need fewer workers because we use machines, robots and computers in factories. Very few people work in primary industries since we are rich enough to buy most of our raw materials from other countries. This helps us to conserve our resources and protect our environment.

Activities

1 Look carefully at map **A**.
  a) List the countries with over:
   • 50 per cent employed in the primary sector
   • 50 per cent employed in the tertiary sector.
  b) Describe the location of the:
   • 'poorer', less economically developed, countries
   • 'richer', more economically developed, countries.
  c) Is graph **D** for a developed or a developing country? Give three reasons for your answer.

2 Copy and complete table **E** by giving the employment structures for India, Brazil and the USA.

E

	Primary (%)	Secondary (%)	Tertiary (%)
India	72		
Brazil			
USA			75

3 As a country becomes more economically developed, why do numbers in:
  a) primary activities decrease
  b) secondary industries increase and then decrease
  c) tertiary activities increase?

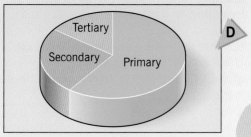

D

Summary There is a link between the employment structure of a country and its level of development. Poorer countries have a larger primary workforce than rich countries.

What is farming?

Farming, or **agriculture**, is the way that people produce food by growing crops and raising animals. In many ways a farm is just like a factory, and as such it can be studied as a **system**. The things that it needs to make it work are called **inputs**. What happens on the farm are its **processes** and what it produces are called **outputs**. A farmer may also **feed back** some of the outputs, such as profits, into the system.

Diagram **A** shows a simple farming system. In reality, farming is much more complex than the diagram suggests, with political, economic and environmental issues causing many problems for farmers. Pages 202–207 look at some of these issues.

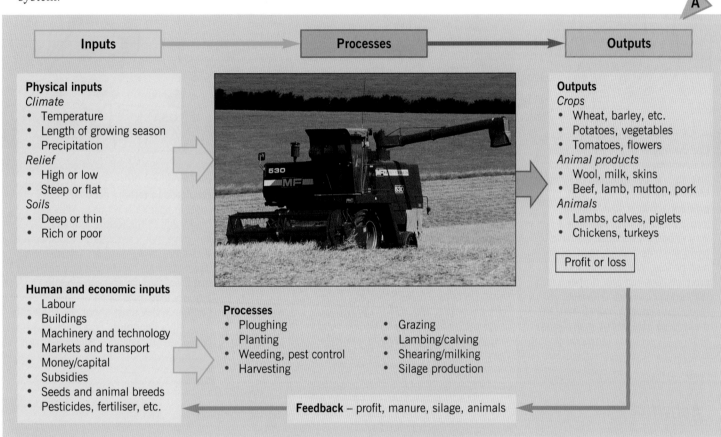

A

| Inputs | → | Processes | → | Outputs |

Physical inputs
Climate
• Temperature
• Length of growing season
• Precipitation
Relief
• High or low
• Steep or flat
Soils
• Deep or thin
• Rich or poor

Human and economic inputs
• Labour
• Buildings
• Machinery and technology
• Markets and transport
• Money/capital
• Subsidies
• Seeds and animal breeds
• Pesticides, fertiliser, etc.

Processes
• Ploughing
• Planting
• Weeding, pest control
• Harvesting
• Grazing
• Lambing/calving
• Shearing/milking
• Silage production

Outputs
Crops
• Wheat, barley, etc.
• Potatoes, vegetables
• Tomatoes, flowers
Animal products
• Wool, milk, skins
• Beef, lamb, mutton, pork
Animals
• Lambs, calves, piglets
• Chickens, turkeys

Profit or loss

Feedback – profit, manure, silage, animals

Farmers are decision makers. They have to choose carefully what crops to grow, what animals to rear and what methods to use. These decisions are dependent on physical, human and economic factors. Of these, physical factors are probably the most important. They include temperature, rainfall, sunshine and soil type, and whether the land is flat or steeply sloping. Economic conditions such as market demand, input costs and transport expenses must also be considered.

Sometimes the farmer may have several choices, so the decision may depend upon individual preferences, traditions and expertise. On other occasions, the farmer's choice may be limited by difficult physical conditions or political pressures.

As needs and conditions vary from place to place, many different types of farming have developed. Geographers find it very useful to sort, or classify, farms into different groups. As you will see later in this unit, it makes studying farming easier if the key features of different farming systems can be identified.

All farms can be classified in three different ways. These put farms in groups depending on their inputs, processes and outputs. Diagram **B** shows the main features of this classification. Notice that the same places keep coming up in each list. For instance, cattle ranching in the USA is extensive, and pastoral, and commercial. It can be described in three different ways.

Classification by inputs

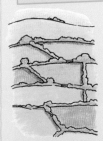

Intensive farming
These farms have large inputs of money, labour or technology to achieve high outputs or yields per hectare. The farms are usually quite small.
Examples:
- Cereal farming in East Anglia
- Greenhouse cultivation in the Netherlands
- Rice farming on the Ganges floodplain

Extensive farming
These farms have comparatively small inputs for large areas of land. They are usually found where conditions are poor, so it is not worth farmers putting a lot of money or work into the land.
Examples:
- Hill sheep farming in upland Britain
- Cattle ranching on the Prairies, USA

Classification by processes

Arable farming is the ploughing of the land and growing of crops.
Examples:
- Wheat and barley in East Anglia
- Rice growing on the Ganges floodplain

Pastoral farming is the leaving of land under grass and the rearing of animals.
Examples:
- Hill sheep farming in upland Britain
- Cattle ranching on the Prairies, USA

Market gardening is when fruit, flowers and vegetables are grown under controlled conditions.
Example:
- Tomatoes in the Netherlands

Mixed farming is when crops are grown and animals are reared in the same area.
Example:
- Dairy farming, with potatoes and turnips, in Cheshire

Classification by outputs

Commercial
The outputs from these farms are mainly or entirely for sale.
Examples:
- Arable farming in East Anglia
- Cattle ranching on the Prairies, USA

Subsistence
Subsistence farmers produce food for themselves and their family. There is rarely any profit or spare food.
Examples:
- Rice growing on the Ganges floodplain
- Shifting cultivation in Amazonia

Activities

1 a) What are farming inputs? Give two examples of economic inputs.
   b) What are farming processes? Give two examples of farming processes for a farm growing wheat.
   c) What are farming outputs? Name two outputs that can be fed back into the system.

2 Copy and complete diagram **C** to show the farming system for a hill sheep farm.

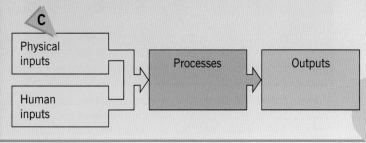

C

3 For a hill sheep farm, list three inputs:
   a) that are outside the farmer's control
   b) that the farmer could alter so as to increase output.
   For each one, suggest how this could be done.

4 What is the difference between:
   a) *intensive farming* and *extensive farming*
   b) *commercial farming* and *subsistence farming*?

5 a) Name a farming type that can be classified as extensive, pastoral *and* commercial.
   b) Give three different ways of classifying rice farming.

Summary

Farming is a system, with inputs, processes and outputs. Farmers have to consider the different inputs when deciding which type of farming is best suited to an area. Farming types may be classified into different groups.

What are the UK's main farming types?

As we have seen on pages 192 and 193, there are several different types of farming around the world. In the UK, as in any other place, climate is the most important factor affecting the type and distribution of farming. Britain's climate is best described as **temperate**, being never too hot or too cold and rarely too wet or too dry. This climate, whilst hardly being suited to growing pineapples or bananas, is good for wheat, barley, potatoes and other such crops. It is also ideal for growing grass on which cattle and sheep can feed.

These climate constraints, along with the other factors explained on page 192, have resulted in there being five main types of farming in the UK.

Arable farms grow crops on prepared land. They include cereal crops such as wheat and barley, and vegetables like potatoes and carrots. These need flat land with a deep fertile soil and a warm and relatively dry climate.

Hill sheep farms produce wool, lamb and mutton. Sheep feed on rough pasture and can graze land that is too steep to raise cattle or grow crops. They are hardy animals and can survive any climate in Britain.

Mixed farms grow crops and raise livestock. Some of the crops grown are used as animal feed, particularly in the winter. Mixed farms need fairly good soil and a climate that is neither too wet nor too dry.

Cattle farms raise cows to produce milk or beef. Raising cows for milk is called **dairy farming**. Cows need land that is not too steep, and a warm, moist, though not too wet, climate for grass to grow.

Market gardening is the growing of fruit, vegetables and flowers, usually on small farms. These use intensive farming methods and are often close to towns where the produce can be sold fresh and at a good price.

It is essential, if a British farmer is to make a **profit** and a livelihood, that farming is as efficient as possible. This means that certain areas will be better suited to one type of farming. Groups of farmers in this area therefore tend to **specialise** in that one type. However, there are likely to be local differences within each specialist area creating some variations.

Map **F** shows the major areas of specialised farming in the UK. It also shows how physical and human inputs into the farming system in places towards the north and west of Britain differ from those in places in the south and east. This helps to determine the pattern of farming across the country. For example, the cool summers, heavy rainfall and hilly conditions of the north and west are most suited to pastoral farming. In contrast, the south and east with its warm summers, drier climate and low-lying flat land are more likely to be arable or mixed.

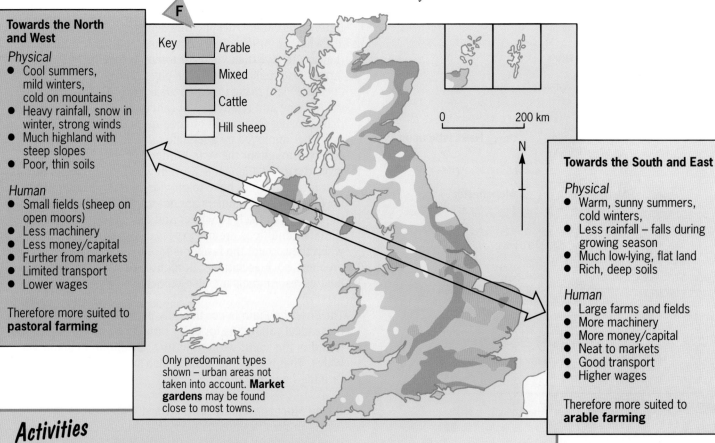

Towards the North and West

Physical
- Cool summers, mild winters, cold on mountains
- Heavy rainfall, snow in winter, strong winds
- Much highland with steep slopes
- Poor, thin soils

Human
- Small fields (sheep on open moors)
- Less machinery
- Less money/capital
- Further from markets
- Limited transport
- Lower wages

Therefore more suited to **pastoral farming**

Towards the South and East

Physical
- Warm, sunny summers, cold winters,
- Less rainfall – falls during growing season
- Much low-lying, flat land
- Rich, deep soils

Human
- Large farms and fields
- More machinery
- More money/capital
- Neat to markets
- Good transport
- Higher wages

Therefore more suited to **arable farming**

Key: Arable, Mixed, Cattle, Hill sheep

Only predominant types shown – urban areas not taken into account. **Market gardens** may be found close to most towns.

Activities

1 Map **G** is a very simplified map of farming in the UK. Make a larger copy of the map and colour in and label the four farming types shown. Add a key to your map and on it, briefly describe the main features of each farming type.

2 a) Which of the farming types shown in **A** to **E** can be described as *pastoral*? Give reasons for your answer.
b) Why are most pastoral farms in the UK found in the north and west?
c) Which of the farming types shown in **A** to **E** can be described as *arable*? Give reasons for your answer.
d) Why are most arable farms in the UK found in the south and east?

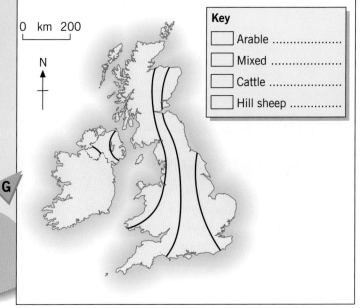

Key: Arable .... Mixed .... Cattle .... Hill sheep ....

Summary There are five main types of farming in the UK. A distinct pattern of farming may be recognised, with the north and west mainly pastoral, and the south and east mainly arable.

195

What are the main features of dairy farming in the UK?

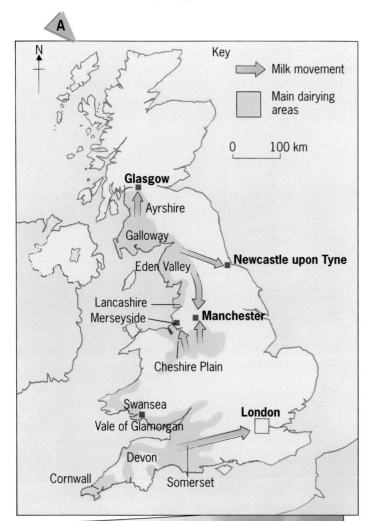

It is hard to imagine a day in which you do not consume milk in one form or another. It may be as a liquid poured onto your breakfast cereal or added to your cup of tea or coffee, or it may be once it has been processed into cheese or butter. Multiply your daily requirements by Britain's 58 million people and you can see the great demand for fresh dairy produce.

Dairy cows cannot, despite this demand, be reared just anywhere in Britain. Dairy cows need certain conditions if they are to give high yields of top-quality milk and if the farmer is to make a profit. The physical conditions of the Cheshire Plain make this area ideal (map **A**). The land here is low-lying and relatively flat (photo **B**). Soils are deep and rich allowing the growth of good-quality grass. The quality is further improved by a reliable rainfall which is spread evenly throughout the year. Winters are quite mild. This gives the grass a longer season in which to grow and means that farmers do not have to provide so much winter fodder for the cows. As summers are not usually very warm then the grass is unlikely to wither and die. Of course these ideal climatic conditions are not expected to occur every year and when they do not the farmer is faced with short-term problems such as summer drought caused by unexpected heatwaves. Animals are also vulnerable to disease.

The Cheshire Plain is not the only place in Britain where these ideal conditions for dairy farming occur. They are also found in the south-western parts of Scotland, England and Wales. That is why those areas also specialise in dairy farming (map **A**).

Gilbert Hitchen's family have farmed 450 hectares (200 acres) on the Cheshire Plain for five generations. The family were encouraged to build up their herd of 100 dairy cows after the Second World War at a time when Britain was very short of agricultural products. Encouragement came initially from the British government and later from the EU which offered **subsidies** on milk production. This meant dairy farmers were guaranteed a fixed price for the milk produced on their farm. They would be given this agreed price even if there was a glut of milk and the market price for the product fell. The government also made efforts to stop other countries 'dumping' their own cheaper dairy products into Britain. The Hitchens were encouraged by their bank to increase their herd to 190 cows. The bank assumed that the cost of buying extra cows and building accommodation for them would soon be repaid by the expected increase in income. The family were unlucky in the 1960s. They were one of many whose herd had to be destroyed after it fell victim to foot-and-mouth disease. The herd was insured but, with so many other farmers also having to rebuild their stock, it had to be replaced with animals costing high prices and which were not always of top quality.

B The Cheshire Plain

During the 1970s and 1980s dairy farmers were helped by improvements in **accessibility** and **technology**. The Hitchens were helped by the building of the M6 and improvements in the local roads. The improvements in accessibility enabled them to get their milk to the urban markets of Manchester and Merseyside quickly and easily. The transport of fresh milk was made easier by the introduction of the refrigerated lorry (photo **C**) and its twice daily collection of milk from the farm. Most milk is bought by the Milk Marque. Another important technological improvement has been the introduction of computers. Amongst other advantages the computer controls the amount and quality of food given to each cow (photo **D**) and punches out cards to record the daily milk yields of each animal. Gilbert Hitchen had invested a lot of money in his farm by increasing his herd to 160 cows, building new barns and introducing computers and modern machinery.

C Refrigerated milk lorry

D Computerised dairy farming

Activities

1 Look carefully at map **A**.
   a) Name eight areas that specialise in dairy farming.
   b) Describe the location of the main dairying areas.

2 Copy and complete table **E** to show the ideal conditions for rearing dairy cows.

E

Temperatures	Precipitation	Relief	Soils

3 Describe and explain how each of the following helped improve the outputs of Mr Hitchen's farm.
   • Subsidies
   • Bank loans
   • Transport improvements
   • Developments in technology

4 Copy and complete diagram **F** to show the farming system for a dairy farm.

F

Physical inputs → Human inputs → Processes → Outputs

Further case studies on UK farms may be found on the National Farmers Union website at:
www.nfu.org.uk/

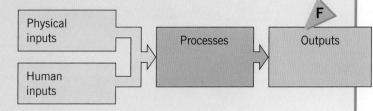

Summary Dairy farms are located mainly in the west of Britain where wetter conditions are ideal for growing good-quality grass. Dairy farming can be affected by improvements in accessibility and technology.

What are the main features of market gardening in the Netherlands?

Market gardening is the intensive cultivation of high-value crops such as fruit, vegetables and flowers. Market gardening has a long history in the Netherlands and accounts for one-quarter of the country's agricultural exports. Its main markets are the EU countries and North America.

Most market gardens are located near to large centres of population and where fast and efficient transport systems are available. This is because fruit, flowers, and to a lesser extent vegetables, perish quickly and are easily damaged. Fertile soils and a mild climate are an advantage but not essential, as in most market gardens, soils and climate are artificially controlled inside greenhouses.

Market gardening in the Netherlands is a **high-tech industry**. Growers are supported by scientific research and government advisory services. Many owners work within groups, or **co-operatives**, to share equipment and access to markets.

The Westland area is a major market gardening area south of The Hague. This is where the Bakker family have a smallholding of less than 2 hectares. The family were originally attracted to farm the area because of its sandy, well-drained soils. Nowadays, however, there are virtually no physical advantages for farming in Westland as almost all production takes place in heated greenhouses with artificial soil.

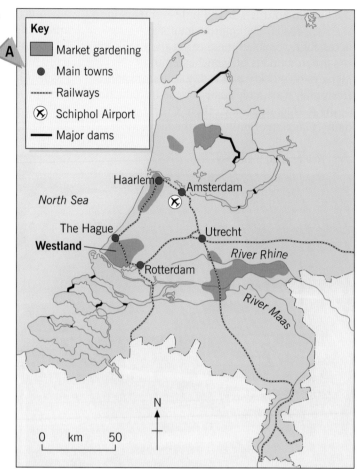

A

Key
- Market gardening
- ● Main towns
- ⋯⋯ Railways
- ⊗ Schiphol Airport
- ▬ Major dams

Haarlem · Amsterdam
North Sea
The Hague
Westland
Utrecht
Rotterdam
River Rhine
River Maas

N

0 km 50

B   The Amsterdam flower market

Market garden greenhouses   C

The Bakkers specialise in salad crops such as tomatoes, lettuce and cucumbers. They also grow flowers and pot plants for the gardening trade. The business is a highly intensive, **commercial** form of farming. Capital investment is very high and the latest technology is used to ensure that maximum use is made of such a small area of land. Heating, ventilation, humidity, watering and fertilising are all computer controlled. The work is **labour intensive**, with six people employed full-time throughout the year.

Market garden products can command a high price if they are fresh and of a high quality. Three times a week, the Bakker co-operative sends produce to Amsterdam on the high-speed rail link. About half is sold in the Amsterdam flower market and the rest distributed to other urban centres in the Netherlands. Every Tuesday a further consignment is sent directly to Schiphol airport. Most of this is air-freighted to London and other European cities. A small amount is flown to Chicago for distribution to the American market.

D

Farming system for a market garden

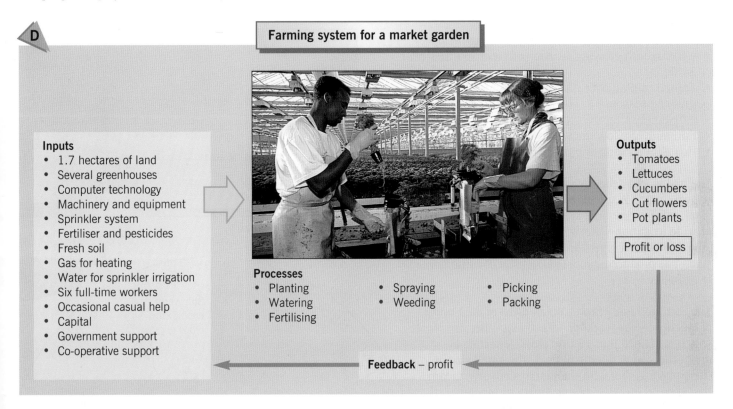

Inputs
- 1.7 hectares of land
- Several greenhouses
- Computer technology
- Machinery and equipment
- Sprinkler system
- Fertiliser and pesticides
- Fresh soil
- Gas for heating
- Water for sprinkler irrigation
- Six full-time workers
- Occasional casual help
- Capital
- Government support
- Co-operative support

Outputs
- Tomatoes
- Lettuces
- Cucumbers
- Cut flowers
- Pot plants

Profit or loss

Processes
- Planting
- Watering
- Fertilising
- Spraying
- Weeding
- Picking
- Packing

Feedback – profit

Activities

1. Give the meaning of the following terms:
   a) *market gardening*   b) *intensive farming*
   c) *commercial farming*   d) *arable farming*.

2. Using map **A**, suggest at least three reasons for Westland being a good location for market gardening.

3. Copy and complete table **E** to show the inputs into the Bakker family's market garden.

4. Why are physical conditions such as relief, soils and climate no longer important to the Bakker family?

5. Briefly describe how the Bakker family's market garden works. Use the headings shown in diagram **F**. You should comment on the inputs into the farm, the processes that take place on the farm, and what the outputs are and where they go to.

E

Original physical	Human	Political	Economic

F

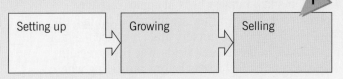

Setting up	→	Growing	→	Selling

Summary   Market gardening is a highly intensive, commercial form of arable farming. Soils and climate are artificially controlled to produce ideal growing conditions and to maximise output.

What are the main features of rice growing in India and Bangladesh?

Rice is the world's most important crop. Estimates suggest that over a third of the world's population rely on it as their **staple**, or main, food. Rice grows best in the monsoon climates of south-east Asia where high temperatures and heavy rainfall provide ideal growing conditions. More than 80 per cent of the world's rice production comes from this region.

The natural inputs required for rice farming include:
- a five-month growing season with temperatures over 21°C
- annual rainfall over 2000 mm with most falling in the growing season
- a dry spell, after the growing season, for harvesting
- flat land, to allow the water to be kept on fields
- rich alluvial soils to provide nutrients.

The plains and delta areas of the lower Ganges valley shown in map **A**, provide these conditions. Temperatures are ideal, rainfall amounts are usually adequate, and the rivers provide irrigation water when needed. The annual floods also deposit rich layers of **alluvium** (silt) which over the years has produced almost perfect soil conditions for growing rice.

The Sandhu family farm a small piece of land north of Calcutta. They work an area of just 1 hectare which is divided into 15 tiny plots. A hectare is about the size of a football field. The small size of the plots, together with the family's poverty, means that there is little use of machinery or modern methods. The farming is **labour intensive**, with much manual effort needed to construct irrigation channels, prepare the fields and plant, weed and harvest the crops. The family have two water buffalo which are used for ploughing. The animals also provide manure which helps supplement soil fertility.

The Sandhu family have learned to manage water carefully, as it is an essential part of rice growing. As photo **D** shows, they grow their rice in flooded fields called padis. Seedlings are first planted by hand in small nursery beds in damp, carefully prepared soil. When they are about 30 cm tall, they are taken out of the nursery and replanted – again by hand – in the flooded fields. The fields remain flooded until a few weeks before harvesting when they are drained. After harvesting, the rice grains are separated from the leafy stems by threshing. They are then dried and stored for later use. Diagram **B** shows the farming year for rice growing in the Ganges valley.

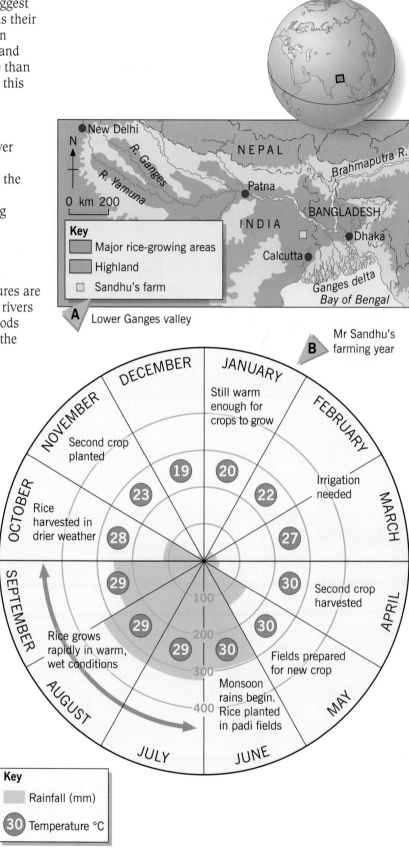

A Lower Ganges valley

Key
- Major rice-growing areas
- Highland
- □ Sandhu's farm

B Mr Sandhu's farming year

Key
- Rainfall (mm)
- ㉚ Temperature °C

Like most people in the region, the Sandhus are **subsistence farmers**. They grow rice mainly for their own consumption and have little left to sell or to trade for other goods. Life can be difficult for these farmers, as they are very poor and are often unable to cope with the many problems that affect farming in the region (diagram **C**). In recent years, new farming methods, irrigation schemes and land reform have been introduced to help reduce the problems. In most cases these have also increased food production and improved standards of living. They are explained on pages 204 and 205.

C

Some problems affecting rice farmers in India

* **Flooding** provides water and fertile silt to grow the rice, but severe flooding can destroy an entire rice crop.

* **Drought** may occur when the monsoon rains fail. With too little water, the rice crop may be ruined.

* A **shortage of land** has caused a reduction in farm size. Some plots are too small to support a family.

* An **increasing population** means that even more food is needed. Food shortages can be a problem.

D Rice cultivation in India

Activities

1 a) Describe five features that make Mr Sandhu's farm ideal for rice cultivation.
   b) Why is irrigation necessary for part of the year?
   c) Why is Mr Sandhu described as a *subsistence* farmer?
   d) Draw a star diagram to show the main difficulties facing subsistence rice farmers like Mr Sandhu.

2 a) Put the terms below into the correct order to describe how rice is grown on a subsistence farm. Add a brief description for each term.

   * weed crop   * plant seedlings   * replant
   * thresh rice   * flood fields   * store crop
   * harvest   * prepare fields

   b) Describe the farm and work being done in photo **D**.

3 Copy and complete diagram **E** to show the farming system for a subsistence rice farm in India.

E

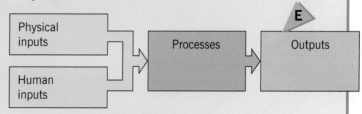

Physical inputs → Processes → Outputs

Human inputs

Summary

Rice growing in India is a highly intensive, subsistence form of farming. The soils and climate of the lower Ganges valley provide ideal growing conditions, but life for most farmers is very difficult.

How is commercial farming changing?

Commercial farming is the growing of crops and raising of animals in order to make a profit. Most commercial farming takes place in the richer, more economically developed regions such as Europe and North America. In recent times, farming in these areas has changed considerably and become increasingly productive.

Many of the changes have been a result of directives issued by the European Union (EU). The EU was founded in 1956 when it was called the European Economic Community (EEC). The organisation was determined to make Europe self-sufficient in food, and the Common Agricultural Policy (CAP) was set up with the main aim of increasing the efficiency of farming. The CAP encouraged farmers to use the latest advances in science and technology to produce more food. Increased investment, a greater use of chemicals and the introduction of **genetically improved** seeds and animals all helped to raise output.

The CAP was very successful in making Europe self-sufficient and providing a source of cheap and varied, quality foodstuffs. However, many people are now concerned that CAP policies have caused serious environmental problems and threatened **fragile environments** (pages 206 and 207). They also worry that as the EU changes its policies to reduce surpluses, some farmers may be forced to change the land use on their farms or go out of business. Some changes in commercial farming are shown in diagram **A** below.

A

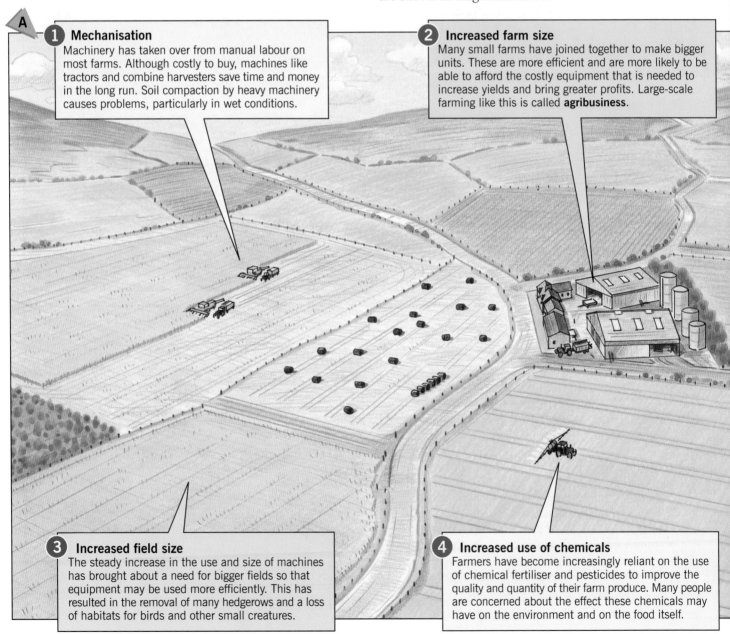

1 Mechanisation
Machinery has taken over from manual labour on most farms. Although costly to buy, machines like tractors and combine harvesters save time and money in the long run. Soil compaction by heavy machinery causes problems, particularly in wet conditions.

2 Increased farm size
Many small farms have joined together to make bigger units. These are more efficient and are more likely to be able to afford the costly equipment that is needed to increase yields and bring greater profits. Large-scale farming like this is called **agribusiness**.

3 Increased field size
The steady increase in the use and size of machines has brought about a need for bigger fields so that equipment may be used more efficiently. This has resulted in the removal of many hedgerows and a loss of habitats for birds and other small creatures.

4 Increased use of chemicals
Farmers have become increasingly reliant on the use of chemical fertiliser and pesticides to improve the quality and quantity of their farm produce. Many people are concerned about the effect these chemicals may have on the environment and on the food itself.

Activities

1 Drawing **B** shows five newspaper headlines that have been torn into two pieces. Match up the torn pieces to give the six original correct headlines.

2 Write two short newspaper articles for the headlines below. In each case describe the change and give its advantages and disadvantages.

| Concerns grow as chemical use increases | Organic farming promises cleaner environment |

3 How can farmers help protect the environment whilst at the same time adding to their income?

B

Set-aside scheme • as machines take over
Caring farmers • as farms get larger
Farmers change jobs • returns natural environment
Profits increase • as tourism takes over
Farming jobs lost • appointed countryside stewards

Summary

There have been many changes to commercial farming in the last 50 years. These changes have increased outputs and reduced food costs. They have also caused problems for the environment.

5 Organic farming
Organic farming is a form of food production that does without chemicals. Much less damage is done to the environment as fewer toxic chemicals are released into the soil or find their way into rivers. Prices are higher and yields are lower than for traditional farming.

6 Natural environments
More efficient farming has led to a surplus of food being produced. 'Set-aside' and 'Farm Woodland' schemes encourage farmers to stop growing crops by offering grants. The land may then return to its natural state, or trees may be planted to enhance the landscape.

7 The Countryside Stewardship Scheme
Through this scheme farmers take on responsibility for the countryside and are paid for their efforts. They may receive up to £300 per hectare for conserving and restoring landscapes at risk such as chalk grasslands, river meadows, coastal dunes and heather moorland.

8 Diversification
Farmers have been encouraged to develop business activities other than farming. The increased demand for rural leisure activities has helped farms become part of the tourist industry. Many now provide holiday accommodation and recreational facilities for visitors.

How is subsistence farming changing?

Subsistence farmers usually produce just enough food for their own needs. In good years there may be a small **surplus** which may be sold to buy other goods. Most subsistence farming takes place in the less economically developed regions of the world like Africa, the Amazon rainforest and India.

The population of many of the poorer countries has been growing rapidly in recent years. With so many extra mouths to feed, it has been essential to increase food production at an equally rapid rate. A variety of methods have been used to try and achieve this. The **Green Revolution** was the name given to the introduction of modern farming methods to the poorer countries of the world in order to increase their food production. Four of the main features of the Green Revolution are described here.

High-yield varieties

In the 1960s, developed countries such as the UK, USA, Germany and Australia provided money to develop high-yield varieties (HYVs) of rice, wheat and maize. In 1965 an improved strain of rice called IR-8 more than trebled yields in India. Improved strains of wheat and maize quickly followed.

The faster-growing HYVs allowed an extra crop to be grown each year and yields have become more reliable as they are more resistant to disease. However, HYVs need large amounts of fertiliser and insecticides as they are more prone to insect attack. The increase in chemical use has meant that farming has become less **sustainable** and, in some cases, has been damaging to the environment.

Advantages
- Yields increased more than three times.
- Possible to grow up to three crops a year.
- Other crops grown which vary the diet.
- Surplus food to sell in cities, creating profit.
- Improved standard of living.

Disadvantages
- Poorest farmers unable to afford machinery, fertiliser and pesticides required.
- HYVs need more water and fertiliser which is expensive.
- Money borrowing put some farmers into debt.
- Maintenance and fuel needed for machines is not always available.

A

B Inundation canal near Calcutta

Irrigation

The monsoon rains are often unreliable and there has always been a need to irrigate the land in these regions. As HYV seeds need much more water than the traditional varieties of rice, wheat and maize, the amount of irrigation required has increased. Irrigation is mainly by wells, canals and reservoirs.
- Wells reach underground water supplies. The water is lifted from the well using a waterwheel or pump. It is then fed along open channels to the fields.
- **Inundation canals** on the river banks fill up as the river floods, and take the water to the fields.
- Nowadays, some areas have a network of canals and channels linked to newly built reservoirs. This provides a larger and more regular supply of water.

Appropriate technology

This is technology suited to the needs, skills, knowledge and wealth of the local people (pages 270–272). Examples include:

- Individual wells with simple, easy-to-maintain pumps.
- Renewable energy sources that use local resources such as wind, solar power and biogas.
- Projects that use labour rather than machinery.
- Low-cost schemes that are **sustainable** and not damaging to the environment.

Land reform

In many poorer countries, farmholdings are very small and broken up into tiny plots that are spread over a wide area. This makes efficient farming difficult. The majority of farmland is also held by a few wealthy landowners. Many of the poorer farm labourers have no land at all and suffer great poverty. The aim of land reform is to:

- increase farm size for small landowners
- set an upper limit on the amount of land owned by the wealthiest landowners
- give surplus land to the landless farm labourers.

Using a simple water pump **C**

The use of new farming methods in the less developed countries has certainly increased production. Fewer people now suffer from food shortages, and more nutritious foods have become available. The resultant changes in diet have helped to improve people's health and raise their **quality of life** (page 274).

There have been problems, however. The Green Revolution brings with it a type of commercial farming that may not be appropriate to the economically developing world. It relies on large inputs of chemicals that damage the environment and which few small farmers can afford. The benefits go to more prosperous farmers who can buy in bulk and produce rice cheaply. There is also concern for small farmers who have been tempted to grow cash crops in order to sell to earn money. This causes local food shortages and a lowering, rather than raising, of living standards.

Activities

1 **a)** Using information in table **D**, draw a line graph to show rice yields in India between 1931 and 2001.
   b) Describe the main features of your graph.
   c) What effect has the Green Revolution had on rice yields?

Year	1931	1961	1981	1991	2001
Yields (kg/ha)	1520	1480	1860	2620	3100 (est.)

D

2 Write a report on the Green Revolution using the headings in drawing **E**.

3 Why is appropriate technology more suited to subsistence farmers in countries like India, than more modern farming methods?

The Green Revolution

- Definition
- Main features
- Successes
- Failures

E

Summary

The Green Revolution has brought many changes to subsistence farming in the world's poorer countries. These changes have increased outputs but have also posed problems.

How has farming affected the environment?

A Open fields where hedges have been removed

Developments in farming, as in other economic activities, lead to changes in the environment. Farmers are continually trying to make their farms more efficient and to improve the quality of their produce. To achieve this there must be careful planning and management. Until recently, improvements to farming were often made with little or no consideration for the management of the environment. It was not realised that developments such as the removal of hedgerows and the increasing use of fertiliser would have such an unintended, adverse effect upon the environment. Today it is a top priority that the costs of developments in farming are balanced against the costs to the environment.

Hedges provide a home for wildlife – birds, animals, insects and plants	**B** Cutting hedges costs the farmer time and money. A hedgecutter costs over £7000
Hedges reduce wind speed	Hedges get in the way of big machinery in fields
Well looked after hedges are attractive	Hedges take up space which could be used for farmland
Hedge roots hold the soil together and reduce erosion	Hedges harbour insect and animal pests as well as weeds

C Soil erosion caused by rainwater run-off

The removal of hedgerows

Although most of Britain's hedgerows were only planted by farming communities in the eighteenth century, they are now often considered to be part of our natural environment. Modern farming, especially in arable areas, uses large machines which are easier to work in large fields. The result was that between 1945 and 1990 over 25 per cent of Britain's hedges were cleared (in parts of East Anglia the figure was over 60 per cent) in order to have larger fields (photo **A**). Figure **B** shows some of the arguments for and against the clearing of hedges. Apart from the loss of a habitat for wildlife, the major and often unintended problem has been the increased risk of **soil erosion**. The leaves and branches of hedges break the fall of heavy rain (page 16), reducing its force before it hits the ground. Roots bind the soil together and slow down the flow of water. Where hedges are removed and the soil is left exposed, then erosion can result either from the soil being washed away during times of heavy rainfall or blown away by the wind under drier conditions (photo **C**).

The use of fertiliser and farm waste

Fertiliser contains nutrients which are necessary for plant growth. Although it is expensive to use, chemical fertiliser replaces nutrients which may have been removed from the soil. This helps to give healthy crops and increasingly higher yields. Farmers also spread slurry, which is animal waste, over their fields. This, being a natural fertiliser, is cheaper to use than bought chemical fertiliser and is just as effective. Both chemical fertiliser and slurry contain nitrate. If the amount of nitrate released into the soil becomes too great for the plants to use, then the excess will find its way into underground water supplies or surface rivers. Excessive nitrate in water supplies can be harmful to human health. Excessive nitrate in rivers acts as a fertiliser and results in the rapid growth of algae and other plants (photo **D**). These use up large amounts of oxygen, leaving too little for fish life.

Farming has also affected the environment by:
- using pesticides to kill insect pests, but this has also reduced numbers of harmless insects such as bees

D Algal growth caused by excess nitrate in the water

- burning straw which releases carbon dioxide into the air and reduces the amounts of nutrient being put back into the soil (this was banned in 1993)
- draining **wetland** wildlife habitats.

Activities

1 **a)** Why have some farmers removed hedgerows?
   b) Why do conservationists object to the removal of hedgerows?

2 **a)** Why do farmers use chemical fertiliser and farm slurry?
   b) How can the use of chemical fertiliser and farm slurry affect the environment?
   c) Copy and complete the flow chart in diagram **E** to show how the use of chemical fertiliser and slurry can kill fish life in rivers. Choose your answers from the following list.
   - Algae and plants use up oxygen
   - Nitrate and slurry reach rivers
   - Fish die due to lack of oxygen
   - Fertiliser and slurry added to field
   - Nitrate makes algae and plants grow
   - Fertiliser and slurry contain nitrate.

E

1	2	3	4	5	6

Summary

Developments in farming need careful planning and management – otherwise they can have unintentional adverse effects upon the environment.

How can farming be made wildlife friendly?

Farming in the UK has become increasingly efficient in recent years, and some areas produce more food than the country needs. This has enabled farmers to become more flexible with their use of land and has given them the opportunity to play a more important part in conserving and caring for the countryside.

The Royal Society for the Protection of Birds (RSPB) is an organisation that works for a healthy environment rich in birds and wildlife. In April 2000, it took over Hope Farm in Cambridgeshire with the aim of researching how farming could be made more wildlife friendly.

Hope Farm is a typical East Anglian arable farm growing mainly cereal crops for the UK market. The previous farmer used large machinery, pesticides and fertiliser in order to maximise outputs and make a profit. The RSPB intends to continue using modern farming methods but to vary planting and chemical use in order to attract and support more wildlife. In this way it hopes to show that by using new ideas and techniques, farming could become more wildlife friendly but still provide a profitable living for the farmer.

The RSPB completed a survey over two years to identify the wildlife present at the farm before changes were introduced. The survey counted 38 different breeding species of birds and a variety of other animals including hares, hedgehogs, stoats, mice and bats. After introducing new farming techniques, further surveys will be completed and results compared.

On purchasing Hope Farm, the aim of the RSPB was mainly to find ways of increasing the variety and numbers of birds but also to encourage other wildlife and promote a sustainable form of farming. It decided to look at five ways of doing that:

1 Improve winter wheat for nesting skylarks.
2 Grow crops in a way that provides winter feed.
3 Use chemicals that leave enough insect food for birds but still increase crop yields.
4 Find ways of sowing crops at different times to provide shelter and food throughout the year.
5 Provide habitats for a variety of wildlife.

Managing hedgerows is one of the easiest and most successful ways of increasing wildlife. At Hope Farm, hedges are cut in late winter after the birds have stripped

A Skylark at Hope Farm

Hope Farm, Cambridgeshire B

the berries. They are then cut just once every three years. This is cheaper than an annual cut and gives wildlife time to establish and increase their numbers.

After two years the small changes so far introduced have already produced an increase in wildlife. Profits are lower than expected but their decline is almost certainly due more to falling wheat prices than changes in farming methods.

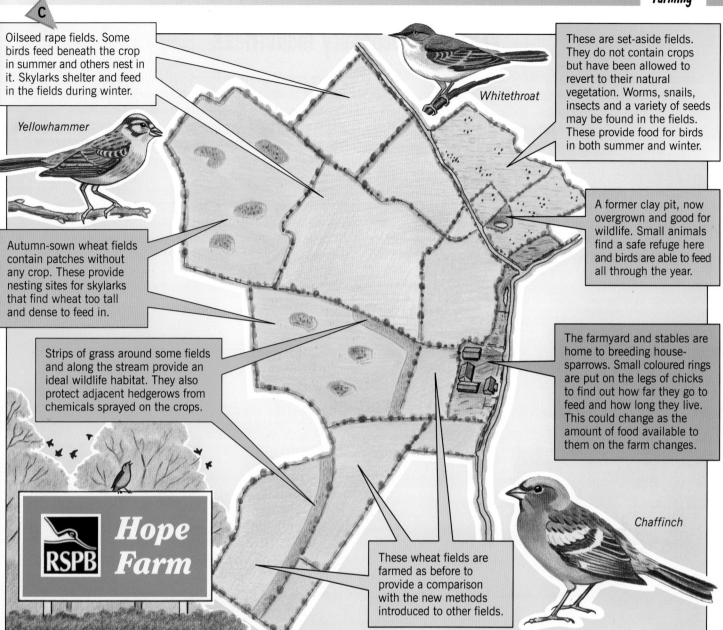

C

Oilseed rape fields. Some birds feed beneath the crop in summer and others nest in it. Skylarks shelter and feed in the fields during winter.

Yellowhammer

These are set-aside fields. They do not contain crops but have been allowed to revert to their natural vegetation. Worms, snails, insects and a variety of seeds may be found in the fields. These provide food for birds in both summer and winter.

Whitethroat

Autumn-sown wheat fields contain patches without any crop. These provide nesting sites for skylarks that find wheat too tall and dense to feed in.

A former clay pit, now overgrown and good for wildlife. Small animals find a safe refuge here and birds are able to feed all through the year.

Strips of grass around some fields and along the stream provide an ideal wildlife habitat. They also protect adjacent hedgerows from chemicals sprayed on the crops.

The farmyard and stables are home to breeding house-sparrows. Small coloured rings are put on the legs of chicks to find out how far they go to feed and how long they live. This could change as the amount of food available to them on the farm changes.

RSPB *Hope Farm*

Chaffinch

These wheat fields are farmed as before to provide a comparison with the new methods introduced to other fields.

Activities

1 Why are farmers now able to play a more important part in conserving and caring for the countryside?

2 Draw a simple farming system diagram for Hope Farm. Include inputs, processes and outputs.

3 The RSPB needs a small information pamphlet to give out to people interested in their work at Hope Farm. Your task is to produce this pamphlet using the headings shown in drawing **D**. You should include diagrams, drawings, a simple, labelled map, and pictures of birds if possible. You might be able to use a computer to word-process your work to make it more professional and attractive. The RSPB website is at: <u>www.rspb.org.uk</u>

D

Hope Farm
A wildlife-friendly farm

- Before RSPB ownership
- Main aims for the farm
- Existing wildlife
- Methods used
- Measuring success

Summary

Modern farming methods have caused damage to the environment. A wildlife-friendly approach can help conserve the countryside whilst still enabling farmers to sustain outputs and make a profit.

What are secondary industries?

Industry as a system

You may remember from page 188 that there are four different types of employment. These are primary, secondary, tertiary and quaternary. This unit looks in detail at secondary activities. Secondary activities are also called manufacturing industries. This is because these industries are a form of employment in which things are made, assembled or produced.

Industry as a whole, or a factory as an individual unit, can be regarded as a **system**. The things that it needs to operate are called **inputs**. What happens in the industry are called **processes**. What it produces are called **outputs**. The finished product is sold and the money earned re-invested in the industry. This money buys more raw materials, pays wages and repays any loans. For an industry to be profitable and remain in business, the value of its outputs must be greater than its inputs.

Diagam **A** shows a simple industrial system. Notice that one of the outputs is 'Waste'. This is a problem for many industries as it may cause pollution and damage the environment. It can be both difficult and expensive to dispose of, or make clean.

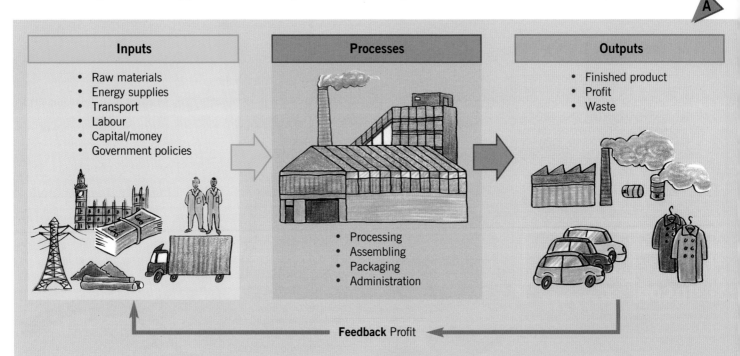

A

Inputs	Processes	Outputs
• Raw materials • Energy supplies • Transport • Labour • Capital/money • Government policies	• Processing • Assembling • Packaging • Administration	• Finished product • Profit • Waste

Feedback Profit

Location of industry

Before building a factory or opening a business, the best possible site has to be found for its location. This will be where the cost of raw materials, energy, labour, land and transport is lowest and where there is a large market for the product (diagram **C**). It is unlikely that all these factors will be available at one particular site, so the correct decision is often very hard to make. A wrong decision could mean failure for the industry, and lead to closure and a loss of jobs.

Not all industries have the same location factors. For some it may be more important to be near raw materials. For others, being near to labour, markets or a good transport system may be more important. Diagram **B** shows some examples of industries needing different location factors.

B

A food processing factory uses fruit and vegetables from nearby farms. These must be fresh so they need to be processed soon after they have been harvested.

A clothing manufacturer The textile industry still employs many people. Factories must be located near to the towns where these people live.

Newspaper printing works Newspapers are now printed in out-of-town locations. They have to reach their markets quickly, so need good transport links.

Bakery Fresh bread is easily damaged and quickly goes stale. Bakeries may be quite small but must be in, or close to, a town where many people live.

Industrial location factors

Physical factors

- **Raw materials**. The factory needs to be close to these if they are heavy and bulky to transport.

- **Energy supply**. This is needed to work the machines in a factory. Early industries were near to coalfields. Nowadays, electricity allows more freedom.

- **Natural routes**. River valleys and flat areas were essential in the days before railways and lorries made the movement of materials easier.

- **Site and land**. Most industries require large areas of cheap, flat land on which to build their factories. Well-drained land is also an advantage.

Human and economic factors

- **Labour**. A suitable labour force is essential. Cost and skill levels are important.

- **Capital**. All industries require some money to set up and start manufacturing. This may be available from banks, government or local authorities.

- **Markets**. An accessible place to sell the goods is needed. This may be in the local area, within the country or abroad as export markets.

- **Transport**. A good transport network helps reduce costs and make the movement of materials easier.

- **Government policies**. Industrial development is encouraged in some areas and restricted in others. Subsidies may be available in development areas.

- **Environment**. Pleasant surroundings with good leisure facilities help attract and retain a workforce.

Oh dear, Now just where shall we go?

Activities

1. **a)** What are *industrial inputs*? Give two examples of economic inputs.
   b) What are *industrial processes*? Give two examples of industrial processes for a factory building cars.
   c) What are *industrial outputs*? Describe some of the problems that waste products may cause.

2. Draw a systems diagram for a factory making chocolate products. Include the following inputs, processes and outputs.

   - cocoa beans • money • chocolate drinks
   - raisins • packaging • sugar • nuts • flavouring
   - wrapping • tasting • chocolate bars
   - chocolate boxes • milk • packing • profit
   - labour • electricity • processing

3. **a)** Look at the information about four industries shown in diagram **B**. Which was the most important location factor for each industry? Choose from:
   • *transport* • *market* • *raw materials* • *labour*.
   b) Look at drawing **D** and choose the best site for each factory in diagram **B**. In each case, write a sentence giving reasons for your choice.

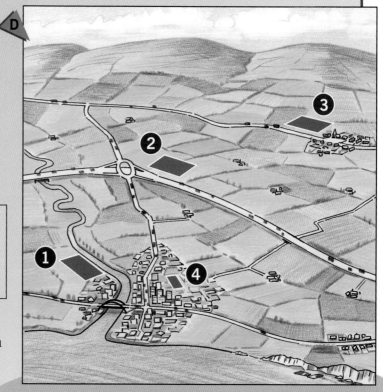

Summary   Industry is a system with inputs, processes and outputs. Many factors need to be considered when choosing a site for a factory. These include transport, and nearness to raw materials, workers and markets for its goods.

211

How has the location of industry in the UK changed?

Industry does not stay the same but changes over the years. New products are developed, new techniques introduced and different location factors become increasingly more or less important. As a result, the type and distribution of industry across the country also changes.

During the Industrial Revolution of the nineteenth century, new industries needed power supplies and raw materials. Power was originally from fast-flowing streams and later from coal. Industry located where these were easily available and gradually a pattern of industrial location developed where different areas **specialised** in industries using local resources. As most natural resources were in the north of England, this became the country's industrial

heartland. Industries at this time were said to have a **raw material location**.

Gradually the use of coal declined as electricity, transmitted through the National Grid, took over as the major power source. Nowadays, industry has a much wider choice of location. Many companies are choosing to locate in areas with an attractive environment, with good transport links and access to a large market for their products. These industries are said to have a **market location**.

Changes in industrial location and the closure of factories and businesses can lead to job losses and unemployment. Some reasons why this may happen are shown in table **B**.

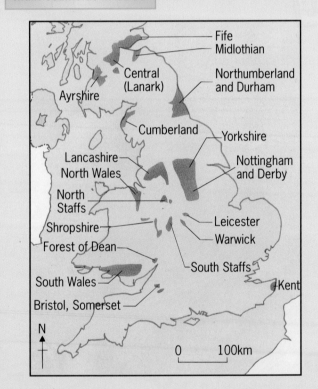

The rapid growth of industry in Britain came in the early nineteenth century with the introduction of steam power. Coal was needed to heat water to produce steam and, as coal was heavy and bulky to move, then coalfields became the ideal location for industry. Industries such as steelmaking and shipbuilding were particularly important. These are called **heavy industries** because they require large amounts of heavy and bulky raw materials like coal and iron ore.

By 1960, industrial development on coalfield sites had reached its peak. Heavy industries still provided most employment in these areas but **light industries** were becoming increasingly important. Light industries make high-value goods like car stereos and fashion clothing. Most regions tended to specialise in one or two types of industry, for example Teesside for chemicals, South Wales for steel and metal working, and the West Midlands for cars and leather goods.

B

Reason for closure	Example
Exhaustion of resources	Pits close as coal runs out
New machinery or new methods	Automated car manufacture needing fewer workers
Fall in demand for product	New synthetic materials cause closure of textile mills
Site needed for other uses	Old factories replaced with shopping complex
Lack of money for investment	Closure of old inefficient steelworks
Competition from overseas	Televisions and electronic equipment from Japan
Political decisions	Government failing to support companies in difficulties

Major industrial regions in 2000

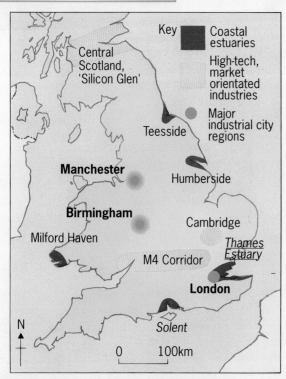

Key
- Coastal estuaries
- High-tech, market orientated industries
- Major industrial city regions

Central Scotland, 'Silicon Glen'

Teesside

Manchester

Humberside

Birmingham

Cambridge

Milford Haven

M4 Corridor

Thames Estuary

London

Solent

N

0   100km

Recently the location of the main centres of industry has again changed. Firstly there has been a shift from coalfields to coastal estuaries as the import of raw materials and export of finished products has become more important. Secondly there has been a movement from the more northerly parts of England to the South mainly by **footloose industries** that are not tied to raw materials. They tend to specialise in very high-value goods like electronics equipment and computers.

Activities

1 Give the meaning of the following terms:
   a) *raw material location*   b) *market location*
   c) *heavy industries*   d) *light industries*
   e) *footloose industries*.

2 Look at the following list of statements about industrial change in the UK. Copy and complete diagram **C** by putting each statement into the correct box.
   - Footloose industries have greater freedom of location.
   - Light industries develop away from coalfields.
   - Most industries have a raw material location.
   - Electric power provides greater freedom of location.
   - Industries move to locations in the South.
   - Coalfield location vital to most industries.
   - Heavy industries located mainly in the North.
   - Many new industries have market locations.
   - Coastal sites favoured for imports and exports.

1900
- _____
- _____
- _____

1960
- _____
- _____
- _____

2000
- _____
- _____
- _____

C

3 Look at the maps in **A** and find an industrial region that is close to you or that you have studied. Name the region and describe how industry there changed between 1900 and 2000. Suggest reasons for the changes.

4 Describe at least three reasons why a business may close down. Give named examples for each one.

Summary

The best site for an industry changes as location factors change. Most UK industry developed on coalfield sites in the North. Nowadays, access to markets and good transport are more important.

213

Why were traditional industries located near to raw materials?

The UK iron and steel industry

The raw materials needed to make iron and steel are iron ore, coal and limestone. Coke, from coal, is used to smelt the iron ore, and limestone is added to help separate the pure iron from impurities.

The nineteenth-century iron industry was concentrated upon coalfields that also contained bands of iron ore. The major producing centres were in the South Wales valleys, north-east England, parts of Central Scotland and around Sheffield. Although later technological developments enabled steel rather than iron to be manufactured, there was no need to change the location of the new steelworks and the distribution pattern remained the same (map **B**).

When a method of smelting low-grade iron ore was discovered in the 1950s, two new steelworks were opened near to the ore deposits at Scunthorpe and Corby (map **B**). Since then much of the UK's accessible coal and iron ore have been used up. As imports of these two raw materials increased, the UK's latest steelworks were built on the coast near to port facilities (map **C**).

A A blast furnace

Competition from overseas has had a severe effect on the UK's steel industry. By 1993 only four **integrated steelworks** remained open. This was reduced to just three in June 2001 when the Llanwern works in South Wales ended its steel production.

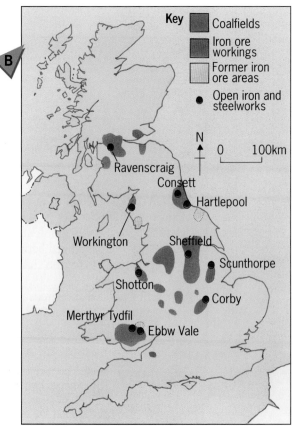

Iron and steelworks in the 1950s **B**

Key
- Coalfields
- Iron ore workings
- Former iron ore areas
- Open iron and steelworks

Ravenscraig
Consett
Hartlepool
Workington
Sheffield
Scunthorpe
Shotton
Corby
Merthyr Tydfil
Ebbw Vale

N
0   100km

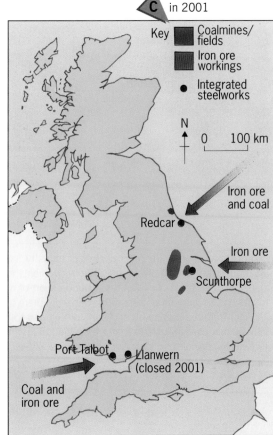

C Integrated steelworks in 2001

Key
- Coalmines/ fields
- Iron ore workings
- Integrated steelworks

N
0   100 km

Redcar — Iron ore and coal
Scunthorpe — Iron ore
Port Talbot
Llanwern (closed 2001)
Coal and iron ore

Iron and steel production in South Wales

The three raw materials needed to make iron and steel – iron ore, coal and limestone – were all available in the valleys of South Wales. More than 150 years ago, the iron industry developed in places like Ebbw Vale and Merthyr Tydfil. After 1860, steelworks slowly began to replace iron foundries. By the 1990s the valley steelworks had all closed because the coal and iron ore once found there had been exhausted.

The government chose sites at Port Talbot and Llanwern for new steelworks. Port Talbot has its own harbour and docks for the import of coal and iron ore. It is an integrated steelworks where all the stages in steel manufacture take place on the same site. The high-quality steel is used locally and exported throughout the world. The closure of Llanwern in 2001 was blamed on competition from overseas, global over-capacity and falling prices.

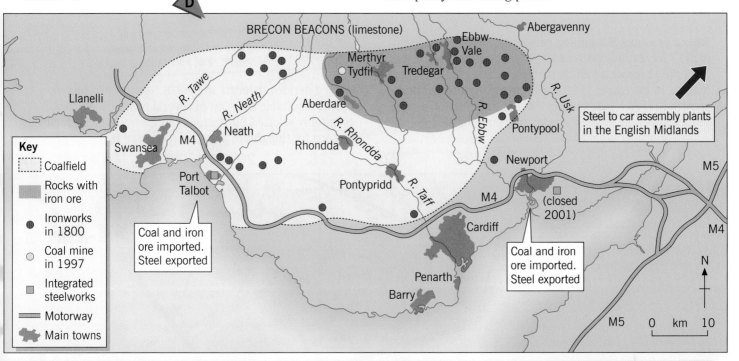

Key
- Coalfield
- Rocks with iron ore
- Ironworks in 1800
- Coal mine in 1997
- Integrated steelworks
- Motorway
- Main towns

BRECON BEACONS (limestone)

Coal and iron ore imported. Steel exported

Coal and iron ore imported. Steel exported

Steel to car assembly plants in the English Midlands

Activities

1 Draw a simple industrial systems diagram for making steel. Include inputs, processes and outputs, and use the terms given below.

- *capital* • *iron ore* • *flat land* • *sheet steel*
- *limestone* • *smelting* • *steel coils*
- *water supply* • *steel ingots* • *energy supplies*
- *shaping* • *labour* • *energy* • *coal*

2 **a)** Describe the distribution of steelworks in the 1950s. Suggest reasons for this distribution.
   b) Describe the distribution of steelworks in 2001.
   c) Why did the pattern change between the 1950s and 2001?

3 Look carefully at map **E**.
   a) Which site would have been best for an ironworks in 1820? Give four reasons for your choice.
   b) Which site would be best for a steelworks in 2001? Give four reasons for your choice.

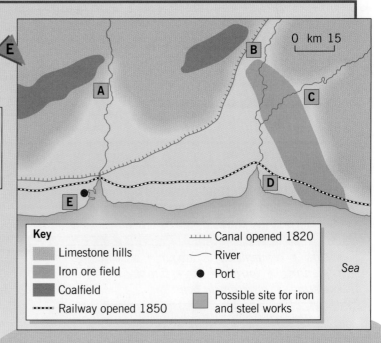

Key
- Limestone hills
- Iron ore field
- Coalfield
- ····· Railway opened 1850
- ⊥⊥⊥⊥ Canal opened 1820
- River
- ● Port
- Possible site for iron and steel works

Sea

Summary   The UK's iron industry grew up near to raw materials. The industry has suffered severe decline, and a better location is now close to coastal ports for importing raw materials and exporting the finished steel.

Why are newer industries located close to markets?

The car industry

Toyota is the world's third largest manufacturer of cars. It has its head office in Japan but operates on a global scale with components factories, assembly plants and selling outlets throughout the world. Huge companies like Toyota, which have offices and factories in several countries, are called **transnational** or **multinational** organisations.

Toyota produces 5.2 million cars a year. As Europe is the world's largest market for cars, Toyota decided that it was essential to locate a vehicle manufacturing plant within that market. In December 1989 the company announced that the plant would be located in the UK. Diagram **A** shows some of the reasons that Toyota gave for choosing the UK.

A

UK location advantages

Toyota – UK

- Large domestic market of its own
- Long tradition of vehicle manufacture
- Skilled and flexible workforce
- Many firms already making car components
- Good communications with the rest of Europe
- Government support at local and national levels

Having decided to locate in the UK, Toyota then had to choose a suitable site. Out of several possibilities, they eventually selected Burnaston near Derby. Some of their reasons for this choice are shown on map **B**.

B

Burnaston location factors

- Large area of flat land available
- Greenfield location with room for expansion
- Central to the UK domestic market
- Many skilled workers living nearby
- Close to West Midlands where many car parts suppliers are located
- Easy access to motorway system
- National rail network nearby
- Attractive countryside, pleasant villages and good amenities nearby
- Government support available

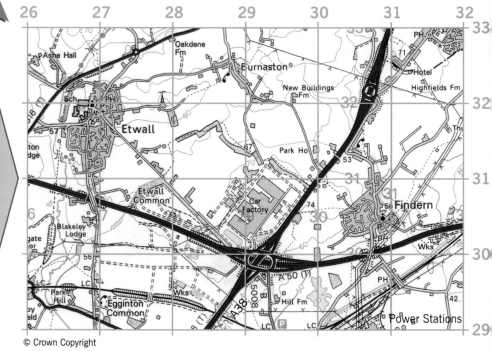

© Crown Copyright

Activities

1 Draw a simple industrial systems diagram for car production. Include inputs, processes and outputs and use the terms given below.

- *cars* • *components* • *labour* • *assembling*
- *painting* • *lorries* • *large, flat site*
- *testing* • *government help*

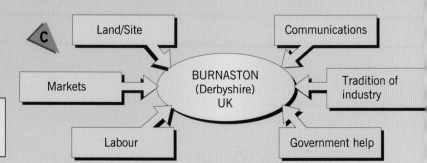

C

Land/Site — Communications — Tradition of industry — Government help — Labour — Markets — **BURNASTON (Derbyshire) UK**

2 **a)** Look carefully at map **B** and identify six location factors that attracted Toyota to Burnaston.
   b) Name each factor and give its map reference.

3 **a)** Make a larger copy of diagram **C**.
   b) Explain why Toyota chose Burnaston as the location for its car assembly plant, by writing a paragraph for each heading.

Toyota today

Toyota Motor Manufacturing (UK) Ltd has two factories – its engine plant in Deeside, North Wales and the vehicle manufacturing plant at Bunaston, Derby. Approximately 70 per cent of production goes to Europe and 10 per cent to 70 countries world-wide. The remaining 20 per cent stays in the UK.

Part of the agreement to allow Toyota to locate in the UK was that 80 per cent of the car must contain local content. To meet this target, Toyota must buy parts from European suppliers. By 2000, Toyota was using over 220 suppliers from 11 European countries, of which over half were UK based (map **E**).

Local suppliers support Toyota's **just-in-time** policy, which is now used by many other firms. With just-in-time, component parts are supplied to the assembly line just minutes before they are needed. Expensive parts do not have to be stored on site so costs are reduced. An efficient and reliable transport system is needed for it to work.

Toyota is a fully integrated operation. It provides technical training and includes the manufacturing of component parts, the assembling of those parts on a production line, and finally the marketing, sales, distribution and post sales service. These features are typical of large modern industries.

D Toyota Motor Manufacturing (UK) Ltd

	Employees		Car production
Location	Engine plant, Deeside	Assembly plant, Burnaston	–
1996	172	2050	110 000 (one assembly plant)
2001	381	2481	179 000 (two assembly plants)

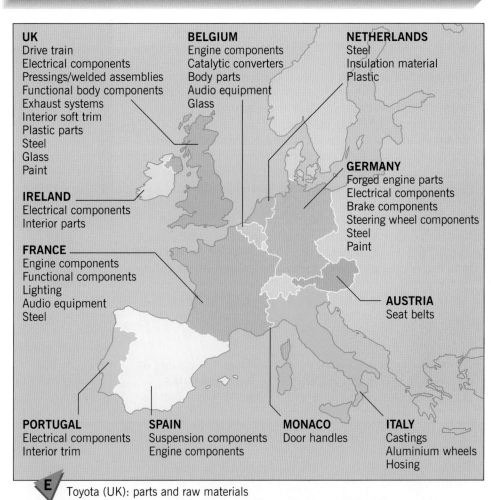

UK
Drive train
Electrical components
Pressings/welded assemblies
Functional body components
Exhaust systems
Interior soft trim
Plastic parts
Steel
Glass
Paint

IRELAND
Electrical components
Interior parts

FRANCE
Engine components
Functional components
Lighting
Audio equipment
Steel

BELGIUM
Engine components
Catalytic converters
Body parts
Audio equipment
Glass

NETHERLANDS
Steel
Insulation material
Plastic

GERMANY
Forged engine parts
Electrical components
Brake components
Steering wheel components
Steel
Paint

AUSTRIA
Seat belts

PORTUGAL
Electrical components
Interior trim

SPAIN
Suspension components
Engine components

MONACO
Door handles

ITALY
Castings
Aluminium wheels
Hosing

E Toyota (UK): parts and raw materials

4 Suggest reasons for each of these slogans:

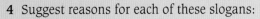

Buy British, buy Toyota

Go foreign, go Toyota

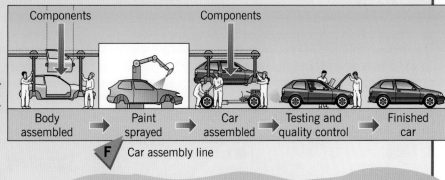

Components   Components

Body assembled → Paint sprayed → Car assembled → Testing and quality control → Finished car

F Car assembly line

5 a) What is an *assembly line*?
  b) Describe how a car is built on the Toyota assembly line.
  c) Why is a good transport system essential to a just-in-time system?

Summary Toyota is an example of a transnational company with factories in several countries. Burnaston, near Derby, was chosen for the company's first European car plant.

What are high-technology industries?

High technology, or **high-tech**, industries make high-value products such as electronic equipment, computers and medical products. They have been the growth industries of recent years and now provide more than 25 per cent of the UK's manufacturing jobs. High-tech industries use the most advanced manufacturing methods and employ a highly skilled and inventive workforce. They put great emphasis on research and development of new products and are always looking to improve existing ones.

Many high-tech companies are divided into two sections. One is involved with product development and the other with product manufacture. The manufacturing section is often a simple assembly plant which puts together **components** that are made elsewhere. These components are usually small, light and easy to transport, so the industry can have a relatively free choice of where to locate. Industries like these are said to be **footloose** (drawing **A**).

As high-tech industries need to attract a highly talented workforce, they are normally located in areas where researchers and operators can enjoy a pleasant environment and a good quality of life. Such areas include Silicon Valley in California and Tsukuba Science City near Tokyo, Japan. Three UK locations are 'Silicon Strip' following the M4 motorway westwards from London, 'Silicon Glen' in central Scotland and 'Silicon Fen' in and around Cambridge.

Firms that make high-tech products often group together on pleasant, newly developed **science** or **business parks**. These parks are often on the edge of the city where there is

A

Footloose industries:
factors that favour a free choice of location

- Tend to use small, light components that are easy to transport to the factory
- End product usually small and easy to transport to markets
- Power source mainly electricity which is available from the national grid
- Small labour force required
- Non-polluting industry so can locate near residential areas
- Suited to locations with good accessibility

plenty of space. The buildings are modern, there is plenty of parking, and the parks are landscaped, often with woodland, gardens, lakes and ponds. All firms on science parks are high-tech and have links with a university. Business parks do not have links with universities and may include hotels and leisure centres.

Activities

1 a) What is meant by the term *high-tech industry*?
  b) What is meant by the term *footloose industry*?
  c) Draw a star diagram to show six reasons why an industry may be footloose.

2 Diagram **B** shows four possible locations for a new science park.
  a) Which of the locations would you choose? Give reasons for your choice.
  b) Why would you reject the other sites?

3 a) Match each of the following features with a number from photo **D**.

• Large, flat site	• Pleasant housing nearby
• Room for expansion	• Landscaped surroundings
• Close to motorway	• Low-density development

  b) List six other factors that make Cambridge an ideal location for a science park.

Summary

High-tech industries use modern technology to produce high-value goods. Most are footloose and locate on edge-of-town sites, often close to universities.

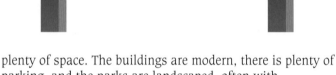

B

Urban area

Motorway

University

CBD

Main road

X Y W Z

Cambridge Science Park

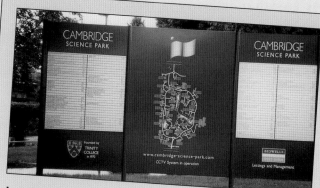

Main features

Cambridge is the site of one of the UK's largest and most successful science parks. It was established in 1970 on 50 hectares of derelict land owned by Trinity College. Since opening, the park has thrived on its close links with the research units of the university.

The work of the companies based at Cambridge ranges from the research and development of medical products such as animal vaccines and kidney dialysis machines, to computers, lasers and information technology products. All companies have a high level of investment and use the latest processing techniques.

As photo **D** shows, the Cambridge Science Park is a low-density development in a park-like setting. The area is landscaped with trees, lakes and ornamental gardens. Nearby are attractive countryside, pretty villages and high-quality residential areas. Over 150 companies have chosen to locate at the park and some 3000 people work there. The largest firm employs 320 people but more than half the companies have fewer than 20 people working for them. Most of the employees are university graduates.

Location factors

- Large, flat greenfield site on edge of city
- Room on site for further expansion
- Attractive site impresses clients and creates a good image
- Highly qualified and skilled workforce available
- Closely linked with excellence of Cambridge University
- Links with university departments and research teams
- Working links with other companies on site
- Close to M11 and M25 motorways
- Near to Stansted Airport for international links
- Good leisure facilities in nearby Cambridge
- Pleasant housing and open space nearby

What are the consequences of industrial decline?

The growth of industry in Teesside

Teesside's industrial development began in the 1850s immediately after the discovery of ironstone at Easton in the nearby Cleveland Hills (map **A**). Within three years there were 35 blast furnaces in the area. Teesside had several advantages for the location of industry.

- Near to the basic raw materials:
  - For early iron production and later, after 1876, steel-making (coal from Durham, iron ore from the Cleveland Hills and limestone from the Pennines).

By 1950, 30 per cent of pig iron produced in the UK came from Teesside.
  - For chemical industry (rock salt from the Tees estuary).
- A tidewater site for importing raw materials (e.g. iron ore when local ores became exhausted, oil for the chemical industry) and the export of finished goods (e.g. ships, bridges, locomotives and machinery).
- Large areas of flat, marshy land which was cheap to buy and easy to reclaim either for docks or as sites for large factories (map **A**).

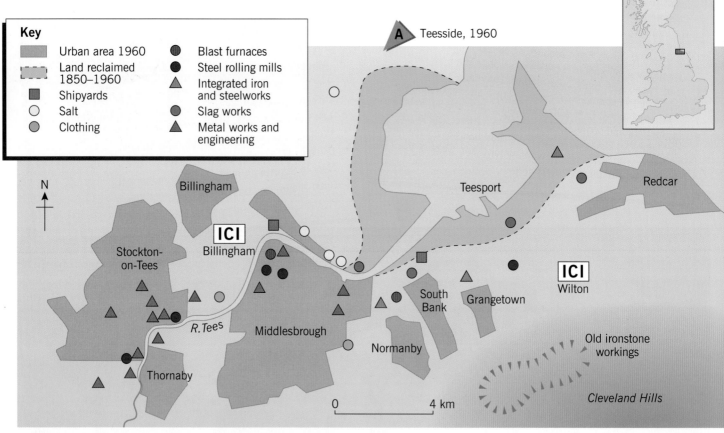

A Teesside, 1960

Key

Urban area 1960	Blast furnaces
Land reclaimed 1850–1960	Steel rolling mills
Shipyards	Integrated iron and steelworks
Salt	Slag works
Clothing	Metal works and engineering

Billingham
ICI Billingham
Stockton-on-Tees
Teesport
Redcar
ICI Wilton
South Bank
Grangetown
R. Tees
Middlesbrough
Normanby
Old ironstone workings
Thornaby
Cleveland Hills

0   4 km

B ICI Wilton Works

During the 1960s, Teesside was regarded as 'the industrial centre of the future'. Successive governments helped modernise both the steel industry (nationalised in 1967) and the chemical industry. A large, modern, integrated iron and steel works was opened at Redcar (map **C** page 214), and the Shell oil refinery at Teesport. At the same time, the expansion of ICI (Imperial Chemical Industries) continued (photo **B**). Indeed, a major reason for creating Kielder Reservoir in Northumberland was the expected demand for water by Teesside. Unemployment fell to 2 per cent and 'full employment' seemed to be guaranteed.

The decline of industry (1980s)

The post-1974 oil price rises led to a world economic recession by the early 1980s. This recession had a major impact upon Teesside's main industries.

- The chemical industry, heavily dependent upon oil, saw job losses and the closure of the Teesport refinery (1984).
- A global fall in the demand for steel and improvements in technology caused British Steel to reduce its workforce from 30 000 in the early 1970s to under 5000 by 1990. Only the large Redcar works remained open.
- Shipbuilding was affected by the fall in steel production, the decline in world trade and competition from shipyards in Asia. Locomotive production ceased as road transport increasingly took over from rail.
- Employment in metal-using industries and in engineering also declined.

By 1987, 25 per cent of the Teesside workforce was unemployed. The traditional manufacturing industries had declined rapidly and the promised growth in service industries had not occurred. The area was suffering from high unemployment (much of it long-term), dereliction of land and buildings caused by industrial decline, and from environmental pollution (photos **C** and **D**).

The British government and the European Regional Development Fund did provide some financial aid. However, new industries that were needed to replace the thousands of jobs lost, the retraining of people for those new types of job, and the conversion of land for the new industries, could not be achieved overnight.

C Teesside in the late 1980s

D Dereliction on Teesside in the late 1980s

E

1966		1987
Primary 1.8%		Primary 2.4%
Chemicals 19.4%		
Services 40.5%	Secondary 57.7%	Services 46.8% — Secondary (industry) 50.8%
Other manufacturing 9.8%	Metal manufacturing 19.0% — Mechanical engineering and shipbuilding 9.5%	
Unemployment 2%		**Unemployment 25%**

Activities

1 **a)** Give six reasons why Teesside became important for the production of iron and steel.
   b) What other industries were attracted to Teesside?

2 **a)** How are the two graphs in **E** different?
   b) What are the main differences between photo **B** and photos **C** and **D**?
   c) Give reasons for your answers to **a)** and **b)**.

Summary

Most industrial regions go through a period of growth and decline. In the 1960s, Teesside was seen as a major European centre of growth. By the late 1980s it had become an unemployment blackspot.

221

What are the consequences of regeneration?

A

Teesside in 2001

The Teesside Development Corporation (TDC) was set up in early 1987. The TDC covered 4500 hectares of land along the banks of the Rives Tees and at Hartlepool. Of this land, almost one-quarter was either derelict or underused. The new emphasis was on two types of development:

1. • private-sector property development
   • public-financed redevelopment of former industrial sites by the TDC
2. projects aimed more towards the leisure and service sector rather than at the traditional work and manufacturing sector.

Fact file **A** lists the achievements of the TDC by 2001. Major projects completed by the TDC by that date included:

- Teesdale – a £500 million development with a mix of offices, housing, shops and leisure facilities (photo **B**). It is built around a series of canals.
- Teesside Park – a major retail and leisure complex now in its second phase of development.
- Tees Offshore Base – the first three phases of development have transformed a derelict shipyard into a centre of excellence in offshore technology with the creation of nearly 1400 jobs (photo **C**).
- Riverside Park – the new home of Middlesbrough Football Club.
- Hartlepool Marina – 80 ha of Hartlepool's South Dock have been converted into a leisure, retail and commercial area.
- A barrage across the River Tees will prevent flooding and improve water quality and water activities (maps **D**).

Fact file – TDC 1987-2001

- ◆ Land reclaimed   386 hectares
- ◆ Housing units completed   1420
- ◆ Roads, built or improved   32 km
- ◆ Gross gain in permanent jobs   23 200
- ◆ Private-sector investment   £1.2 billion
- ◆ Drop in unemployment   8%

2001

Primary 0.4%
Secondary 20.3%
Services 79.3%

Unemployment 17%

B Teesdale

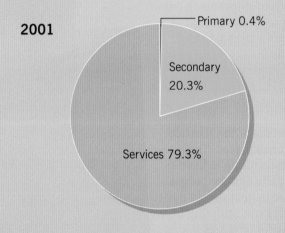

C Tees Offshore Base

A cleaner Tees

A major clean-up of the River Tees and a boost to industrial development on Teesside will result from the Tees Estuary Environmental Scheme (TEES). TEES is a response by Northumbria Water, ICI and the industries of Wilton and Seal Sands, to the increasingly high EU environmental standards (maps **D**). Under the European Urban Waste Water Treatment Directive:

- Northumbria Water must provide further treatment for wastewater at present being discharged from its three sewage treatment works on Teesside

- major industries on Teesside must reduce their effluent flowing into the Tees.

New pipelines will be laid on both the north and south sides of the river. These will take sewage and industrial effluent to a combined new treatment works on land provided by ICI. The scheme, completed in 2002, builds upon improvements in water quality of the Tees already made in the last 25 years (maps **D**).

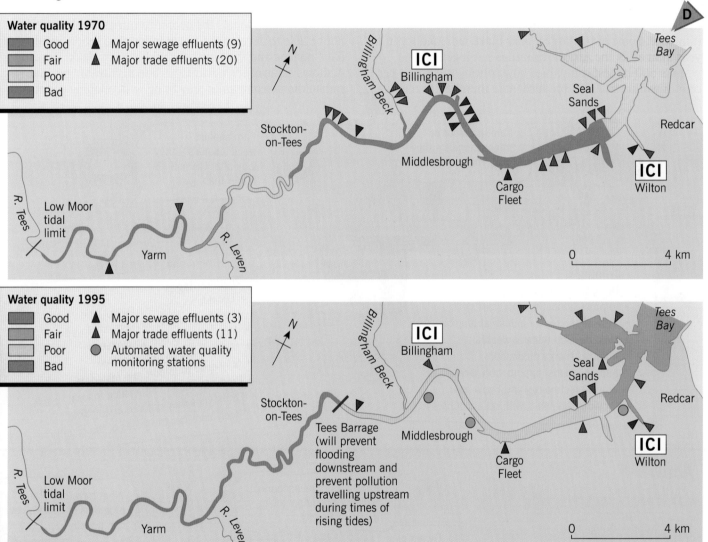

Water quality 1970

Good	▲ Major sewage effluents (9)
Fair	▲ Major trade effluents (20)
Poor	
Bad	

Water quality 1995

Good	▲ Major sewage effluents (3)
Fair	▲ Major trade effluents (11)
Poor	● Automated water quality monitoring stations
Bad	

Tees Barrage (will prevent flooding downstream and prevent pollution travelling upstream during times of rising tides)

Activities

1 a) Why was the Teesside Development Corporation set up?

b) In what ways has it been successful?

c) What problems still remain?

2 a) Why was much of the water in the lower Tees of poor or bad quality in 1970?

b) What attempts have been made, and are being made, to improve water quality?

c) How successful have these attempts been so far?

Summary

A regeneration programme can help a declining industrial area to recover. Although there have been recent improvements in employment and in the environment in Teesside, changes have been slow and are still not sufficient.

What are transnational corporations?

Transnational corporations, or multinational companies as they are often called, are very large businesses that have offices and factories in several different countries. The headquarters and main factory are usually located in developed countries, particularly the USA and Japan. Smaller offices and factories tend to be in the developing countries where labour is cheap and production costs are low.

In the past 30 years, transnationals have grown in size and influence. Some of the largest ones make more money in a year than all of the African countries put together. The world's 500 largest companies now control at least 70 per cent of world trade and produce more than half of the

world's manufactured goods. Being so large, they also influence consumer tastes and lifestyles and are responsible for many of today's scientific and technological breakthroughs.

Many people are concerned about the effects of transnationals. They argue that they locate in poorer countries just to make a profit, and pay low wages, particularly to women and young children. Others say that without transnationals the poorer countries would simply not be able to develop their own industries. People would have no jobs and their future would be very bleak. Drawing **A** shows some of the advantages and disadvantages that transnational companies may bring to developing countries.

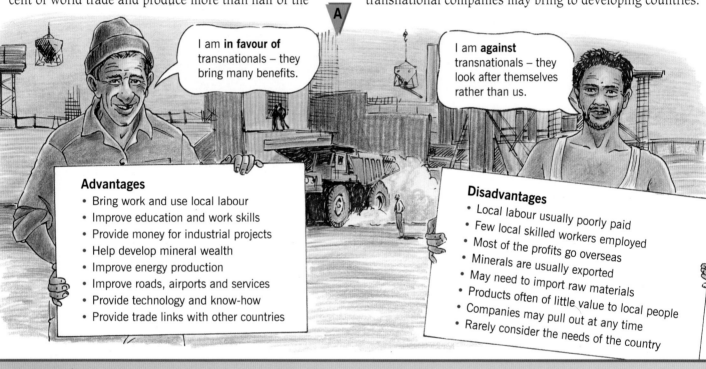

A

I am **in favour of** transnationals – they bring many benefits.

I am **against** transnationals – they look after themselves rather than us.

Advantages
- Bring work and use local labour
- Improve education and work skills
- Provide money for industrial projects
- Help develop mineral wealth
- Improve energy production
- Improve roads, airports and services
- Provide technology and know-how
- Provide trade links with other countries

Disadvantages
- Local labour usually poorly paid
- Few local skilled workers employed
- Most of the profits go overseas
- Minerals are usually exported
- May need to import raw materials
- Products often of little value to local people
- Companies may pull out at any time
- Rarely consider the needs of the country

Activities

1 **a)** What are *transnational corporations*?
   b) Give ten facts about transnationals.

2 Answer the four questions posed by the people in drawing **B**.

3 Do you think that transnationals are good or bad for developing countries? Give reasons for your answer.

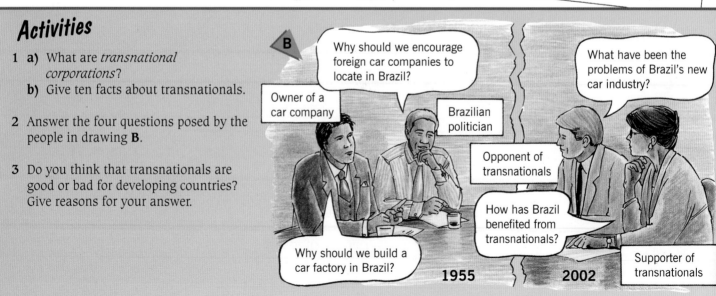

B

Why should we encourage foreign car companies to locate in Brazil?

Owner of a car company

Brazilian politician

What have been the problems of Brazil's new car industry?

Opponent of transnationals

How has Brazil benefited from transnationals?

Why should we build a car factory in Brazil?

1955

2002

Supporter of transnationals

Transnationals in Brazil

In the mid-1950s Brazil was a very poor country and the government decided to try and attract foreign companies to help development. They thought that car manufacturing in particular would encourage the growth of support industries, bring in thousands of jobs and help raise the living standards of Brazilians. They also hoped that many new industrial skills would be learned by workers and management which would help other areas of Brazil's economy.

But why should the giant car companies of the USA, Europe and Japan want to locate in Brazil? Four good reasons were put forward:

1 They could build and operate modern factories cheaply.
2 They would have a guaranteed market in Brazil and could supply other South American countries.
3 The wages in Brazil were very low, so production costs could be reduced.
4 The strong military government of Brazil meant there would be little strike action.

By the mid-1960s, several transnational companies had built car works in Brazil. Ford of the USA and Volkswagen of Germany opened factories in São Paulo, and Fiat opened a plant in Belo Horizonte.

The companies quickly profited from the growing demand for cars in Brazil. They were able to produce cars more cheaply than in Europe and the USA, and now export cars all over the world. For the companies involved, locating in Brazil had been a success. But what of Brazil? Certainly the arrival of transnationals increased job opportunities and brought wealth to the country. Vehicle exports now make up 5 per cent of Brazil's export earnings. There have been problems, however, and some of these are shown in drawing **D**.

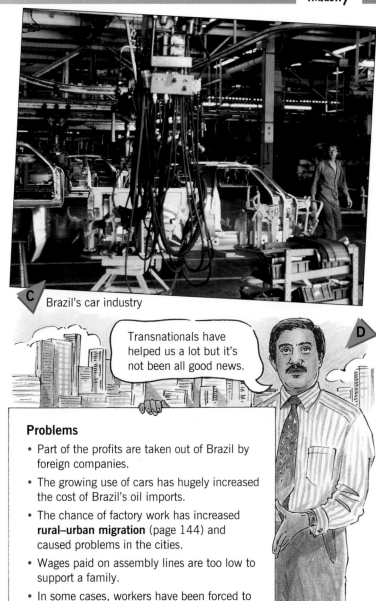

C Brazil's car industry

Transnationals have helped us a lot but it's not been all good news.

D

Problems

• Part of the profits are taken out of Brazil by foreign companies.
• The growing use of cars has hugely increased the cost of Brazil's oil imports.
• The chance of factory work has increased **rural–urban migration** (page 144) and caused problems in the cities.
• Wages paid on assembly lines are too low to support a family.
• In some cases, workers have been forced to work very long hours in difficult conditions.

4 a) Make a copy of diagram **E**.
  b) Add the following labels to the correct positions marked 1 to 6.
   • New firms start up to make parts.
   • Training helps develop a skilled workforce.
   • Workers earn money to spend on products from other industries.
   • Government invests in other industries and education and health services.
   • Exports earn foreign currency.
   • Iron ore mining develops.
  c) Briefly describe how the setting up of car assembly plants in Brazil helped the economy in other ways.

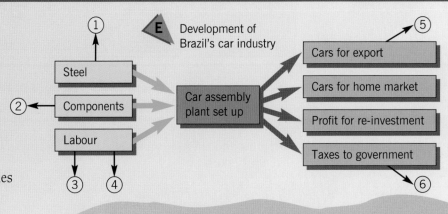

E Development of Brazil's car industry

① Steel

② Components

Labour

③ ④

Car assembly plant set up

Cars for export ⑤

Cars for home market

Profit for re-investment

Taxes to government ⑥

Summary

Transnational corporations are large businesses that have offices and factories all over the world. They can bring many benefits to developing countries but can also cause problems.

What are newly industrialised countries?

In the past, most of the world's industries were located in Europe and North America. The situation now is very different, as many developing countries have started to catch up and even overtake the large industrialised nations. Countries like Japan, Singapore, Thailand and Malaysia have all rapidly developed a large manufacturing industry, and as a result have seen their economies grow and their countries become richer. Countries that have undergone rapid and successful industrialisation are called **newly industrialised countries** or **NICs**.

The countries of eastern Asia have enjoyed the most spectacular growth. The greatest increase has been in South Korea, Taiwan and Singapore, which became known as the three 'Tigers'. Hong Kong was known as the fourth Tiger, but is now part of China and no longer independent.

Like Japan, these Tiger economies lacked basic raw materials but each had a strong government and a highly motivated workforce. The workforce were initially willing to work long hours for little pay. This helped their companies produce cheap goods and slowly take over world markets. Since the late 1970s, Malaysia, Thailand and Indonesia have begun to join the list of NICs. The next to emerge, and certainly the largest, is likely to be China.

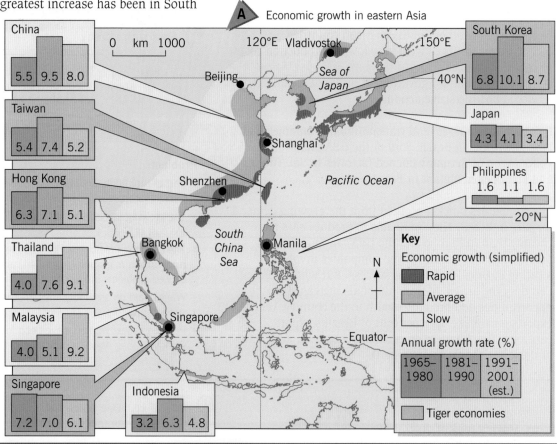

A Economic growth in eastern Asia

China		
5.5	9.5	8.0

South Korea		
6.8	10.1	8.7

Taiwan		
5.4	7.4	5.2

Japan		
4.3	4.1	3.4

Hong Kong		
6.3	7.1	5.1

Philippines		
1.6	1.1	1.6

Thailand		
4.0	7.6	9.1

Malaysia		
4.0	5.1	9.2

Singapore		
7.2	7.0	6.1

Indonesia		
3.2	6.3	4.8

Key

Economic growth (simplified)
- Rapid
- Average
- Slow

Annual growth rate (%)

1965–1980	1981–1990	1991–2001 (est.)

- Tiger economies

Activities

1 a) What is meant by the term *newly industrialised country (NIC)*?
   b) Using the 1991–2001 figures on map **A**, list the NICs in order of growth. Give the highest first.

2 a) From graph **B**, give Thailand's GDP per capita in 1980, 1985, 1990 and 2000.
   b) Describe the pattern of growth shown by the graph.
   c) During which period was growth greatest?

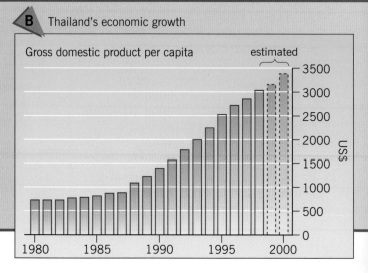

B Thailand's economic growth

Gross domestic product per capita

Thailand – a newly industrialised country

Thailand is one of the world's fastest-growing economies. In less than 30 years it has changed from an economy based on agriculture to one dominated by manufactured goods. This has resulted in a huge improvement in living standards and the development of the country's capital, Bangkok, as a prosperous industrial region.

Change began in the 1970s when industries such as oil refining, motor vehicle assembly, steelmaking and chemicals were developed. This was followed in the 1980s and early 1990s by the rapid expansion of the electronics industry when foreign companies were invited to open factories in Thailand. Japanese and US firms moved in, attracted by low wages and a strong government. In the mid-1990s the government further speeded up development by moving towards high-tech industries producing high-value, high-profit goods.

Thailand's rapid growth has brought many benefits:
- The country's GDP (wealth) has tripled in 20 years.
- There are more jobs available.
- Average wages have increased rapidly.
- More people now own TVs, hi-fi's, refrigerators, etc.
- Education and training has been improved.
- New hospitals have been built.
- Improvements have been made to roads and railways.
- Improvements in Bangkok include a modern international airport and new rapid transit system with a Skytrain and underground rail network.

Rapid growth has brought problems, however. Many people are concerned that growth has been centred on Bangkok, whilst the rest of the country has gained little benefit. Wages, for example, are ten times higher in Bangkok than they are in the rural north-east of the country. Some other problems are shown in diagram **D**.

C Bangkok's air-conditioned Skytrain

D

Growth has brought us many benefits but it's certainly caused some problems!

Bangkok
- Seriously overcrowded and congested
- Said to have the world's worst traffic jams
- Badly polluted air with very high lead levels
- Many buildings poorly constructed
- Only 2% of residents with sewerage facilities

Countrywide
- Rapid rural–urban migration (page 144) has left both cities and rural areas with big problems.
- Working conditions in some factories are very poor, with long working hours and very low pay.
- The sea and many rivers and canals have become polluted with industrial waste.

---

3 Describe Thailand's emergence as a newly industrialised country, using the headings in diagram **E**.

4 Thailand has been described as one of 'the Young Tigers of Asia'. Suggest reasons for this.

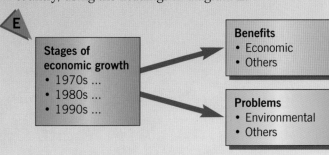

E

Stages of economic growth	Benefits
• 1970s ...	• Economic
• 1980s ...	• Others
• 1990s ...	
	Problems
	• Environmental
	• Others

Summary

Many countries in eastern Asia have seen dramatic levels of industrialisation and economic growth in the last few decades. They are now called 'newly industrialised countries' or NICs.

What is informal employment?

Working in the poorer countries of the world can be very different from working in the UK. Most people work long hours and earn low wages. Few people have a proper, full-time job and there is little government support for those families with no money. The situation in the cities is particularly bad. Here there are many more people than jobs. With the rapid growth of these cities due to inward migration from the countryside, the job situation is continually worsening (pages 180–181).

Some people manage to gain employment in the **formal sector**. These jobs provide a regular income and may be in an office, shop or organised factory. Most, however, have to find work in the **informal sector** (drawing **A**). Here they may work in jobs such as street-trading, shoe-shining or luggage-carrying (photo **F** page 185). There is little or no security in the informal sector. People live from day to day and, as diagram **B** shows, they are trapped by a lack of opportunity to improve their positions.

A

We have created our own employment to meet local demand. We work long hours but have no security or regular wage.

Our prices are negotiable but we earn little money. We don't pay taxes and our work is often illegal.

C Selling food on a beach in Bali, Indonesia

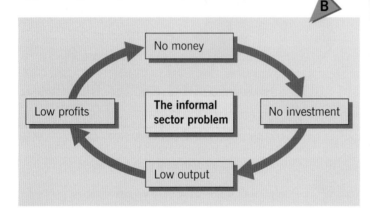

B

No money

Low profits

The informal sector problem

No investment

Low output

Informal employment in Kenya

In some cities there are schemes to support informal sector workers and to improve their working conditions. In Nairobi, for example, there are several small **Jua Kali** workshops (diagram **E**). In these workshops, scrap metal is collected and hammered into an assortment of products which are sold cheaply and used locally. The government supports Jua Kali businesses by providing electricity and basic skills training.

Many people consider that schemes like this are ideal for developing countries. This is largely because they use **appropriate technology** that meets the needs of the local people and the environment in which they live (diagram **F**). They are also **sustainable** because they can go on for year after year without using up resources or harming the environment. Indeed the Jua Kali scheme actually recycles waste material so it is particularly resource friendly.

Six year old boy selling crockery in Cuzco, Peru D

E

Jua Kali metal workshops, Nairobi	
Type of industry	Usually small-scale family enterprises. Long and irregular hours. Low and uncertain wages. Use simple tools and basic equipment. A market location. **Informal employment**.
Raw materials	Cheap scrap metal is recycled by melting it down and hammering it into various shapes. Products include boxes, cooking utensils and water barrels. Charcoal is used to heat the scrap metal.
Energy	Electricity for lighting. Scrap wood for burning as charcoal. Manual labour.
Technology	**Appropriate technology**, which is **sustainable** and suited to the skills of the people, the availability of raw materials and capital. Highly labour intensive.
Investment	Very limited. The government has supported this Jua Kali scheme by supplying electricity and providing roofs to protect the workers from the weather. Jua Kali means 'under the hot sun'.
Trade	All the products are designed and made for people in the surrounding area. They are sold and used locally.
Workforce	Over 1000 workers in an area of about 300 x 100 metres. Workers use traditional skills or learn new ones. Large numbers of workers are needed since production is labour intensive.
Advantages to Kenya	Estimates suggest that 600 000 people are employed in 350 000 small-scale Jua Kali enterprises in Kenya. The workers have an enterprising and hard-working spirit. Firms needs little capital, recycle materials that otherwise would be wasted, provide low-cost training, and can react quickly to market changes. They provide the backbone to Kenya's industrial development

F

Appropriate technology ...
* uses methods that are appropriate to the area where they are used
* is usually small-scale
* uses cheap and simple equipment
* has a small demand for energy
* uses local resources
* involves traditional skills
* helps ordinary people improve their quality of life.

Activities

1 What is the difference between *formal employment* and *informal employment*? Give three examples of each.

2 What do you think are the advantages and disadvantages of selling food as shown in photo **C**? Consider the seller, the customer and the local authorities.

3 Give the meaning of the terms *appropriate technology* and *sustainable development*.

4 Describe and give examples to show how Nairobi's Jua Kali workshops use appropriate technology. Use the headings in drawing **G**.

5 Briefly describe how informal activities are just as important to countries like Kenya, as formal activities.

G

Jua Kali industries and appropriate technology

* Methods and equipment   * Size of enterprises

* Energy needs   * Resource needs

* Skills required   * Sustainable approach

Summary

The lack of regular, waged employment in the poorer countries of the world has led many people to seek work in the informal sector. Here, they work long hours, usually for themselves, but earn little money.

16 Tourism

What is the tourist industry?

Tourists are people who travel for pleasure, whilst the **tourist industry** looks after the needs of tourists and provides the things that help them get to places where they can relax and enjoy themselves. In the last 50 years the demand for tourism has soared as the popularity of long annual holidays, weekend breaks and day visits to tourist centres has increased. As a result, tourism is now the world's fastest-growing industry and employs more people worldwide than any other activity. It is an important factor in the economy of most developed countries and is seen by many developing countries as the one possible way to obtain income and create jobs.

Tourism, like any other industry, relies on both natural and human resources. The Mediterranean countries, for example Greece and Spain, have warm, sunny climates and sandy beaches. The Alps and Rockies have stunning mountain scenery, snow and lakes, while East Africa is the home of some of the world's most spectacular wildlife. These are all examples of natural attractions. Some resources for tourism are, however, man-made. People visiting places like Paris and Rome are attracted by the cities' buildings, culture, history and night life. The variety of holidays now available to tourists is enormous. Some examples are shown below.

Coastal tourism needs sun, sand and sea and involves large amounts of relaxation.

Winter tourism takes place in mountain regions and includes skiing and snowboarding.

Adventure tourism is a recent development and includes climbing, trekking and cycling.

Cultural tourism includes visits to historic sites, interesting cities and different cultures.

Graph **E** shows the increase in tourist numbers. At present there are no signs of the increase slowing down. Indeed, as more places develop their tourist attractions, and as travel becomes easier, tourist numbers are expected to increase for many years to come.

Tourism has changed considerably in the last 50 years. Not only has there been a dramatic increase in tourist numbers, but there is now a wider range of holiday destinations and attractions. People also seem to be travelling further, and more often than ever before. There are several reasons for these changes, and some are shown in drawing **F**.

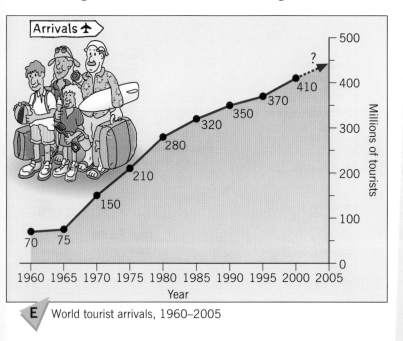

E World tourist arrivals, 1960–2005

Whilst the increase in tourism has brought many benefits it has also caused problems. With so many people travelling around the world, tourism has changed places and cultures. In some cases it has spoilt the very environment that attracted people there in the first place. Careful management and a **sustainable** approach to tourism is needed if this problem is to be overcome (page 242).

F

Reasons for the growth in tourism

Greater affluence – people are generally wealthier now than in the past. They have more money available to spend on luxury items such as holidays and leisure pursuits.

More leisure time – people now have a shorter working week and longer paid holidays, and tend to retire earlier. This gives them more time for leisure pursuits.

Greater mobility and accessibility – the increase in car ownership and improvements in road, rail and air travel have reduced journey times and made travel easier.

Package holidays – cheaper holidays are possible with block bookings and deals where travel, accommodation and meals are all included in the price.

Advertising – holiday programmes and brochures promote new and different places and activities. People are now more aware of the holiday opportunities available to them.

Activities

1 a) Copy table **G** and for each of the photographs **A**, **B**, **C** and **D** describe the main features of the holiday.
  b) Add to your table descriptions of the two holidays shown in drawing **H**.

G

Holiday description	Physical attractions	Human attractions

2 Describe your favourite holiday, giving its location, main features and both physical and human attractions.

3 Look at drawing **F**. Which reasons do you think have been most important? Explain your answer.

H

Visit Disneyland Paris – Europe's greatest leisure resort

Take a luxury cruise and enjoy the best of the Caribbean

Summary

Tourism employs over 200 million people worldwide and is the world's largest and fastest-growing industry. Greater wealth, improved accessibility and more leisure time are the main reasons for growth.

What are National Parks?

National Parks are large areas of attractive countryside where scenery and wildlife are protected so that everyone can enjoy them. Britain's first National Parks were created in the 1950s when the government of the time felt there was a real danger that some of the country's wildest scenery would be damaged or destroyed. The main idea of the Parks was one of **conservation** and a need to protect the environment. It was also hoped that they would help to look after the way of life and livelihood of people already living there. These included people working on farms, in forestry and in various other activities.

National Parks in Britain have two main aims:
1 To preserve and enhance the natural beauty of the landscape.
2 To provide a place for recreation and enjoyment.

To these can be added another aim:
3 To protect the social and economic well-being of people who live and/or work in the National Park.

To help fulfill these aims, each National Park in Britain is managed by a National Park Authority (NPA) which is controlled by a committee, or board, made up of local people and representatives of the government. Each NPA has a number of roles or duties, some of which are shown in drawing **B**.

The term 'National Park' can be misleading. They are not 'parks' in the sense of an urban park, and the public do not have complete freedom to wander where they would like. This is because, unlike most National Parks in Europe and America, they are not 'national' in the sense that they are owned by the nation. They are, though, considered to be 'national' and belonging to the nation because their beauty and leisure opportunities are vital to the country. These two misconceptions can lead to conflicts between different landowners and different land users.

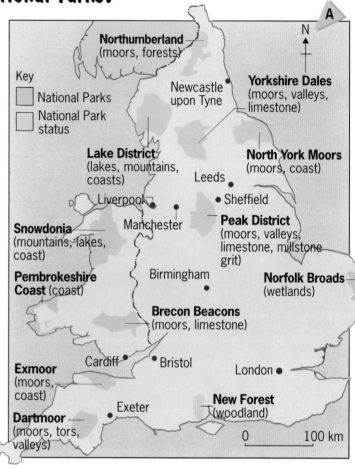

A

Key
☐ National Parks
☐ National Park status

Northumberland (moors, forests)
Newcastle upon Tyne
Yorkshire Dales (moors, valleys, limestone)
Lake District (lakes, mountains, coasts)
Leeds
North York Moors (moors, coast)
Liverpool
Sheffield
Snowdonia (mountains, lakes, coast)
Manchester
Peak District (moors, valleys, limestone, millstone grit)
Pembrokeshire Coast (coast)
Birmingham
Norfolk Broads (wetlands)
Brecon Beacons (moors, limestone)
Cardiff
Bristol
London
Exmoor (moors, coast)
New Forest (woodland)
Exeter
Dartmoor (moors, tors, valleys)
0   100 km

B National Park Authorities
* provide information services
* manage traffic and provide car parks
* signpost and maintain footpaths for walkers
* provide wardens to help visitors
* buy land to preserve the landscape
* control the design of new buildings
* manage and plant woodland
* advise farmers and landowners
* encourage job creation for local people

Activities

1 a) Why were National Parks set up?
 b) What is meant by the term *conservation*?

2 a) Name the Parks that have upland features.
 b) Name the Parks that have coastal features.
 c) Describe the distribution of the UK's National Parks.

3 Make a copy of table **C** and sort the duties from drawing **B** into the correct columns. Some might come under both headings.

C

Ways in which National Park Authorities...	
...help protect the countryside	...help visitors

D

The Lake District National Park

The Lake District is the largest and most popular National Park in Britain. It has mountains, moors and woodland as well as lakes and a windswept coastline. The National Park Authority describes the area as having the most spectacular and varied landscape in the country. Over 14 million people visit the Park each year and as survey **E** shows, most are attracted by the fine scenery and peaceful nature of the area.

Over 42 000 people live in the Lake District National Park and for many of them tourism provides an important source of income and employment. Estimates suggest that about a third of all jobs in the area are tourist related. In addition to direct employment, tourism supports local services such as the bus and rail network, village shops, public houses and various recreational amenities. The importance of tourism to the Lake District was shown in 2001 when foot-and-mouth disease reduced visitor numbers by over a third and cost the area an estimated £230 million in lost income.

E

Lake District National Park Authority

Reasons for visiting the Park:

Spectacular scenery	75%
Clean air	68%
Peace and quiet	64%
Not too crowded	60%
Good walking	58%
Wildlife and plant life	42%
Attractive villages	38%
Shopping	22%
Water sports	16%
Climbing or scrambling	14%
Visiting restaurants or cafés	12%

[The survey asked visitors what aspects of the Lake District National Park were most important to them.]

4 Draw a horizontal bar graph to show what attracts visitors to the Lake District. Colour the physical attractions *green* and the human attractions *red*.

5 Write a short article for a travel magazine describing the attractions of the Lake District National Park as a holiday destination.

6 Make a list of tourist jobs that may be available to school leavers in the Lake District. Try to give more than fifteen.

Summary

National Parks help to protect the countryside. National Park Authorities care for the landscape and help to look after the needs of visitors and people who live and work there.

How can tourism be managed in the National Parks?

The problem

The number of people visiting National Parks has increased rapidly in recent years. This has caused problems for the Parks, where overcrowding, damage to the environment and conflict between users is now much more common than in the past.

Bowness-on-Windermere, or simply Bowness, was once a quiet village where a few tourists relaxed and admired the unspoilt landscape of woodland, lake and mountains. Over the years, for the reasons shown in drawing **A**, Bowness has become increasingly popular. Today, especially at weekends in summer, the village has become a **'honeypot'** that is swarming with tourists and jammed with traffic (photo **B**). The problem is how to preserve the natural beauty of the area – the reason for it attracting so many people – while still providing facilities for the numerous peak-time visitors. This is a problem not just for Bowness but for several other honeypot locations in the Lakes.

The large number of people wishing to use the National Park for a wide range of activities sometimes leads to **conflict** or disagreement. As diagram **C** shows, this conflict is often between different groups of people each interested in a particular activity. Examples of conflict include farmers complaining about visitors' dogs roaming free during lambing time, and powerboat users disturbing the peace and tranquillity preferred by anglers on the same lake.

Reducing conflict and protecting the environment from damage and overuse can be very difficult. The National Park Authority has only limited powers and has to encourage landowners and the county council to make what they consider are the right decisions for the Park. The Lake District National Park Authority has developed a number of carefully considered plans to cope with the problems of overuse.

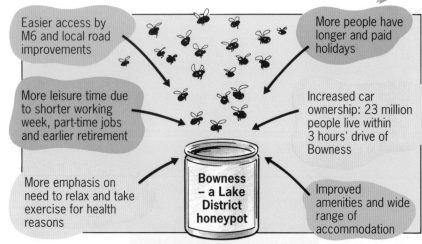

A

Easier access by M6 and local road improvements

More people have longer and paid holidays

More leisure time due to shorter working week, part-time jobs and earlier retirement

Increased car ownership: 23 million people live within 3 hours' drive of Bowness

More emphasis on need to relax and take exercise for health reasons

Bowness – a Lake District honeypot

Improved amenities and wide range of accommodation

A honeypot is a place of attractive scenery or special interest which attracts people in large numbers

B Bowness-on-Windermere

C Some conflicts between different groups in National Parks

Farmers and visitors

Walls broken down

Crops trampled by walkers

Gates left open

Dogs worry sheep

Farmers

Residents and visitors

House prices increase

Roads are congested

Houses bought up

Litter left behind

Residents

Different groups of visitors

Noisy power boats

Mountain bikers

Water skiers

Motorbike users

Visitors looking for peace and quiet

Management plans in the Lake District National Park

1 Three main areas of quiet have been established within the Park. These will retain the traditional characteristics of the area, and facilities that encourage large numbers of visitors will not be allowed. For example, parking will be restricted, roads downgraded to 'lakeland lanes', and caravan sites discouraged. These methods are examples of **negative planning** techniques.

2 Along the two main access corridors between Penrith, Keswick and Kendal, **positive methods** of management have been employed. Roads here have been upgraded, parking improved and tourist facilities provided. Caravan sites are concentrated within this area so that they are easily accessible and do not cause congestion on narrow roads.

3 Certain lakes have been identified as 'free access' lakes. Windermere, for example, will retain its 'honeypot' role as a centre for most water activities. Restrictions on powerboat use will reduce noise pollution, visitor numbers and overcrowding both on and along the lake. Other lakes such as Wastwater and Haweswater are designated 'natural' lakes, allowing only lakeside owners to use them. These lakes are in the quiet zones of leisure activities.

Despite these approaches, it is expected that the high levels of tourist activity in the so-called honeypot areas will continue to be a problem. However, the intention is that these areas will be contained and controlled and will not be allowed to spread to the quieter areas of the National Park.

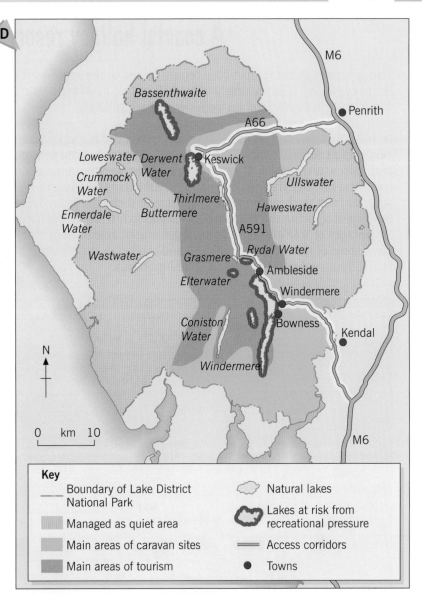

Key

—·—·— Boundary of Lake District National Park	Natural lakes
Managed as quiet area	Lakes at risk from recreational pressure
Main areas of caravan sites	Access corridors
Main areas of tourism	● Towns

Activities

1 Why are visitors to Bowness-on-Windermere increasing in number?

2 In what ways has the increase in leisure activities harmed the very environment that was the source of its attraction? (**Hint:** Look at survey **E** on page 233.)

3 Table **E** is the result of a survey of tourists who were making a return visit to Bowness. Do you think Bowness has been completely spoilt, or not? Give reasons for your opinion.

4 For each group of people shown in diagram **C**, give an example of a conflict that affects them and explain the problems that it may cause.

5 Describe the main features of the management scheme developed by the Lake District National Park Authority. Use the following headings:
   a) Quiet areas   **b)** Busy areas
   c) Lake environments   **d)** Honeypots.

E

74% still liked Bowness
26% didn't like it much
10.2% thought it had become too commercialised
15.3% thought it had become too crowded

Summary

As National Parks have become more popular, overcrowding, damage to the environment and conflict between users has steadily increased. Careful management by National Park Authorities can help reduce the problem.

A coastal holiday resort — Tolo, Greece

Tolo is located in the Peloponnese peninsula and to the south-west of Athens (map **A**). It has a long, sandy, sheltered and curving beach which stretches over 3 km between two rocky headlands.

In 1979 Tolo was still an unspoilt village with a linear shape. Many small shops selling local produce extended either side of the one long and narrow main street. The major occupations were farming (mainly fruit) and fishing. The only tourists were Athenians escaping in summer from the heat, noise and pollution of their city. Today Tolo is the largest resort in the Peloponnese. Landsketch **B** shows the natural advantages of Tolo and some of its added tourist amenities. It also gives some of the causes of pollution and conflict in the present resort. Tourism brings both benefits and problems.

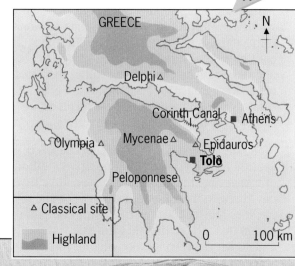

A

GREECE

N

Delphi △

Corinth Canal ■ Athens

Olympia △   Mycenae △ △ Epidauros

Tolo

Peloponnese

△ Classical site

Highland

0   100 km

B

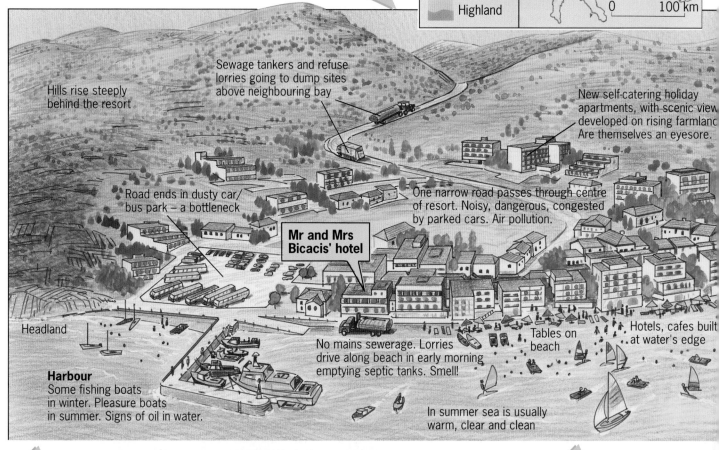

Hills rise steeply behind the resort

Sewage tankers and refuse lorries going to dump sites above neighbouring bay

New self-catering holiday apartments, with scenic view developed on rising farmland Are themselves an eyesore.

Road ends in dusty car/ bus park – a bottleneck

One narrow road passes through centre of resort. Noisy, dangerous, congested by parked cars. Air pollution.

Mr and Mrs Bicacis' hotel

Headland

Tables on beach

Hotels, cafes built at water's edge

Harbour
Some fishing boats in winter. Pleasure boats in summer. Signs of oil in water.

No mains sewerage. Lorries drive along beach in early morning emptying septic tanks. Smell!

In summer sea is usually warm, clear and clean

C   Tolo – the natural setting

D   Tolo – added amenities

E

Nikos and Katarina Bicacis

Mr and Mrs Bicacis live behind the hotel which they run together. As the holiday season is so short, Mr Bicacis also runs a farm 20 km away. Their working year is:

* **March** Hotel opens; very quiet.

* **May to August** The hotel is very busy and extra domestic help is needed. While tourists like the hot, dry weather, it does not help the farm where crops (mainly fruit) need to be watered each day.

* **September** Hotel is quiet.

* **October/November** Domestic staff are laid off. Repairs and redecorating jobs are done in the hotel.

* **December to February** Hotel is closed. This is the busiest time on the farm for oranges, the major crop.

Hot, dry sunny summers. Increased risk of water shortage for new hotels and apartments, and for farming. Winter rainfall is not sufficient to replace water taken from ground in summer. Farmers get water from boreholes which are becoming salty.

Only one narrow road into resort. Busy in summer, noisy, little room to pass

New apartments built on farmland

Hotels, cafes, tavernas and craft shops all close for several months in winter

Former local shops turned into craft and souvenir shops. Tavernas and bars are noisy at night.

Remains of a Mycenaean settlement 3500 years old

Sand dunes

Thin sandy beach, now only 3 m wide as new hotels are built at the tideless water's edge

Sea and sand dunes polluted with litter from tourists

Sea full of swimmers, pedalos, windsurfers and yachts, all competing for the same space

Headland

Activities

1 a) Describe the appearance of Tolo in 1979.
   b) What natural advantages did Tolo possess to attract tourists?
   c) What amenities had been added by 2002 to attract more tourists to the resort?

2 Tourism can brings advantages and can create problems and conflicts. Use the information on these two pages to answer these questions.
   a) What advantages has tourism brought to Tolo?
   b) What problems has tourism created?
   c) What conflicts have been created between:
     • groups of local people
     • local people and tourists
     • groups of tourists?

Summary

People are attracted to Mediterranean resorts because of the hot, dry summer weather, the spectacular scenery and the added tourist amenities. Local residents, often poor by EU standards, see tourism as an opportunity to improve their standard of living even if it also changes their way of life and spoils their environment.

Mountain holidays — Nepal

Nepal has become increasingly popular with tourists. It has some of the world's most spectacular mountain scenery as well as friendly people and a fascinating culture. The Himalayas, which form the backbone to the country, are the main attraction. As explained on page 62, they are an example of **fold mountains** and were formed when two **tectonic plates** pushed together causing the rock layers to bend and fold as they were forced upwards.

Visitors to Nepal have a wide choice of holiday. Some base themselves in Kathmandu and enjoy the rich history and unique culture that a country squeezed between India and Tibet has to offer. Others choose to visit the mountains on trekking, cycling or climbing holidays which are offered by many specialist tour operators. The holiday described below is a medium-grade trek which takes visitors to the base of Mount Everest (8850 m), the world's highest mountain.

A

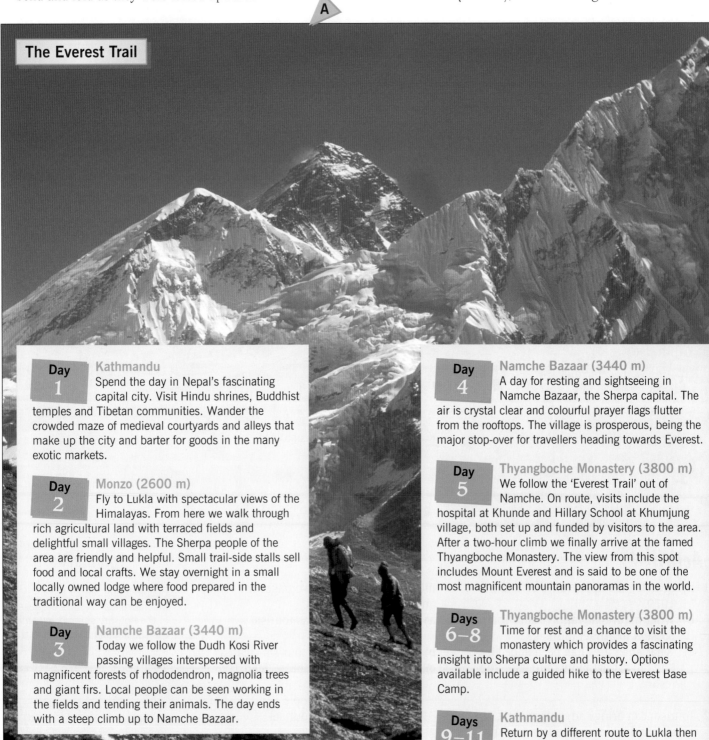

The Everest Trail

Day 1 — Kathmandu
Spend the day in Nepal's fascinating capital city. Visit Hindu shrines, Buddhist temples and Tibetan communities. Wander the crowded maze of medieval courtyards and alleys that make up the city and barter for goods in the many exotic markets.

Day 2 — Monzo (2600 m)
Fly to Lukla with spectacular views of the Himalayas. From here we walk through rich agricultural land with terraced fields and delightful small villages. The Sherpa people of the area are friendly and helpful. Small trail-side stalls sell food and local crafts. We stay overnight in a small locally owned lodge where food prepared in the traditional way can be enjoyed.

Day 3 — Namche Bazaar (3440 m)
Today we follow the Dudh Kosi River passing villages interspersed with magnificent forests of rhododendron, magnolia trees and giant firs. Local people can be seen working in the fields and tending their animals. The day ends with a steep climb up to Namche Bazaar.

Day 4 — Namche Bazaar (3440 m)
A day for resting and sightseeing in Namche Bazaar, the Sherpa capital. The air is crystal clear and colourful prayer flags flutter from the rooftops. The village is prosperous, being the major stop-over for travellers heading towards Everest.

Day 5 — Thyangboche Monastery (3800 m)
We follow the 'Everest Trail' out of Namche. On route, visits include the hospital at Khunde and Hillary School at Khumjung village, both set up and funded by visitors to the area. After a two-hour climb we finally arrive at the famed Thyangboche Monastery. The view from this spot includes Mount Everest and is said to be one of the most magnificent mountain panoramas in the world.

Days 6–8 — Thyangboche Monastery (3800 m)
Time for rest and a chance to visit the monastery which provides a fascinating insight into Sherpa culture and history. Options available include a guided hike to the Everest Base Camp.

Days 9–11 — Kathmandu
Return by a different route to Lukla then fly to Kathmandu.

Nepal is one of the poorest countries in the world and most of its people earn little money and have very difficult lives. The country appreciates the value of its scenery and varied culture. These are seen as a major source of income for a country that is desperately short of money and employment opportunities. Income from tourism can be used to pay for improving services, building more houses and roads, and creating new jobs.

Unfortunately, tourism can also cause problems. Large numbers of people going on holiday can damage the very environment that attracted them in the first place. Pressure is put on the environment and local people may struggle to retain their traditional way of life. In Nepal the worst problems may be found in the more popular valleys where trekkers can sometimes outnumber the local population.

The effects of tourism in Nepal

Benefits

Problems

Much of our forest has gone. The trees have been used as firewood.

The loss of forest has increased soil erosion and made it more difficult to farm.

Tourism can create many new jobs. We are very short of work.

Farmers can sell their produce to tour operators.

Soil eroded from the hillside has blocked rivers and caused flooding.

Tourism brings a lot of overseas money into Nepal.

Money from tourism can pay for new schools and hospitals.

I hope to become a mountain guide or porter when I leave school.

Most tour companies are foreign-owned, so Nepal receives little tourist money.

Our traditional way of life has been changed by the increase in tourists.

Our valleys and villages are full of litter and rubbish left by tourists.

Activities

1 Read the Everest Trail itinerary and list the attractions of the holiday under the headings of **Physical attractions** and **Human attractions**.

2 Write a paragraph to explain how the development of tourism in Nepal has helped:
   a) the local people and
   b) the country as a whole.

3 Copy and complete flow chart **C** by putting the statements from list **D** into the appropriate box.

4 Suggest what could be done to increase the amount of money that Nepal receives from tourism.

D
- Rivers blocked
- Farming made more difficult
- Flooding increases
- Food production reduced
- Hillsides cleared of trees
- Firewood needed for tourists
- Soil erosion increases

Summary

Mountain areas are becoming increasingly popular with tourists. Tourism can bring jobs and earn money for less developed countries like Nepal. However, it can also cause problems for the local people and their environment.

How can tourism change the environment?

Many tourists are attracted to areas of great scenic beauty or where there is abundant wildlife. Many of these environments are fragile, and so they can easily be changed and damaged by the large number of tourists they will attract. Therefore it is essential that such environments are protected. To do this needs careful planning and management. One example which illustrates these points is the east coast of Kenya (transect **A**).

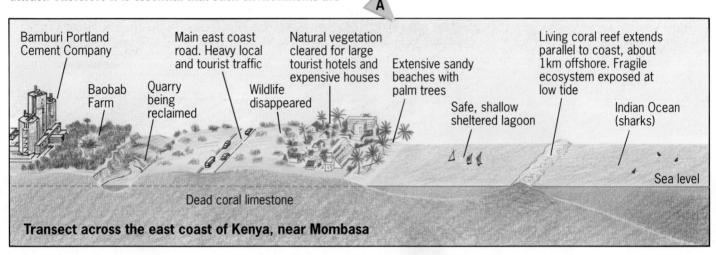

Bamburi Portland Cement Company

Baobab Farm

Quarry being reclaimed

Main east coast road. Heavy local and tourist traffic

Wildlife disappeared

Natural vegetation cleared for large tourist hotels and expensive houses

Extensive sandy beaches with palm trees

Safe, shallow sheltered lagoon

Living coral reef extends parallel to coast, about 1km offshore. Fragile ecosystem exposed at low tide

Indian Ocean (sharks)

Sea level

Dead coral limestone

Transect across the east coast of Kenya, near Mombasa

The coral reef – a fragile environment

Coral is an organism which can only live in very warm, clear and clean seas. It produces attractive shapes and colours and provides a habitat for an abundance of marine life.

C Hotel near Mombasa

B Coral reef, Kenya

Unfortunately it can easily be damaged or destroyed by even the lightest contact, and suffocated if covered by sand or silt. The reef off the coast of Kenya is uncovered at low tide (photo **B**). In places it is possible either to wade out across the shallow lagoon, sheltered from large waves and shark attack, or to take a boat to reach the reef. The reef is destroyed when these boats drop anchor, or when tourists trample over the living coral. Swimmers and snorklers disturb sand on the lagoon floor and this, together with increased pollution from the tourist resorts, suffocates the coral. Several marine parks and reserves have been created where it is illegal to spear fish, to collect coral and cowrie shells, or to disturb the highly sensitive ecosystem. It is still possible, for a small fee, to scuba-dive to see underwater coral.

The beach hotels – tourism changing the natural environment

The coast is, arguably, Kenya's most important natural resource. Lying on the Equator, temperatures on the Kenyan coast reach 30°C every day. There is plenty of sunshine for most of the year and there are only two short rainy seasons. The beaches are long, clean and sandy. The coral reef extends along the whole of the Kenyan coast apart from a gap at Mombasa, Kenya's only port. Behind the coral reef the shallow lagoon is warm and sheltered, making it safe for swimming and other water sports. The result has been the construction of an almost unbroken line of holiday hotels and beach complexes extending for many kilometres on either side of the port of Mombasa (photo **C**). The building of these hotels has led to the removal of any remaining natural vegetation, the disappearance of wildlife, and the pushing inland of local farmers. The result has been a complete change in the local landscape and habitat.

Baobab Farm – restoring a damaged environment and creating a wildlife habitat

Behind the coastal hotels, the Bamburi Portland Cement Company quarries coral limestone. By 1971, 25 million tonnes of coral limestone had been removed, leaving only a huge scar. The environment had been totally damaged – there was no soil, no vegetation and no wildlife. The Swiss transnational cement company then appointed Dr Haller to try to restore the environment from what he called 'a lunar landscape filled with saline ponds'. He found that one species of tree, the Casuarina, could grow in the coral rubble, the saline soil and in temperatures reaching 40°C. The Casuarina tree (photo **D**) grew by 3 metres a year. Thousands of red millipedes were then added to break down the large number of fallen leaves to produce humus. Earthworms and indigenous grasses and trees were added, but no chemicals were used. Within 20 years, 15 cm of soil had formed and a tropical rainforest had been created (photo **E**). The ecosystem was completed by the addition of insects, birds, and herbivores.

Yet the project is not just an environmental success. It has become a sustainable commercial venture with income from 46 enterprises including a fish farm, a crocodile hatchery and the sale of fruit and rice. The Baobab Farm, the name of the restored area, is open to school parties every morning, while in the afternoon it attracts other visitors. The near 120 000 visitors in 2001 made it the largest tourist attraction in the Mombasa area.

D Casuarina trees planted by the Bamburi Portland Cement Company

E Baobab Farm restored environment

Activities

1 **a)** Why do coral reefs form a 'fragile environment'?
   b) How is tourism affecting this fragile environment?
   c) What has been done to try to protect this fragile environment?
   d) How has tourism changed the coastal landscape of Kenya?

2 The Baobab Farm has become one of the largest and most successful environmental reclamation schemes in the world. Copy out and complete flow chart **F** by putting the correct phrases from the list A–H into the appropriate box.

A Thousands of red millipedes added
B Insects, birds and herbivores added
C Indigenous grasses and trees planted
D Bare lunar landscape
E Tropical rainforest ecosystem
F No soil, vegetation or wildlife
G Humus produced and 15 cm soil formed
H Casuarina trees planted

Summary

Many environments in the world are fragile and can easily be damaged or changed by human activities such as tourism. It is important to try to protect, plan and manage those environments that have not yet been damaged, and to restore those that have.

F

D Bare lunar landscape → → → → → → → →

What is ecotourism?

Tourism has brought many benefits to the poorer countries of the world. For many it is the only way to earn money and improve living standards. Unfortunately, tourism can also be damaging. Without planning and careful management it does little to benefit local people and is often harmful to their culture and environment.

Many developing countries are now looking at **ecotourism**, or 'green tourism' as it is sometimes called, as a way of overcoming these problems. Ecotourism is a **sustainable** form of tourism which aims to protect the environment and respect the local culture and customs. Ecotourists usually travel in small groups and share special interests. They include wildlife enthusiasts, birdwatchers and photographers. They often visit National Parks and game reserves where the wildlife and scenery is protected and managed. An example of ecotourism in Kenya is shown below.

Ecotourism in Kenya

Tourism is Kenya's fastest-growing industry and largest earner of money from overseas. Most visitors are attracted by the wildlife and scenery, which they can view in one of the many National Parks or game reserves.

Amboseli National Park lies in the south of the country and has a great variety of wildlife, and spectacular views of Mount Kilimanjaro. The land here was used by the Maasai people who were, by tradition, pastoral herders. In 1970 when the Park opened, the Maasai were forced to move off their land and live elsewhere. They gained no benefit from tourism and no longer had an interest in protecting wildlife in the Park. As tourism grew in popularity, so the problems of the Park increased. Vehicles disturbed animals, vegetation was damaged and wildlife numbers declined. Gradually, fewer visitors came to the Park.

Eventually in 1996 the problem was recognised and a Wildlife Conservation Area was set up with the aim of both protecting the environment and bringing greater benefits to local people. The idea is that small areas outside the National Park will be leased to ecotourist companies, each of which will be allowed to build just one hotel in a specified location. They will have to employ local people, buy local food when possible and use local resources in a sustainable way. This will give the local community a greater interest in protecting wildlife because it will also be the source of its income.

Several sites have been identified for development and the first one, the Kimana Wildlife Refuge, opened in 1997. It is already a success, attracting visitors who are keen to view wildlife in its natural and unspoilt setting whilst at the same time benefiting the local community.

A Mount Kilimanjaro and a herd of elephants

B Maasai herdsmen near Amboseli National Park

Whilst tour companies are largely responsible for setting up eco-friendly tourism, tourists themselves can play an important role in reducing the negative effects of their visits. As drawing **C** shows, this can be done mainly by showing a respect for the environment and appreciating the needs and wishes of the local people.

Ecotourism has enjoyed much success but is, perhaps, facing a difficult future. Some operators are using the 'eco' label simply as a selling point and are not always ensuring environmentally friendly practices. In addition, as more people take this type of holiday it will become increasingly difficult to maintain the small numbers and low impact of trips.

C

Traveller's Code

Friends of Conservation

To enjoy your visit abroad whilst reducing the environmental and social pressures that tourism can bring, please observe these guidelines.

When buying holiday souvenirs, remember that local crafts make unusual gifts and help support the local community.

Resist buying or collecting souvenirs from reefs such as coral, shells and starfish. It contributes to the degradation of the reefs and marine life.

Buying exotic souvenirs like ivory, spotted furs and certain plants is illegal and can threaten the most endangered species.

Buying local clothing, shopping in local outlets and eating local food is a great way of enjoying a country.

Remember to watch out for wildlife and avoid breeding areas during breeding seasons. Respect wildlife and marine reserves.

Remember that you are on holiday on someone else's doorstep. Respect the people, culture and natural surroundings of the countries you visit.

Leave no litter. As well as being unattractive it can damage fragile environments and have serious consequences on wildlife.

Adapted from 'Travellers Code' by the Friends of Conservation at **www.foc-uk.com**

Activities

1 Define in your own words the meaning of:
   a) *ecotourism* and   **b)** *sustainable tourism*.

2 Describe the problems caused by tourist development in Kenya's Amboseli National Park.

3 Describe the main features of the ecotourist development at Kimana Wildlife Sanctuary, using the headings shown in drawing **D**.

4 Describe how the Traveller's Code can help:
   a) to reduce damage to the environment
   b) to support the local economy.

D

Kimana Wildlife Sanctuary

1 Location
2 Aims
3 Methods used
4 Benefits for:
   a) local people
   b) wildlife
   c) tourists

Summary

Ecotourism is a sustainable form of tourism which tries not to damage the environment and respects and brings benefits to the local community.

What are resources?

Resources can be defined as any material or product that people find useful. Stone, for example, becomes a resource when people use it to build houses. Grass is a resource when cows eat it and produce meat and milk. Attractive countryside is a resource for people relaxing in their leisure time. Resources may be either natural or human. **Natural resources**, which this unit is mainly concerned with, are physical features like climate, vegetation, soils, and raw materials such as minerals and fuel. **Human resources** include the workforce, skilled labour, machinery and money. Natural resources are usually described as **non-renewable** or **renewable**. Definitions and examples of these are shown in diagram **A** below.

A Types of resource

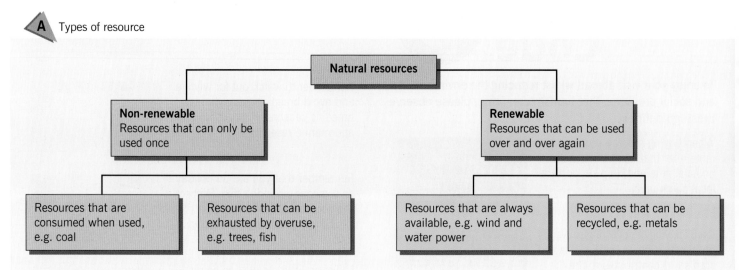

Natural resources

Non-renewable
Resources that can only be used once

Renewable
Resources that can be used over and over again

Resources that are consumed when used, e.g. coal

Resources that can be exhausted by overuse, e.g. trees, fish

Resources that are always available, e.g. wind and water power

Resources that can be recycled, e.g. metals

Managing resources

The demand for and use of the world's resources continue to grow at an increasingly rapid rate. This is mainly due to:

1 **Population growth** – there is a rapid increase in the number of people on the Earth. There are now 6 billion of us compared with 5 billion just 12 years ago (page 130).
2 **Economic development** – more countries are trying to develop their industry and wealth. As they do this, they use up more and more of the Earth's resources.

As the demand for resources grows, there is an increasing need to manage their use and look after them more carefully. If this is not done, two problems may result. The first is that some resources will simply run out. The second is that the environment will become even more damaged and polluted.

Diagram **B** shows some methods of managing resources in a **sustainable** way. Sustainable methods use resources sensibly and in a way that does not waste them or cause damage to the environment (page 270).

B

Sustainable resource management

Using renewable resources – such as wind and wave power which can be used over and over again.

Recycling – turning waste into something useful that can be used again.

Increasing efficiency – using resources in a less wasteful way, e.g. low-energy light bulbs.

Controlling pollution – by reducing emissions from vehicles and power stations.

Conservation – protecting and preserving wildlife and scenery.

Energy resources

Energy is very important to us. It provides lighting and heating in our homes and offices. It powers computers, televisions, and factory machinery. It fuels transport and has helped us develop and improve the quality of life on our planet. Graph **C** shows the world's main sources of energy. Some of these are non-renewable resources and some are renewable.

So far, **non-renewable energy resources** have been easy and quite cheap to use. Coal, oil and natural gas are called **fossil fuels** because they come from the fossil remains of plants and animals. Unfortunately fossil fuels create a lot of pollution and are also thought to be causing changes in the world's climate through **global warming** (pages 104–105). **Renewable**, or **alternative, energy resources** are mainly forces of nature, like water, wind and the sun, which can be used over and over again. They tend to be difficult and expensive to use but cause little pollution and are a sustainable form of energy.

Energy and development

Although only a quarter of the world's population live in industrialised countries, they use over three-quarters of the world's energy. As graph **D** shows, there is a close link between the wealth of a country and the amount of energy that it uses. The relatively few countries with a high standard of living consume large amounts of energy to operate machines in industry, transport and the home.

The first countries to become industrialised were those that had large amounts of their own easily accessible fossil fuels. Britain, for example, had large reserves of coal which helped power its early industrial growth. Today, some of these countries no longer have sufficient resources of their own but they usually have enough money to import what they need.

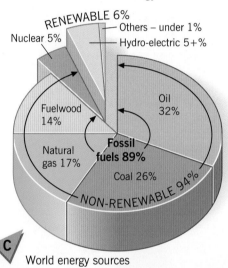

C World energy sources

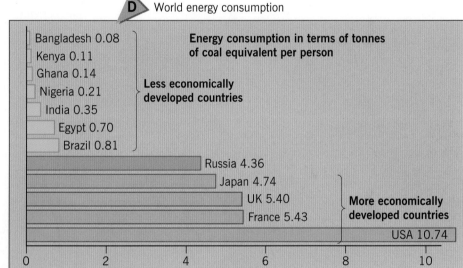

D World energy consumption

Energy consumption in terms of tonnes of coal equivalent per person

Less economically developed countries

Bangladesh	0.08
Kenya	0.11
Ghana	0.14
Nigeria	0.21
India	0.35
Egypt	0.70
Brazil	0.81

More economically developed countries

Russia	4.36
Japan	4.74
UK	5.40
France	5.43
USA	10.74

Activities

1 Give the meaning of the following terms. Include at least three examples for each.
   a) *Natural resources*   b) *Human resources*
   c) *Renewable resources*   d) *Non-renewable resources*
   e) *Fossil fuels*.

2 a) Copy and complete drawing **E** using these labels:
   • Already using large amounts of energy
   • Large and rapid increase in energy use
   • Small increase in energy use
   • Using small amounts of energy
   • Poorer less developed countries (LEDCs)
   • Richer industrialised countries (MEDCs)
   b) Describe the effects on energy demand shown by the drawing. Suggest reasons for this.

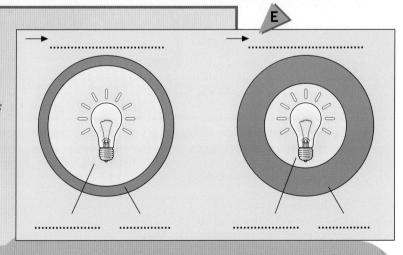

E

Summary

The increased demand for resources is due mainly to world population growth, increased wealth and economic development. Resource use is greatest in the industrialised countries of the developed world.

245

How can the production of electricity affect the environment?

In Britain our lives seem to depend upon electricity. Whether at work, play or at home, electricity is usually taken for granted. All it needs is the touch of a switch. Yet in many parts of the world, electricity is either unavailable or its provision is unreliable. Places without electricity often tend to remain less economically developed and their people are likely to have a poorer quality of life.

There are many different sources of energy (page 245). All have advantages for the economy and people of a place. They all affect the environment in one way or another.

Non-renewable types of energy are used mainly in industrialised countries. These resources are usually very efficient but they create considerable environmental problems. Renewable resources cause fewer environmental problems but we often still lack the technology needed for their economic development. Most of Britain's electricity is produced in **thermal power stations** using coal, oil or natural gas. The remainder comes either from **nuclear** or **hydro-electric power stations**. Fact files **A**, **B** and **C** explain how each of these three types of power station affect the environment.

Fact file A: thermal (coal-fired)

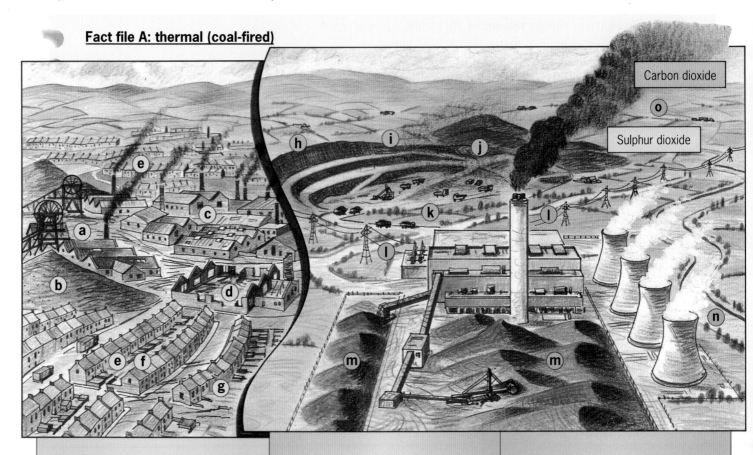

Mining village dating from the nineteenth century	Present-day open-cast mine	Thermal power station
• Coal-mine creates noise and dust (**a**) • Spoil tips from waste mined along with the coal (**b**) • Coal-using industries are located nearby – noise, smoke (**c**) • Unsightly, derelict buildings if coal-mine or industries close (**d**) • Tightly packed terraced houses blackened by smoke (**e**) • Little open space left (**f**) • Roads and buildings affected by subsidence due to underground workings (**g**)	• Top soil is cleared and farmland is lost (**h**) • Hedges, wetlands and wildlife habitats are lost (**i**) • Dust and noise (**j**) • Heavy traffic on narrow roads (**k**)	• Unsightly power station, cooling towers, chimneys and pylons (**l**) • Stock piles of unused coal; ash from used coal (**m**) • Cold water from the river is used to condence the steam in the power station back to water. Warm water is then returned to the river and has harmful effects on fish and plant life (**n**) • Waste gases released into air. Sulphur dioxide helps cause acid rain. Carbon dioxide contributes towards greenhouse effect and global warming (**o**)

Fact file B: nuclear

Advantages

- Very little natural resource (uranium) is needed.
- Nuclear waste is limited and can be stored underground.
- Cleaner than fossil fuels. There is less air pollution and it is not a major contributor to acid rain or global warming.

Disadvantages

- Safety risk. It is feared that an accident at a power station could kill many people.
- Accidental releases of radio-active material can contaminate land and sea areas.
- Health risk, e.g. possible links with leukaemia and other serious illnesses.
- Nuclear waste is dangerous and can remain radioactive for a long time.
- Safety fears in transporting and storing nuclear waste.

Nuclear power station

Fact file C: hydro-electricity

Advantages

- Very clean form of energy. No contamination of the air, land or water.
- A renewable form of energy so it is not using up the Earth's resources.
- Usually produced in highland areas that have heavy, reliable rainfall. These places rarely have high population densities.
- Dams built to store water also reduce the risks of flooding and water shortages.

Disadvantages

- Where storage lakes (reservoirs) have formed behind dams, large areas of farmland and wildlife habitats may have been flooded.
- People and animals are forced to move as lake levels rise.
- Unsightly pylons take electricity from power station.
- Silt previously spread over farmland is now deposited in the lake.

Hydro-electric dam, Scotland

Activities

1 How has the environment been affected by the:
   a) mining of coal in the nineteenth century
   b) present-day open-cast mining?

2 a) How can the environment be affected by the production of electricity in:
   i) thermal power stations
   ii) nuclear power stations
   iii) hydro-electric power stations?

   b) In groups of two or three, compile a newspaper report describing which of the three methods of producing electricity has the:
   i) most harmful effect
   ii) least harmful effect
   upon the environment.
   Give reasons for your opinions.

Summary

Electricity can be produced from a wide range of energy sources. While all these sources affect the environment, some have a much more harmful effect than others.

Can wind help provide our energy needs?

Renewable resources can be used over and over again without being used up. Sunshine, waves and wind can be used to produce electricity and are renewable energy resources because they will never run out. At present, less than 2 per cent of energy used in Britain comes from renewable resources. The government has said that it wants to increase this to 10 per cent by the year 2010. Using more wind power would help reduce air pollution and conserve non-renewable resources like coal, oil and natural gas.

In the last decade, Britain has put more money into developing wind power than any other form of renewable energy. Wind turbines, to be at their most efficient, need to be in areas with high and regular wind speeds. As map **A** shows, such sites are usually found on exposed coasts or in upland areas of western Britain. As 30 metre high turbines are expensive to build and maintain, it is usual to group several together to form a 'wind farm'.

Britain's first wind farm was opened in 1991 near Camelford in Cornwall. The farm is located on moorland 250 metres above sea level where average wind speeds are 27 km/hr. It generates enough electricity for 3000 homes. In 2000 the first offshore wind turbines started to produce electricity in the North Sea near Blyth, Northumberland. By 2002, there were almost 80 wind farms either working or planned in the UK. They contributed about 0.37 per cent of the UK's total energy needs.

Many other countries are also developing wind power. Denmark, for example, relies on wind power more than any other country, and intends to install 2000 offshore turbines by 2009. The USA has also invested in wind power on a large scale. California alone has more than 23 000 turbines which together produce enough electricity to supply a city the size of San Francisco. One of the largest wind farms is at Palm Springs near Los Angeles. Here there are 4600 high-tech turbines producing power for privately owned companies. The site is ideal because the winds are strongest at the time of year when power demands are greatest.

A Some UK wind farms and possible future sites

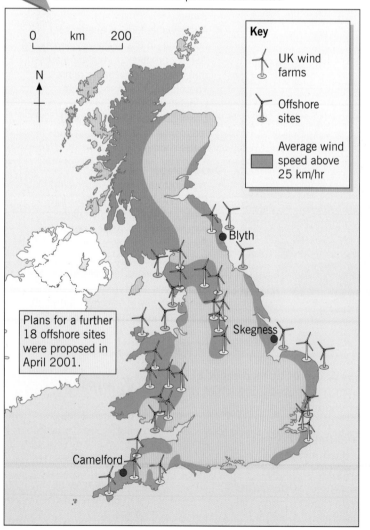

Key

- UK wind farms
- Offshore sites
- Average wind speed above 25 km/hr

Plans for a further 18 offshore sites were proposed in April 2001.

Blyth

Skegness

Camelford

B Features of a typical wind turbine

Turbine produces electricity

15 metre blades made from high-tech carbon fibre

Computer adjusts turbine to face wind

Blades may be 'feathered' in storm conditions

Access ladder for servicing

Strong steel column

Columns up to 50 metres high

Underground power cable to control unit

Whilst wind power has many advantages as a source of energy, many people wonder if it can ever be a realistic alternative to fossil fuels. It has been estimated that it would take more than 7000 wind turbines to replace one nuclear power station, and 200 000 would be needed to supply all of Britain's electricity. A wind farm that size, if located in the UK's South West region, would cover all of Cornwall and much of Devon!

C

Advantages
- Wind turbines do not cause air pollution.
- Winds are stronger in winter when demand for electricity is greatest.
- Electricity production is relatively cheap.
- Farmers paid rent for allowing turbines on their land.
- Wind could generate 10% of the UK's electricity.

Disadvantages
- Wind does not blow all of the time.
- Wind farms spoil the look of the countryside.
- Large numbers of turbines are needed.
- Wind turbines are noisy and can interrupt radio and TV reception.

Activities

1 In your opinion, what are the two most important advantages, and the two most important disadvantages, of wind power? Give reasons for your answer.

2 a) Study sketch map **D** and match locations ①, ②, ③, ④ and ⑤ with the following advantages (some sites may be used more than once):
- Away from main settlements
- Near existing industry
- In a windy location
- Easily accessible.

b) Give a disadvantage for each location.

c) Which two of the sites do you think would be the best location for a wind farm? Give reasons for your answer.

3 Offshore locations are becoming increasingly popular as wind farm sites. Describe the advantages and disadvantages of these sites.

D

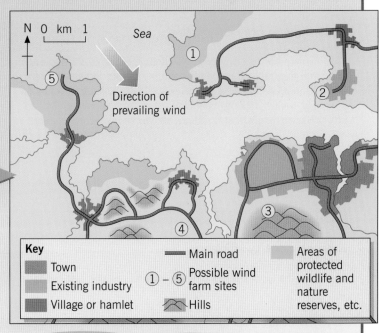

N 0 km 1   *Sea*

Direction of prevailing wind

Key
Town	—— Main road
Existing industry	①–⑤ Possible wind farm sites
Village or hamlet	〰 Hills

Areas of protected wildlife and nature reserves, etc.

Summary

Wind power is a renewable energy resource that is clean, safe and relatively cheap. It could contribute to the UK's energy needs and help reduce the use of fossil fuels.

How can resources be conserved?

The Earth's resources are being used up at an increasing rate because of population increase and economic development. Some non-renewable resources such as oil, for example, could run out before we have developed replacements. Our increased use of resources is also causing pollution and serious damage to the environment at local, national and global scales.

Most people agree that a way of solving the problem is through **sustainable development**. Sustainable development does not waste resources or damage the environment. It is progress that can go on for year after year and helps improve our quality of life today, but does not spoil our chances in the future.

Sustainable development requires careful planning and, because it affects us all and extends across national frontiers, needs everyone to be involved. For example, there is little benefit in just one house in a street using energy-saving light bulbs, or one country reducing the amount of fossil fuel it burns. What is needed is an approach where ordinary people, national governments and global organisations work together and try to conserve resources and improve the quality of life for everyone. This is called **global citizenship**.

Some ways in which non-renewable resources can be conserved are shown in **A** below. Look carefully at the conservation methods used in drawing **B**. How does your classroom compare?

A

Recycling

Waste materials can be re-used by making them into new products. This may be done for scrap metals, aluminium cans, electric cable, glass bottles, paper and old clothes, for example. Recycling reduces the amount of non-renewable resources such as iron ore, bauxite and copper that has to be mined. Page 229 shows an example of recycling waste materials in Kenya.

Reducing resource consumption

This involves saving electricity by switching off lights, computers and other electrical equipment when they are not in use. Turning down central heating by a degree or two also saves large amounts of energy, whilst walking, cycling or using the bus or train instead of a car saves fuel. Other resources should be used carefully and not wasted.

Increasing energy efficiency

Improving the quality of buildings so that less heat is lost is one way of doing this. Heat already given off by machinery can be used – for example houses could be heated by the warm water from power stations, or more fuel-efficient planes, buses and cars could be produced.

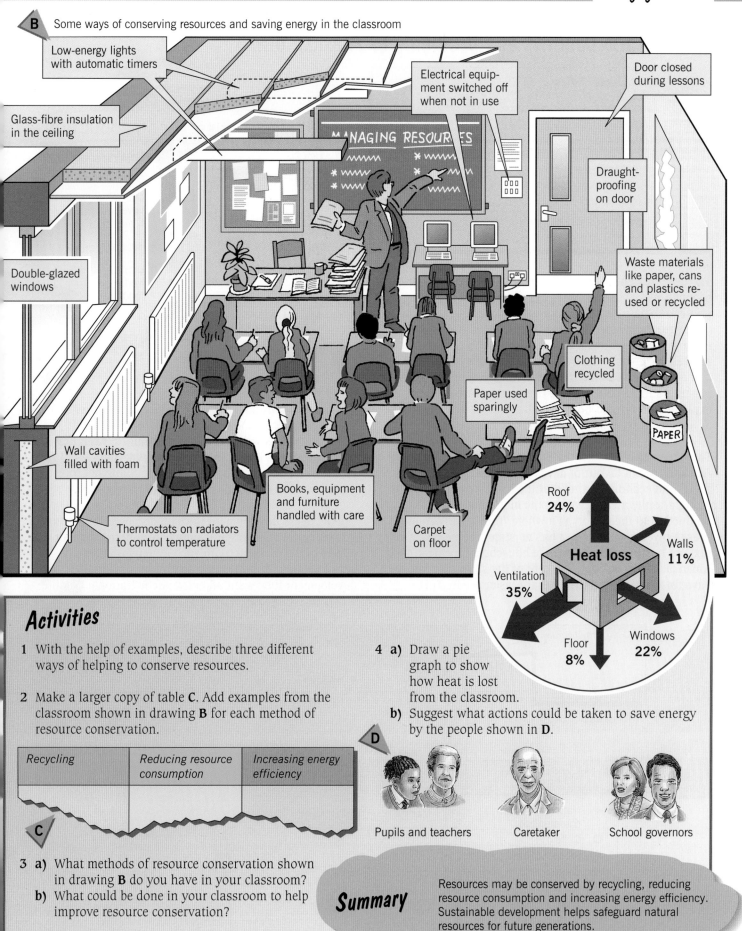

B Some ways of conserving resources and saving energy in the classroom

Low-energy lights with automatic timers

Glass-fibre insulation in the ceiling

Electrical equipment switched off when not in use

MANAGING RESOURCES

Door closed during lessons

Draught-proofing on door

Double-glazed windows

Waste materials like paper, cans and plastics re-used or recycled

Clothing recycled

Paper used sparingly

PAPER

Wall cavities filled with foam

Books, equipment and furniture handled with care

Carpet on floor

Thermostats on radiators to control temperature

Heat loss

Roof **24%**

Walls **11%**

Ventilation **35%**

Windows **22%**

Floor **8%**

Activities

1 With the help of examples, describe three different ways of helping to conserve resources.

2 Make a larger copy of table **C**. Add examples from the classroom shown in drawing **B** for each method of resource conservation.

Recycling	Reducing resource consumption	Increasing energy efficiency

C

3 **a)** What methods of resource conservation shown in drawing **B** do you have in your classroom?

b) What could be done in your classroom to help improve resource conservation?

4 **a)** Draw a pie graph to show how heat is lost from the classroom.

b) Suggest what actions could be taken to save energy by the people shown in **D**.

D

Pupils and teachers

Caretaker

School governors

Summary

Resources may be conserved by recycling, reducing resource consumption and increasing energy efficiency. Sustainable development helps safeguard natural resources for future generations.

What are the benefits and problems of Alaskan oil?

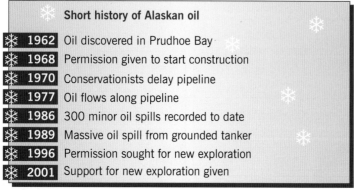

Short history of Alaskan oil	
1962	Oil discovered in Prudhoe Bay
1968	Permission given to start construction
1970	Conservationists delay pipeline
1977	Oil flows along pipeline
1986	300 minor oil spills recorded to date
1989	Massive oil spill from grounded tanker
1996	Permission sought for new exploration
2001	Support for new exploration given

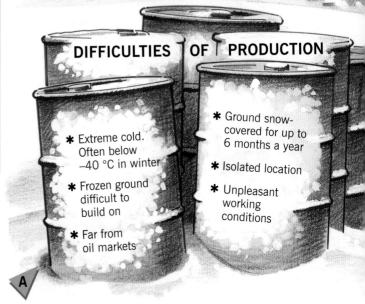

DIFFICULTIES OF PRODUCTION

* Extreme cold. Often below −40 °C in winter
* Frozen ground difficult to build on
* Far from oil markets
* Ground snow-covered for up to 6 months a year
* Isolated location
* Unpleasant working conditions

A

The development of the oil industry in northern Alaska has brought many benefits but it has also caused problems. It is now more than 20 years since the first oil began to flow out of the ground near Prudhoe Bay. Since then Alaska has become rich, a serious oil spill has occurred, and environmentalists have continued to voice concerns about the damaging effects of the industry. Now, with reserves beginning to run out and the need for new fields to be explored and opened up, the arguments for and against development have become stronger.

The governor of Alaska, the State Congress and a majority of Alaskans are in favour of development. They point to the social and economic gains that the industry has already brought to the area as well as the importance of oil production to the nation as a whole. They rightly claim that Alaska owes much to oil. It has no state income tax or sales tax and has the highest average household income in all of America. Oil provides 85 per cent of the state's income and pays for virtually everything that Alaska needs. This

includes schools, hospitals, public buildings, water and sewerage schemes, new roads and improved communications with isolated communities. Even now that oil production is beginning to slow down, Alaska still produces a quarter of the nation's output and helps to provide over 700 000 jobs throughout the country.

Opponents to development have other views. Environmental groups and wildlife organisations are concerned about the effects of the oil industry in such an environmentally sensitive area. To protect the area from industrial development they want all of the Arctic National Wildlife

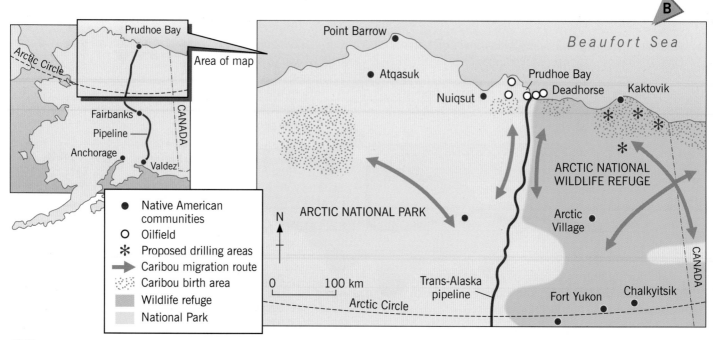

B

Beaufort Sea

Point Barrow

Prudhoe Bay

Atqasuk

Nuiqsut

Deadhorse

Kaktovik

ARCTIC NATIONAL PARK

ARCTIC NATIONAL WILDLIFE REFUGE

Arctic Village

Trans-Alaska pipeline

Fort Yukon

Chalkyitsik

Arctic Circle

N

0   100 km

Legend:
- ● Native American communities
- ○ Oilfield
- ✳ Proposed drilling areas
- → Caribou migration route
- ⠂ Caribou birth area
- ▨ Wildlife refuge
- ▨ National Park

Prudhoe Bay
Area of map
Arctic Circle
Fairbanks
Pipeline
Anchorage
Valdez
CANADA

Prudhoe Bay **D**

C

CONCERNS ABOUT THE OIL INDUSTRY

* Possible damage to fragile tundra vegetation and wildlife

* Loss of wilderness area

* Disruption of caribou migration routes

* Caribou forced away from sites where they give birth

* Local people may have to move away

* Only 5% of local people given oil-based jobs

* Native American people wish to retain control of their land

* Destruction of traditional lifestyles

* Fear of an oil spill

Refuge to be designated as a statutory wilderness. Local Native American groups, including the Inupiat and Gwich'in, are also actively opposed to the scheme, arguing that they do not want oil money but wish only to continue their unspoilt way of life. At first the US government was against drilling and showed a reluctance to support further exploration. In 2001, however, President Bush, aware of the benefits that new oilfield discoveries would bring to the nation, announced his support for further development. The decision angered many local people and environmental groups.

Activities

1 Draw a timeline to show the history of Alaskan oil.

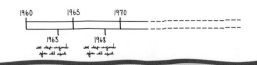

2 Describe the main features of Alaska's oil industry by completing fact file **E**.

3 Imagine that you work for the United States Department of the Interior and have been asked to write a report on the future of oil production in northern Alaska. Your report should be short, factual, and give both sides of the argument. It should provide reasoned recommendations as to whether further exploration and development should go ahead.

Fact file: Oil production in Alaska

E

1 **Location**
   Maps and a written description will be needed here to show Alaska and the oilfield area. An atlas and map **B** will help.

2 **Main features**
   You could describe the features shown in photo **D** or draw a labelled sketch. Add any other relevant geographical information.

3 **Benefits**
   Describe or list these under the headings 'Social' and 'Economic'. Remember to mention who has benefited.

4 **Problems**
   Describe or list these under the headings: Environment, People, Transport, Climate. Try to be specific and give examples.

Summary

Industrial activity in Alaska has brought social and economic gains. Many people are concerned about the damaging effects it may have on local communities and the environment.

What problems were there building the Alaskan pipeline?

When oil was first discovered at Prudhoe Bay, reserves were estimated to be in the region of 25 billion barrels. This made it the largest oilfield in America with a life expectancy of over 25 years. Unfortunately, the main demand for the oil came from thousands of kilometres away to the south, in America's great industrial cities. A major problem was how to transport the oil so far – a problem that was made even worse by the difficult physical and environmental conditions of the region.

The oil companies suggested three different routes.

1 Ship the oil directly by tanker from Prudhoe Bay. This was both difficult and dangerous as the Beaufort Sea is frozen most of the year.

2 Construct a Trans-Canadian pipeline to the east coast. This was an unpopular option because of its great length, danger to the environment and enormous cost.

3 Construct a pipeline south across Alaska to the ice-free port of Valdez on the Pacific coast. This was the most difficult route and attracted great opposition from conservation groups.

After much debate and a huge amount of technological research and expense, the third option was chosen and the 1285 km Trans-Alaska pipeline finally opened in June 1977.

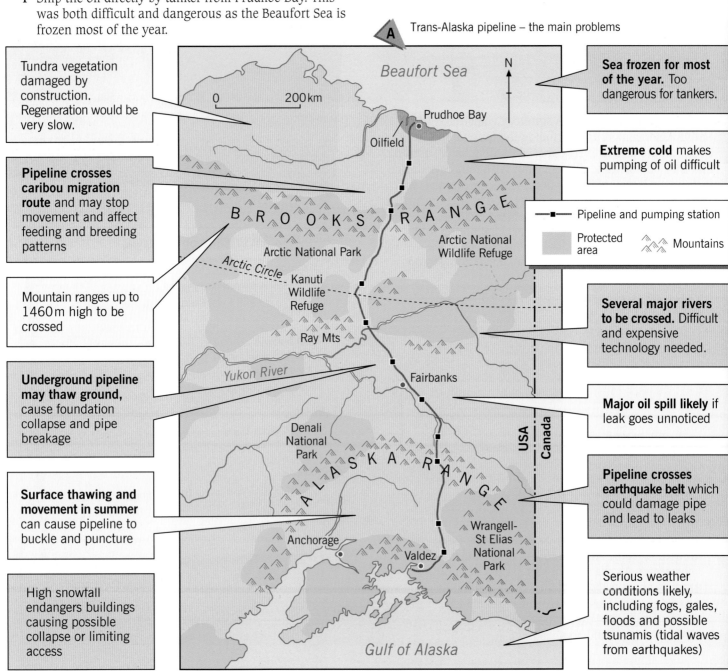

A Trans-Alaska pipeline – the main problems

Tundra vegetation damaged by construction. Regeneration would be very slow.

Pipeline crosses caribou migration route and may stop movement and affect feeding and breeding patterns

Mountain ranges up to 1460 m high to be crossed

Underground pipeline may thaw ground, cause foundation collapse and pipe breakage

Surface thawing and movement in summer can cause pipeline to buckle and puncture

High snowfall endangers buildings causing possible collapse or limiting access

Sea frozen for most of the year. Too dangerous for tankers.

Extreme cold makes pumping of oil difficult

Pipeline and pumping station

Protected area

Mountains

Several major rivers to be crossed. Difficult and expensive technology needed.

Major oil spill likely if leak goes unnoticed

Pipeline crosses earthquake belt which could damage pipe and lead to leaks

Serious weather conditions likely, including fogs, gales, floods and possible tsunamis (tidal waves from earthquakes)

Beaufort Sea

0   200 km

Prudhoe Bay

Oilfield

B R O O K S   R A N G E

Arctic National Park

Arctic National Wildlife Refuge

Arctic Circle

Kanuti Wildlife Refuge

Ray Mts

Yukon River

Fairbanks

Denali National Park

A L A S K A   R A N G E

Anchorage

Valdez

Wrangell-St Elias National Park

USA   Canada

Gulf of Alaska

Trans-Alaska pipeline – attempted solutions

Permafrost causes many problems for builders in polar regions. Permafrost is permanently frozen ground and covers much of the pipeline route. It is difficult to build any structure on it because the warmth of the structure melts the permafrost and can cause movement and subsidence.

Above the permafrost is the active layer. The active layer is up to 3 metres thick and causes further problems as it thaws in summer but freezes in winter.

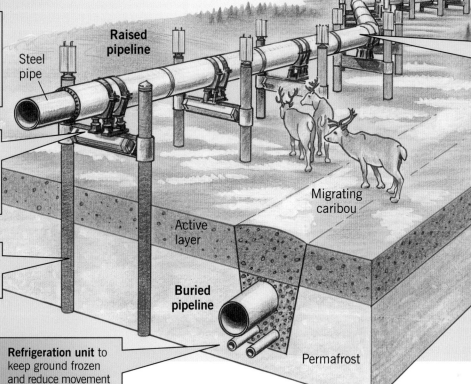

Pipeline built on stilts 3 metres above ground to reduce melting of permafrost and allow animal movement underneath

Raised pipeline

Steel pipe

Pipeline insulated and oil pumped at 80 °C

Sliding shoes allow sideways movement in event of ground shaking from major earth movements

Trans-Alaska pipeline built from Prudhoe Bay to ice-free Valdez

Migrating caribou

Pumping stations can cut down oil flow in damaged sections

Support pylons sunk through active layer to give stability

Active layer

Buried pipeline

Suspension bridges built to carry pipeline across rivers

Refrigeration unit to keep ground frozen and reduce movement

Permafrost

Activities

1 **a)** Using map **A** and cross-section **C** below, describe the route of the Trans-Alaska pipeline. Include any towns, mountain ranges and protected areas, and mention distances.
   b) How might the route that you have described, cause problems for pipeline builders?

2 **a)** Make a larger copy of cross-section **C** below.
   b) Write the following in the correct places and draw an arrow to each feature:
   • Brooks Range • Alaska Range • Yukon River
   • Beaufort Sea • Gulf of Alaska • 1460 metres.

3 **a)** Make a larger copy of table **D**.
   b) Match each of the eight locations from the cross-section with a problem taken from map **A**. (Only write in the **bold** words.)
   c) Write a solution from drawing **B** for each of the problems. (Again, use only the **bold** words.)

Location	Problem	Attempted solution
①		
②		

D

Summary

Difficult physical conditions have hindered oil production in Alaska. Builders of the pipeline had to overcome rugged countryside, extreme climatic conditions and complex environmental problems.

C

Prudhoe Bay

Valdez

① ② ③ ④ ⑤ ⑥ ⑦ ⑧

How has Alaskan oil damaged the environment?

24 March 1989 ... 12.28 a.m. ... *Exxon Valdez* ... 25 miles out of Valdez ... carrying 11 million gallons of oil ... off course ... run aground on reef ... badly holed ... oil pouring out ... need help immediately ...

A short radio message on the coastguard's emergency wavelength and it was everyone's worst nightmare come true: a huge tanker pouring oil into the calm, crystal-clear waters of Prince William Sound. One of America's great wilderness environments perhaps damaged for ever by a single mistake. Coastguards could not understand why the ship was so far off course until they found out that the captain was drunk and an inexperienced third mate had been left in charge.

The effects of the spill were massive, deadly and long-lasting. An oil slick 10 cm thick quickly spread out from the stricken tanker, eventually reaching 1000 km to the south-west. An estimated 1700 km of coastline was oiled, and within days more than 10 million birds were reported dead and much of the local marine life poisoned and virtually wiped out (figure **C**).

The major problem was – and still is – how to clean up the oil and reduce the impact of the spill. Four main methods were used in Alaska.

1 Encircle the spill with booms and skim off the oil. This helped protect some areas and was successful on small patches of oil.

2 Disperse with chemicals. This was partially successful although the fumes are poisonous and kill wildlife.

3 Burning off the oil. This cleared some patches but also produced poisonous fumes and became dangerous in high winds.

4 Hand cleaning. A large team of workers scrubbed the rocks by hand and used high-pressure hot water to wash down the shoreline. This was a slow process and research now suggests that it may have washed the oil deeper into the beaches and caused more permanent damage.

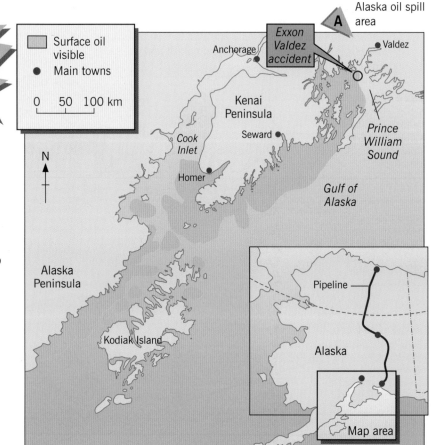

A

Alaska oil spill area

Surface oil visible

● Main towns

0 50 100 km

B

The process of cleaning up has gone on for years, and a 1997 report showed that large areas of coastline remained contaminated. Some wildlife including harbor seals and several species of duck are still not showing any signs of recovery.

C Effects of oil spill

| Destruction of fragile ecosystem |

| 10 million shore birds and water fowl dead |

| Commercial salmon hatcheries polluted |

| Land-based animals such as caribou poisoned by eating seaweed |

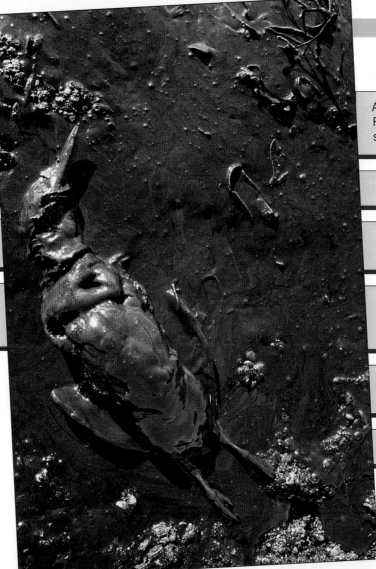

| At least 13 whales killed. Reduction in number of births since disaster |

| 1700 km of coast covered in oil |

| Loss of business for tourist industry |

| Over 5000 sea otters killed. Recovery by 1997 reported as very low |

| Serious loss of coastline vegetation and seaweeds |

| Local fishing industry ruined |

The cost of the clean-up has been considerable and not totally successful. In the first year after the disaster, Exxon, the company found to be at fault, spent £843 million and employed 11 000 people to clear oil from the beaches. Exxon also agreed to pay £600 million over the next ten years to complete the clean-up and restore the environment as far as possible to its original state. Sadly, this has not yet happened. Oil remains on the coastline, and wildlife and its habitat have still not fully recovered. Many doubt if they ever will.

The failure to prevent a disaster or even to significantly reduce its impact has worried many people. Most worrying, perhaps, is the fact that this happened despite the most up-to-date technology, the latest advanced safety systems and the most expensive clean-up scheme ever undertaken.

The incident has made people think carefully about the impact of economic activity on the environment. There is no doubt that the oil industry has brought many social and economic benefits to Alaska and the United States as a whole. It has also, however, caused serious and irreparable damage to a fragile environment. People have to choose which is more important – to use these resources or to protect and conserve the environment.

Activities

1 Write a newsflash to be read out on television giving news about the *Exxon Valdez* disaster. The newsflash will be broadcast on the day of the incident. Write between 50 and 80 words.

2 Draw a star diagram to show the methods used to disperse the oil.

3 Produce two posters, one to show the benefits of the oil industry in Alaska and the other to show the problems and possible damage that it may cause.
   • Make the posters colourful and attractive.
   • Give facts and figures where possible.
   • Include headlines, maps and sketches for impact.

4 Do you think oil production should continue in places like Alaska? Give reasons for your answer.

Summary

Industrial activity may bring benefits but can also cause problems. The *Exxon Valdez* oil spill was the worst in the history of the United States, and caused extensive long-term damage to the environment.

What are the main features of international trade?

No country can provide everything that its inhabitants want or need. To provide these needs, a country has to **trade** with other countries. It buys goods and services that it is either short of, or which it can obtain more cheaply from elsewhere. These are referred to as **imports**. To pay for these goods, a country must sell things of which it has a surplus, or which it can produce more cheaply than other countries. These are known as **exports**. The imports and exports of a country are likely to include raw materials, energy resources and manufactured goods.

Ideally a country hopes to have a **trade surplus**. This means that it will earn more money from the goods that it exports than it needs to spend on imports. A country with a trade surplus will become richer and, if the extra money is used sensibly, it can help to provide services and to improve the standard of living and quality of life of its inhabitants. Unfortunately not every country has a trade surplus. A country that has to spend more on imports than it earns from exports has a **trade deficit**. As a result it is likely to remain poor, will have insufficient money to develop new industries or improve services, and is likely to fall into debt.

Countries that trade with other countries are said to be **interdependent**.

Patterns of world trade

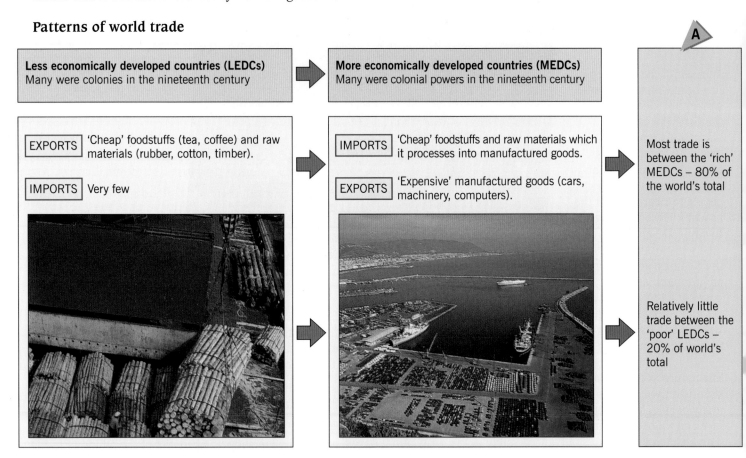

Less economically developed countries (LEDCs)
Many were colonies in the nineteenth century

More economically developed countries (MEDCs)
Many were colonial powers in the nineteenth century

| EXPORTS | 'Cheap' foodstuffs (tea, coffee) and raw materials (rubber, cotton, timber). |

| IMPORTS | Very few |

| IMPORTS | 'Cheap' foodstuffs and raw materials which it processes into manufactured goods. |

| EXPORTS | 'Expensive' manufactured goods (cars, machinery, computers). |

Most trade is between the 'rich' MEDCs – 80% of the world's total

Relatively little trade between the 'poor' LEDCs – 20% of world's total

There is, however, a wide imbalance in the trade between the LEDCs (the less economically developed countries) and the MEDCs (the more economically developed countries). As diagram **A** shows, this is mainly because:
- the LEDCs provide **primary goods**, such as foodstuffs and raw materials, which they sell to the MEDCs – primary goods are usually sold at a low and often fluctuating price
- the MEDCs process primary goods, either bought from LEDCs or available within their own country, into **manufactured goods** – these are sold at a high and usually a steady price (see graph **D**).

The result has been:
- a trade deficit for many LEDCs and a trade surplus for most MEDCs
- an increase in the share of world trade by the MEDCs and a decrease by the LEDCs.

Over a period of time, the trade imbalance between the richer countries of the 'North' and the poorer countries of the 'South' has widened (map **B** page 266).

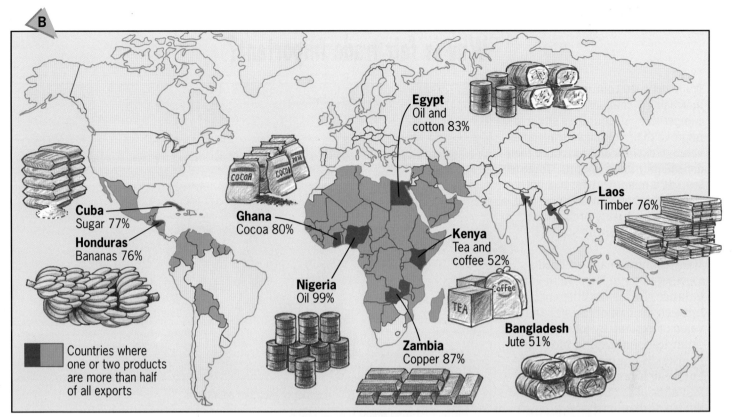

B

Cuba — Sugar 77%
Honduras — Bananas 76%
Ghana — Cocoa 80%
Egypt — Oil and cotton 83%
Nigeria — Oil 99%
Kenya — Tea and coffee 52%
Zambia — Copper 87%
Bangladesh — Jute 51%
Laos — Timber 76%

Countries where one or two products are more than half of all exports

A further problem for many LEDCs is their reliance upon a limited range of primary products that they are able to export. Map **B** shows that many LEDCs, especially those in Africa, rely heavily on just one or two commodities. The economy of a developing country can be seriously affected if there is:

- a crop failure, which may follow a natural disaster, or a mineral is used up
- there is a decline in demand for the product, especially during a global economic recession
- competition from a rival producer which may sell goods more cheaply.

C

INTERNATIONAL TRADE		
	Advantages	*Disadvantages*
Richer countries (MEDCs)		
Socio-economic	Cheap imports of foodstuffs and raw materials. Expensive exports of manufactured goods. Trade surplus.	Often obtainable from considerable distances: high transport costs.
Environmental	Limited mining and deforestation. Money to improve environment.	Manufacturing goods can create air, water, noise and visual pollution.
Political	Can exert pressure on LEDCs.	–
Poorer countries (LEDCs)		
Socio-economic	Raw materials have a ready market in MEDCs. Source of work.	Limited range of exports. Trade deficit as imports cost more than exports.
Environmental	–	Problems created by mining, deforestation and overgrazing.
Political	May be able to obtain overseas aid.	Often tied to/dominated by MEDCs.

Activities

1 **a)** What is the difference between:
   - *imports* and *exports*
   - *a trade surplus* and *a trade deficit*?
   b) What types of goods are exported by countries in:
   - the 'South' (LEDCs)   • the 'North' (MEDCs)?
   c) How does graph **D** help to explain why the gap between the rich and poor countries is getting wider?

2 **a)** Name six countries that rely heavily upon the export of one or two commodities (map **B**).
   b) Why do these countries:
   - usually have a trade deficit
   - find it difficult to improve their standard of living and quality of life?

D

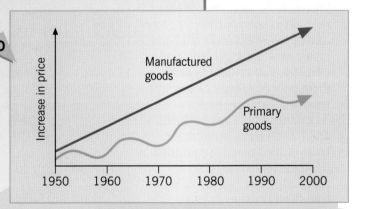

Summary

Countries trade in order to share out resources and to earn money. Countries that rely upon the export of raw materials do not develop economically as quickly as those that export manufactured goods.

Why is fair trade important?

Trade between countries has become increasingly complex and competitive. All countries strive to improve their volume of trade and value of exports and to reduce their dependency upon imports. Trade is seen as a major way for a country to improve its standard of living and, as a result, the quality of life of its inhabitants.

Many countries have grouped together to try to improve their **trade balance**. This is the difference between the cost of imports and the value of exports. By joining together, countries form **trading blocs** (map **A**). The UK is a member of the European Union (EU). One of the first aims of the EU (or EEC as it was known at that time) was to try to improve trading links between its members. This was achieved by eliminating customs duties previously paid on goods that were moved between member countries. This lowered the prices of goods, making them cheaper and more competitive against goods from non-EU countries. Also, as the number of EU member countries has grown, so too has its internal market. The larger the internal market, the greater the number of potential customers. This is even more important when, as in the case of the EU, many of these potential customers are wealthy and able to purchase a greater number of higher-value goods.

Two points should be noted:
1 World trade is not shared out equally between countries. Whereas, for example, the EU and the USA are responsible for 52 per cent of the world trade total and Japan a further 8 per cent, the developing countries between them only contribute 20 per cent (map **B**).
2 As trading blocs such as the EU and NAFTA try to increase the trade of their member countries, the trade of non-members, especially those in developing countries, is reduced. This apparent unfairness is a cause of the widening gap between the rich countries of the 'North' and the poorer countries of the 'South'.

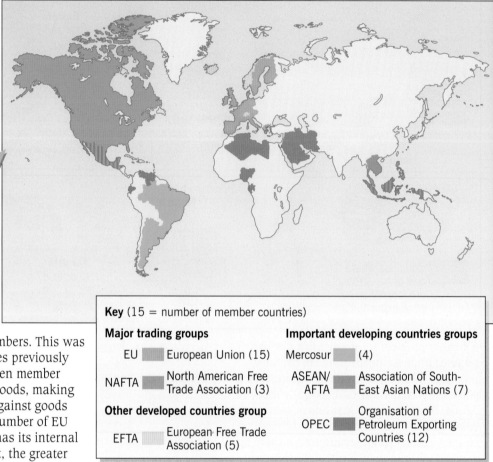

A

Key (15 = number of member countries)

Major trading groups

EU — European Union (15)

NAFTA — North American Free Trade Association (3)

Other developed countries group

EFTA — European Free Trade Association (5)

Important developing countries groups

Mercosur — (4)

ASEAN/AFTA — Association of South-East Asian Nations (7)

OPEC — Organisation of Petroleum Exporting Countries (12)

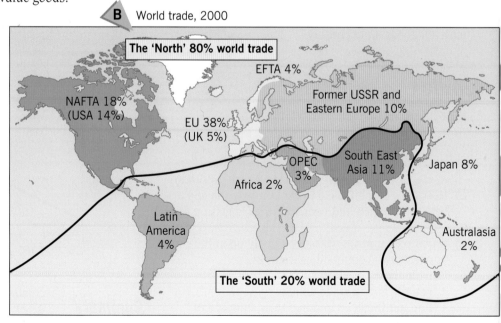

B   World trade, 2000

The 'North' 80% world trade

EFTA 4%

Former USSR and Eastern Europe 10%

NAFTA 18% (USA 14%)

EU 38% (UK 5%)

OPEC 3%

South East Asia 11%

Japan 8%

Africa 2%

Latin America 4%

Australasia 2%

The 'South' 20% world trade

Free trade, tariffs and quotas

In an ideal world there should be 'free trade' between all countries (diagram **C**). In the real world this rarely happens. Virtually all governments, especially those of the MEDCs, have tried to regulate overseas trade. They have achieved this by creating trade barriers which governments hope will protect jobs and industries within their own country. The most common methods of affecting the levels and patterns of international trade are through **tariffs** and import **quotas** (diagram **C**). Tariffs can either reduce the cost of imports (helping the trade balance) or protect home-made products. Quotas tend to be restricted to primary goods and so work against LEDCs.

Free trade is when governments neither restrict nor encourage the movement of goods.

Tariffs are taxes or customs duties paid on imports. The exporter has to pay a percentage of the value of the goods to the importer. Importers may add tariffs just to raise money, but usually the idea is to put up the price of imported goods in order to make them more expensive and therefore harder to sell.

Quotas limit the amount of goods that can be imported. They tend to be restricted to primary goods and so work against the LEDCs.

C

The World Trade Organisation (WTO)

Attempts began in 1986 to try to replace the various protectionist blocs and self-interest groups with one large global free trade area. A breakthrough occurred in 1994 when the major trading countries agreed to reduce tariffs on many industrial products. This decision benefited the newly industrialised countries (page 226) and, as it increased competition and lowered prices, consumers. However, farm subsidies and tariffs on agricultural goods were only to be reduced over a period of time. This delay, caused by strong agricultural pressure groups in the EU and the USA, was to the detriment of the many LEDCs that rely upon the export of agricultural products.

Activities

1 **a)** Map **D** shows four trading blocs. For each trading bloc:
   i) give its initials
   ii) find out the names of three member countries.
   b) Give two reasons why countries group together to form trading blocs.

2 Diagram **E** shows the volume of trade between the EU, the USA and Japan.
   a) What percentage of world trade is shared by these three?
   b) Of the three, which has:
   i) a trade surplus with the other two
   ii) a trade deficit with the other two
   iii) a trade surplus with one and a trade deficit with the other?

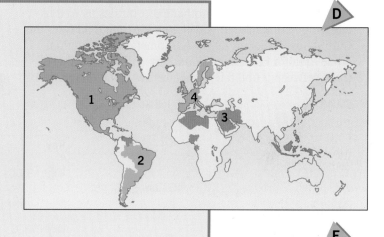

D

E

USA to Japan 48.6	EU to Japan 28.9	Japan to EU 53.9
USA to EU 98.1	EU to USA 97.4	Japan to USA 90.0

Exports in US$ billion (1993)

3 Study figure **F**.
   a) What are: i) *tariffs* ii) *quotas*?
   b) How do tariffs and quotas work in favour of rich countries and against poorer countries?

F

You have most of the world's trade, yet you impose tariffs and quotas to prevent us from selling our goods

We must protect our jobs and industries by limiting your cheap imports

Economically less developed countries = 20% of world trade

Economically more developed countries = 80% of world trade

Summary Governments have tried to influence patterns of international trade by grouping together to form trading blocs or by imposing tariffs and quotas.

How interdependent are Kenya and Japan?

Kenya

As diagram **A** shows, Kenya's trade is typical of most LEDCs. Exactly half of the country's exports consist of three relatively low-value agricultural products: tea, coffee, and horticultural crops (mainly peas, beans and flowers). Although **manufactured goods** appear to be significant, most of these are made from either low-value or **recycled** raw materials (page 229). These products, many of which need only limited technology, tend to be sold to surrounding and equally poor African countries. In contrast, Kenya's imports are dominated by high-value manufactured goods and oil. The result is that Kenya has:

- a large **trade deficit** – over US$ 1100 in the late 1990s
- a low amount of trade per person – US$ 110
- to borrow money from more wealthy countries.

Kenya's most important trading partner is the UK (diagram **B**). This is because Kenya was once a British colony and, although independent since 1963, it still provides raw materials for its former colonial power. Kenya also now provides raw materials for other EU countries as well as receiving manufactured goods from Japan and oil from the Gulf States. Although Kenya has joined the Preferential Trade Area for Eastern and Southern Africa, the value of trade between it and fellow members is small (diagram **B**).

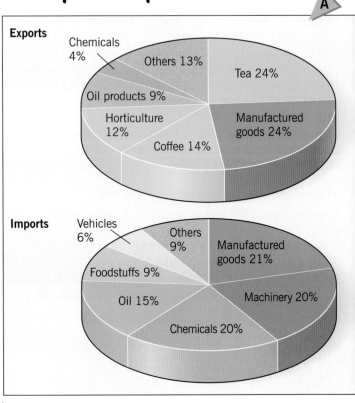

A

Exports: Tea 24%, Manufactured goods 24%, Coffee 14%, Horticulture 12%, Oil products 9%, Chemicals 4%, Others 13%

Imports: Manufactured goods 21%, Machinery 20%, Chemicals 20%, Oil 15%, Foodstuffs 9%, Vehicles 6%, Others 9%

B

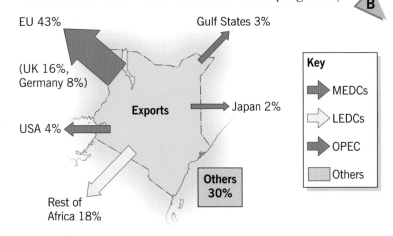

Exports
- EU 43% (UK 16%, Germany 8%)
- Gulf States 3%
- Japan 2%
- USA 4%
- Rest of Africa 18%
- Others 30%

Key
- MEDCs
- LEDCs
- OPEC
- Others

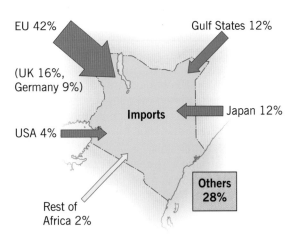

Imports
- EU 42% (UK 16%, Germany 9%)
- Gulf States 12%
- Japan 12%
- USA 4%
- Rest of Africa 2%
- Others 28%

Activities

1 **a)** Draw four bar graphs like those in diagram **E**, then complete:
  - the first to show Kenya's exports
  - the second to show Kenya's imports
  - the third to show Japan's exports
  - the fourth to show Japan's imports.
  b) On each graph, colour raw materials (minerals, fuel resources and foodstuffs) in *green* and manufactured goods in *red*.
  c) Suggest how your completed graphs account for:
  - Kenya's trade deficit
  - Japan's trade surplus.

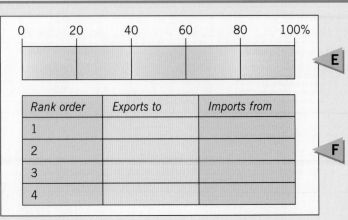

E

0   20   40   60   80   100%

F

Rank order	Exports to	Imports from
1		
2		
3		
4		

Japan

As diagram **C** shows, Japan's trade is typical of most MEDCs. Japan exports large amounts of high-value manufactured goods such as electrical goods, cars and high-technology products. The Japanese have achieved this by working long hours, introducing modern machinery, developing high levels of technology and producing high-quality, reliable goods. Japan, which has few natural resources, limited flat land and a big population, has to import large amounts of raw materials, foodstuffs and energy resources. With the exception of oil, most of these imports are relatively cheap in value. Having become a wealthy nation, Japan can also afford to import expensive manufactured goods from other MEDCs. The result is that Japan has:

- a large trade surplus – over US$ 110 000 in the late 1990s (although the country has since then experienced its first economic recession)
- a large amount of trade per person – US$ 4849.

It is the world's second largest trading country after the USA, and has been able to give economic aid to less wealthy countries (page 264).

Unlike all other MEDCs, Japan has not grouped itself with any major world trading bloc. However, its need to import raw materials and its desire to export large volumes of manufactured goods, makes it as interdependent as any other country in the world (diagram **D**).

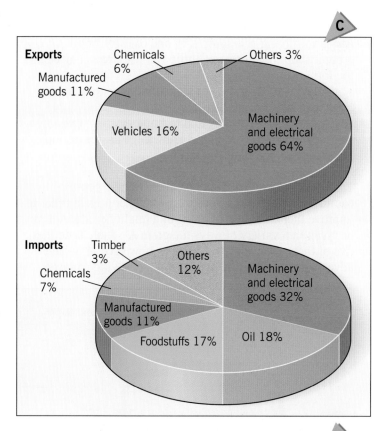

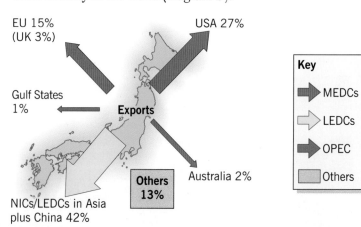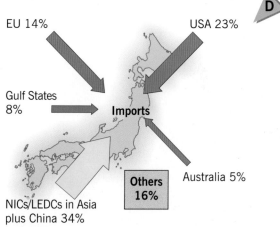

2 **a)** Make two copies of table **F**, the first for Kenya and the second for Japan.

  b) Complete the two tables by ranking in order:
  - the places to which most exports are sent
  - the places from which most goods are imported.

3 If Japan can buy one tree from a developing country for £2 and then sells a table, made from that tree, back to the developing country for £20, what effect will that have on the trade balance and standard of living in:
  - Japan
  - the developing country?

Summary

Countries need to be interdependent if they are to share in the world's resources. Countries like Kenya which export primary goods do not develop as quickly economically as those like Japan which export manufactured goods.

Why is aid needed?

Aid is the giving of resources by one country, or an organisation, to another country. The resources may be in the form of:
- money, often as grants or loans that have to be repaid even if interest rates are low
- goods, food, machinery and technology
- people who have skills and knowledge as, for example, teachers, nurses and engineers.

The basic aim in giving aid should be to help poorer countries to develop their economy and services in order to improve their standard of living and quality of life. Aid can be given in a variety of ways (diagram **A**). Table **D** defines five of these methods and describes how each may have advantages and disadvantages to the receiving country.

Aid is needed by many LEDCs:
- because there is an imbalance in trade giving a trade deficit
- because differences in development have led to global inequalities
- because there is a need to improve their basic amenities (water supply, electricity) and infrastructure (new roads, schools and hospitals)
- to encourage and promote self-help schemes and sustainable development
- to combat the effects of environmental hazards such as earthquakes, tropical storms or drought
- following human-created disasters such as civil war.

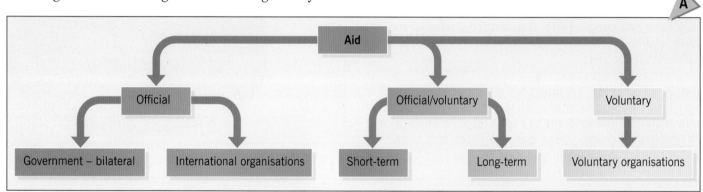

A

| Aid |
| Official | Official/voluntary | Voluntary |
| Government – bilateral | International organisations | Short-term | Long-term | Voluntary organisations |

B Lake Kariba, Zimbabwe

C Emergency aid being given following an earthquake

Unfortunately foreign aid does not always reach the places or people that are most in need. It can be directed into over-ambitious schemes or end up in the pockets of politicians and corrupt officials. There are also strong arguments as to whether aid should be given at all. Some people think that aid discourages the development of LEDCs and makes them dependent upon rich donor countries and organisations. Others believe that as we all live in the same world we must help each other and try to improve the quality of life for everyone – this is the concept of **global citizenship**.

Type of aid	Definition	Disadvantages and advantages
Government (bilateral)	Given directly by a richer country (donor) to a poorer country – often **tied** with 'strings attached'	• 'Tied' meaning LEDC has to buy goods from donor, e.g. arms, manufactured goods • Money often has to be spent on prestigious schemes such as dams (photo **B**) and international airports • Large schemes take up land belonging to local people • Aid often encourages corruption – money rarely reaches poorer people living in more remote areas • LEDC unable to repay money – gets further into debt • LEDC becomes increasingly dependent on donor country • Can provide grants for students to study in MEDCs
International organisations (multilateral)	Given by organisations such as the World Bank and the IMF (International Monetary Fund)	• Not meant to be tied, but less likely to be given to countries with unfavourable economic and political systems • Encourages farming and industry but products are sent to MEDCs rather than consumed in LEDCs • LEDCs become increasingly dependent on aid, and often fall increasingly into debt • Helps LEDCs to develop new crops, raw materials and industry
Voluntary	Organisations such as Oxfam and ActionAid which collect money and receive gifts for people in LEDCs	• Not tied • Deals with emergencies • Encourages low-cost self-help schemes • Money more likely to reach poorer people in more remote areas • Dependent on charity's ability to collect money • Annual amounts uncertain – requires longer-term planning
Short-term/ Emergency	Needed to cope with the effects of environmental hazards such as earthquakes and tropical storms	• Immediate help – provides food, clothes, medical supplies and shelter (photo **C**) • Goes to places and people most in need • Not tied, and less chance of corruption • Also helps refugees (photo **E**)
Long-term/ Sustainable	Organisations such as Intermediate Technology that help people in LEDCs to support themselves	• Encourages development of local skills and use of local raw materials • Trains local people to be teachers, nurses, health workers • Helps equip schools and development of local agriculture and small-scale industry • LEDCs do not fall into debt (see photo **F**)

D

E Refugees in Africa

F

A village well helps people to help themselves

Activities

1 a) What is *aid*? Why is it needed by many LEDCs?
b) What is the difference between:
   • *official aid* and *voluntary aid*
   • *short-term aid* and *long-term aid*?

2 a) With the help of table **D**, list:
   • five advantages of receiving aid
   • five disadvantages of receiving aid.
b) Which of the five types of aid do you consider to be best for an LEDC? Give reasons for your answer

Summary   Aid can bring many benefits but it can also create problems in receiving countries. Global citizenship is when countries and people try to help each other.

How can contrasts in development be measured?

Geographers are concerned with:
- differences in levels of development between places, both between countries and within countries
- mapping these differences to see if there are recognisable patterns in development
- trying to explain why these differences have occurred and suggesting how they may be evened out.

The term **development** can be defined, and interpreted, in many different ways. It is, therefore, difficult for geographers and other groups of people to find methods of measuring development.

The traditional, and arguably the easiest, method of comparing different places is to measure their wealth. The wealth of a country is measured in terms of its **gross national product** or **GNP**. The GNP per capita is the total value of goods produced and services provided by a country in a given year divided by the number of people living in that country. In order to make comparisons between countries easier, GNP is given in US dollars (US$). Figure **A** shows that every person living in the UK, regardless of their age, should have received the equivalent of US$ 18 700 had the country's wealth been shared out evenly. Although table **C** shows differences between countries, it hides differences between places and groups of people within those countries.

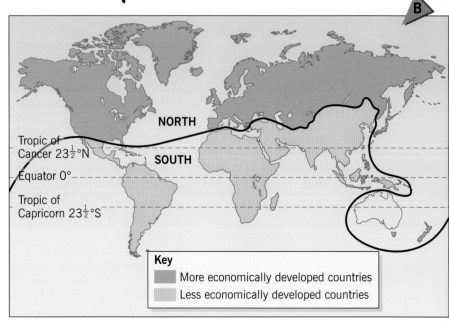

Key
  More economically developed countries
  Less economically developed countries

Map **B** shows how, based on GNP, the world can be divided into two economic groups:
1 **More economically developed countries (MEDCs)**, which have the highest GNPs and include the richer, more industrialised countries of the 'North'.
2 **Less economically developed countries (LEDCs)**, which have the lowest GNPs and include the poorer, less industrialised countries of the 'South'.

To many people living in a western, industrialised country such as the UK, economic development has been associated with a growth in wealth. This suggested that the GNP of a country had to increase if the standard of living and quality of life of its inhabitants were to improve. More recently the meaning of development has been widened to include various social, health and education indicators (table **C**). However, it might be argued that these indicators themselves are often dependent upon a country's wealth.

As summarised on table **D**, LEDCs, when compared with MEDCs, usually have:
- higher birth, death and infant mortality rates, a greater natural increase in population, and a shorter life expectancy (pages 131–134)
- a lower level of literacy and fewer doctors per size of population
- more jobs in the primary sector and fewer in the secondary and tertiary sectors (pages 188–191)
- a smaller volume of trade and, probably, a trade deficit (page 258).

Latest figures show that in one year the UK earned $1 094 000 million. If you divide this by the UK's total population of 58.3 million, this gives a GNP per capita of $18 700.

	Country	GNP	Population indicators						Other indicators				
			Birth rate	Death rate	Natural increase	% population 0–14 yrs	Infant mortality	Life expectancy	% in primary sector	Energy used per person	% urban population	People per doctor	Adult literacy
M E D Cs	Japan	39 640	10	8	2	18	4	80	7	498	77	608	99
	USA	26 980	15	9	6	21	7	76	3	1139	78	421	98
	UK	18 700	13	11	2	19	6	76	2	533	89	350	99
N I Cs	Malaysia	3890	26	5	21	38	23	70	26	229	51	2564	82
	Brazil	3640	20	9	11	35	53	62	25	85	81	844	81
L E D Cs	India	340	25	9	16	37	69	60	62	37	29	2459	50
	Kenya	280	32	11	21	50	55	54	81	12	25	21 970	75
	Bangladesh	240	30	11	19	44	100	55	59	9	21	12 884	36

		Developing countries	Developed countries
Population (birth rate, death rate and natural increase, page 131; infant mortality rate and life expectancy, page 134)		• High birth rate (falling). • Relatively high death rate (falling). • Rapid natural increase. • High infant mortality rate (falling slightly). • Short life expectancy (increasing).	• Low birth rate (steady). • Low death rate (steady). • Slow natural increase. • Low infant mortality rate (steady). • Long life expectancy (increasing).
Health		• Few doctors, nurses and hospitals. • Each doctor may have several thousand patients.	• More doctors, nurses and hospitals. • Each doctor may have several hundred patients.
Education (literacy: the % of adults who can read and write)		• Insufficient money for full-time education. • Low percentage of literate adults, especially among women.	• Full-time education. • High percentage of literate adults, including women.
Jobs (employment structures, page 188)		• Most jobs in primary sector (highest percentage in farming). • Relatively few in secondary and tertiary sectors.	• Few jobs in the primary sector. • Larger number in the secondary sector. • Highest percentage in the tertiary sector.
Trade (page 258)		• Small in volume and value. • Mainly raw materials (minerals and foodstuffs that are cheap to buy). • Often a trade deficit.	• Large volume and value. • Mainly manufactured goods (expensive to buy). • Usually a trade surplus.

Activities

1 a) What do you understand by the term *development*?

b) Why is development hard to define?

c) What is *gross national product (GNP)*?

d) What is the GNP for each of the USA, the UK, Brazil, Kenya and India?

e) Why do the USA and the UK have a higher GNP than Kenya and India?

f) Describe the location and distribution of the less economically developed countries.

2 Name six indicators, other than GNP, that can be used to show differences in levels of development between regions and countries.

Summary

GNP is the most frequent method used to show differences in development between places. Development can also be measured using social, health and educational indicators.

How else may development be measured?

It has already been pointed out that GNP provides the easiest method of measuring, and the simplest way to compare, different levels of development. However, it is now widely accepted that the term 'development' has a wider meaning than just 'wealth'.

In 1990, the United Nations published the first of what is now an annual report using the **Human Development Index (HDI)**. Using the HDI, the UN rank countries according to the quality of life of their citizens rather than using traditional economic figures. The HDI is a composite of three variables: life expectancy, education, and income per capita.

- **Life expectancy** is regarded as the best measure of a country's health and safety.
- **Education** attainment is obtained by combining adult literacy rates and the average number of years spent at school.
- **Income per capita** is adjusted to actual purchasing power, i.e. what an income will actually buy in a country.

Each variable is given a score ranging from 1.000 (the best) to 0.000 (the poorest). The HDI is the average of the three scores. The latest UN report, published in 2001, puts Norway at the top with a score of 0.992 and Sierra Leone at the bottom with a score of 0.042 (table **A**). The HDI shows countries that have a better or worse quality of life than might be expected by their GNP. Countries that are:

- better off include Australia, Sri Lanka and India
- worse off include several wealthy oil-producing countries in the Middle East such as Saudi Arabia.

Even so, the world map showing HDI (map **B**) reveals that:

- countries with scores that exceed 0.900 correspond very closely with the MEDCs in the 'North' (map **B** on page 266)
- countries with scores below 0.250 equate equally closely with the LEDCs in the 'South' (in 2001, the bottom 20 countries were all located in Africa south of the Sahara).

In 2001, data was only available for 162 countries and for some of these its reliability might be questioned. The last time data was given for Afghanistan, it was ranked second from the bottom (1993).

A

	Rank order	Country	Life expectancy at birth (years)	Adult literacy rate (%)	Years of schooling (average)	Real GNP per capita (PPP$*)	HDI
Top 4	1	Norway	78	99.5	11.8	24940	0.939
	2	Australia	80	99.5	11.6	18950	0.938
	3	Canada	79.5	99.5	12.1	21130	0.936
	4	Sweden	78.5	99.5	11.4	18540	0.935
Selected countries	6	USA	76	99	12.3	26980	0.934
	8	Netherlands	78	99	10.8	19950	0.931
	9	Japan	80.5	98.5	10.7	21110	0.928
	13	France	79	99	11.6	21030	0.924
	14	UK	77	99	11.6	19260	0.923
	17	Germany	76	99	11.2	20070	0.921
	20	Italy	78.5	97.5	10.3	19870	0.909
	21	Spain	78.5	98	9.2	14520	0.908
	42	Poland	72.5	97.5	8.4	5400	0.828
	45	Mexico	74	87	6.2	6400	0.790
	48	Malaysia	70	82	6.1	9020	0.774
	61	Brazil	66.5	81	4.9	5400	0.750
	67	Saudi Arabia	69.5	61	4.0	10020	0.740
	108	China	70.5	70	4.8	2920	0.718
	118	Egypt	62	49	2.8	3820	0.635
	122	Kenya	54.5	75	3.4	1380	0.514
	127	Pakistan	59	36	2.3	2230	0.498
	129	India	60.5	50	2.8	1400	0.471
	136	Nigeria	54.5	53	1.5	1220	0.455
	139	Bangladesh	56	36	2.2	1380	0.440
Bottom 4	159	Burkina Faso	42	18	0.4	780	0.320
	160	Burundi	49	50	0.2	620	0.309
	161	Niger	41	53	0.2	750	0.274
	162	Sierra Leone	48	29	0.5	580	0.258

Human Development Index

*Purchasing power parity in US dollars (200

B Human Development Index (HDI)

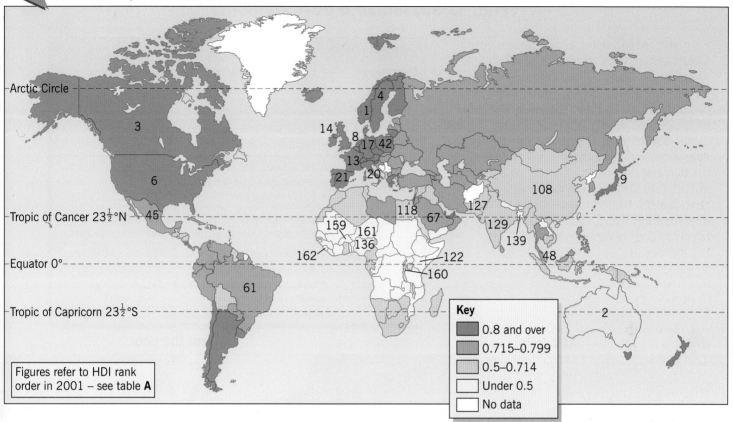

Figures refer to HDI rank order in 2001 – see table **A**

Key
- 0.8 and over
- 0.715–0.799
- 0.5–0.714
- Under 0.5
- No data

The HDI can also:
- highlight where poverty is worst, both within a country (diagram **C**) and between countries
- act as a measure to show how far a country has developed
- help that country to set targets that can lead to improvements in the quality of life if its citizens, e.g. better health care to increase life expectancy, and better education to improve literacy.

C

If the HDI was only applied to white Americans, then the USA would be top of the ranking table.

If the HDI was only applied to black Americans, then the USA's ranking would fall to the lower 30s.

Activities

1 a) Why did the United Nations suggest the Human Development Index (HDI)?
  b) Which three variables are used to determine the HDI?
  c) How is the HDI worked out?
  d) In what ways is the HDI a better guide to development than the more traditional use of GNP?

2 Describe carefully the location of those countries with an HDI score of:
  i) over 0.90
  ii) between 0.50 and 0.99
  iii) under 0.50.

3 What might a country do to try to raise its HDI score and to improve its world ranking?

Summary

The Human Development Index (HDI) has extended the meaning of development to include real income, education and life expectancy. Even so there are considerable similarities between the GNP and HDI maps.

What is sustainable development?

Economic, social, political and environmental factors can all have an impact on development (diagram **A**). Many of the rich countries of the 'North' (the MEDCs) became 'developed' by creating industries that used raw materials which they obtained from their former colonies in the 'South' (the LEDCs). This development, apart from widening the gap in wealth between the rich and the poor (photos **B**), paid little attention to the effect it was having upon the environment. During the 1990s, **sustainable development** was increasingly used to try to improve levels of development.

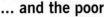

Development may be hindered by:	
Economic	• Lack of raw materials and energy reserves • Inefficient farming and unreliable food supplies • Lack of jobs • Lack of money for investment • Trade debt and growing international debt • Limited technology
Social	• Insufficient money to improve education and health • Lack of an efficient transport system • Poor housing conditions and amenities
Political	• Lack of government aid or involvement • Effects of civil war, or receiving numerous refugees
Environmental	• Effects of natural disasters such as drought, storms, flooding, earthquakes and volcanic eruptions • Effects of disease (malaria, AIDS) • Pests (e.g. locusts) • Unreliable supplies of clean water

The rich ...　　　　　　　　　　　**... and the poor**

Sustainable development

Sustainable development involves a sensible use of resources, especially those that are non-renewable, and **appropriate technology**. It should lead to an improvement in people's:

• **quality of life** – allowing them to become more content with their way of life and the environment in which they live
• **standard of living** – enabling them, and future generations, to become better off economically.

Sustainable development needs careful planning, a commitment to conservation and the co-operation of different countries.

Sustainable development in LEDCs

In LEDCs, whose inhabitants are poor, alternative forms of technology to those used in the MEDCs need to be adopted. These alternative forms, collectively referred to as **appropriate technology**, are summarised in diagram **C**.

The Intermediate Technology Development Group (ITDG) is a British charitable organisation that works with people in developing countries. It:

• helps local people acquire the tools and techniques they need in order to work themselves out of poverty
• helps people to meet their basic needs of food, clothes, housing, energy and jobs
• uses, and adds to, local knowledge by providing technical advice, training, equipment and financial support so that local people can become more self-suffcient and independent.

C APPROPRIATE TECHNOLOGY BY ITDG

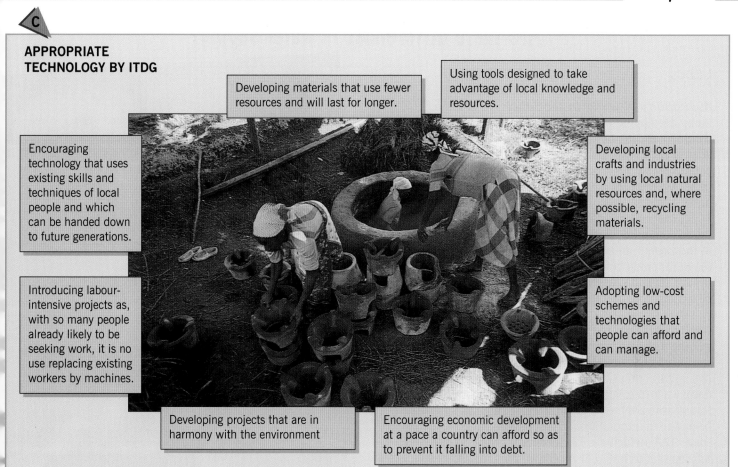

Developing materials that use fewer resources and will last for longer.

Using tools designed to take advantage of local knowledge and resources.

Encouraging technology that uses existing skills and techniques of local people and which can be handed down to future generations.

Developing local crafts and industries by using local natural resources and, where possible, recycling materials.

Introducing labour-intensive projects as, with so many people already likely to be seeking work, it is no use replacing existing workers by machines.

Adopting low-cost schemes and technologies that people can afford and can manage.

Developing projects that are in harmony with the environment

Encouraging economic development at a pace a country can afford so as to prevent it falling into debt.

Sustainable development in MEDCs

It is often mistakenly believed that sustainable development is not appropriate for MEDCs. This is not the case, as it is the MEDCs that:

- consume most of the world's resources, such as fossil fuels, minerals and timber
- are the greatest contributors to global pollution, for example global warming and acid rain.

An appropriate technology can contribute to a more sustainable way of life for people who are rich or poor, living in places that are considered to be either developed or developing. The only difference is that for those living in the MEDCs, the appropriate technology is likely to be high-tech.

Activities

1 How can environmental conditions and hazards contribute to different levels of development?

2 **a)** What is *sustainable development*?
   b) Why is it important that future development should be sustainable?
   c) Make a copy of diagram **D**. Complete it by adding six advantages of sustainable development.

3 **a)** What are the aims of the Intermediate Technology Development Group (ITDG)?
   b) Why is its support often more valuable to an LEDC than large loans given by the World Bank or an MEDC?

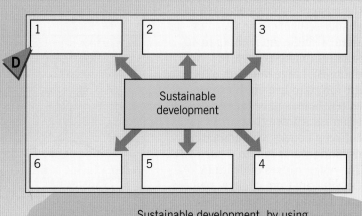

D

| 1 | 2 | 3 |

Sustainable development

| 6 | 5 | 4 |

Summary

Sustainable development, by using technologies that are appropriate, should improve the quality of life and standard of living of people across the world.

Are there differences in water supply?

LEDCs

Availability of water in the Sahel

Many developing countries in Africa have a tropical continental climate where there is a pronounced wet and dry season each year (map **C** page 93). These include the Sahel countries that lie immediately to the south of the Sahara Desert (map **A** page 126). Here, not only is the wet season short, but annual rainfall amounts are unreliable (page 96) and, since 1970, have even decreased. Even when rain does fall, it may fall in torrential downpours that cause flash floods, and is lost to local communities. If the rains fail during one year, the result for crops, animals and people can be disastrous. If they fail for several years, the result is increased soil erosion and desertification, as shown on page 126.

Like most LEDCs, the Sahel countries have neither the money nor the technology to build dams to store water. Relatively few people have piped water and, as most of the inhabitants live in rural areas, there is often a long, time-consuming daily walk to collect water from the nearest well or river. Recent improvements in rural water supplies have mainly been due to international charity organisations, such as ITDG (page 270), which have helped local people to introduce appropriate technology. As shown on photo **A,** this includes self-help schemes such as digging wells and using modern pumps.

Clean water in Calcutta

In many LEDCs where rainfall is usually available, the supply of water is not clean (map **C**). This may be due to the large numbers of people who live in shanty settlements in large urban areas such as Nairobi (page 184), Calcutta (pages 180–181) and São Paulo, who have no adequate, safe method of disposing of sewage. One-quarter of Calcutta's population of 12 million has no access to piped water. It is not uncommon for people living in the bustees – the name given to shanty settlements in Calcutta – to share a single tap with up to 40 families (photo **D** page 181, and photo **B**). Large numbers of people are forced to live on the streets and, as few homes have toilets, sewage often runs down the narrow lanes between houses, contaminating drinking water (photo **C** page 181). The major causes of death in Calcutta are from cholera, typhoid and dysentery – diseases linked to poor sanitary conditions.

A

B

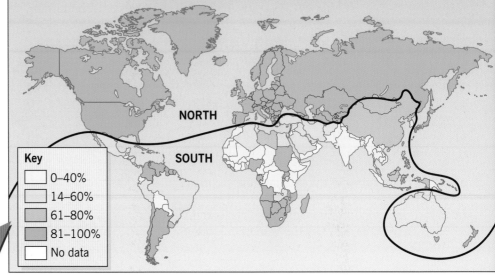

NORTH

SOUTH

Percentage of the population with access to safe water C

Key	
	0–40%
	14–60%
	61–80%
	81–100%
	No data

MEDCs

Water supply in the UK

In Britain it is taken for granted that it will rain every few days, and that rain will fall fairly evenly throughout the year. This means that most of the UK's water supply can be obtained from surface water in rivers and lakes, and the remainder from underground water-bearing rocks. When, in previous years, the demand for water increased as the country's population and industry grew, the UK had the capital and technology to build storage dams and later to provide piped water for everyone (except for just a few living in isolated rural areas). Today most householders are linked to the main sewerage system. This means that most people have both a guaranteed and a clean supply of water. The UK's only problems are that:

- most rain falls in the north and west, and most people live in the south and east (map **D**) – this means that surplus water has to be collected and stored in the north and west and transferred when it is needed to the south and east
- in occasional years there is a lengthy drought – as in 1975/76 (photo **E**) and 1995/96.

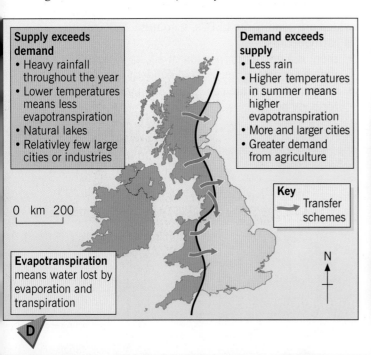

Supply exceeds demand
- Heavy rainfall throughout the year
- Lower temperatures means less evapotranspiration
- Natural lakes
- Relativley few large cities or industries

Demand exceeds supply
- Less rain
- Higher temperatures in summer means higher evapotranspiration
- More and larger cities
- Greater demand from agriculture

Key
→ Transfer schemes

0 km 200

Evapotranspiration means water lost by evaporation and transpiration

N

D

E Haweswater Reservoir, Lake District during the 1976 drought

Activities

1 **a)** Why is the availability of water a problem in the Sahel countries of Africa?
   b) Why is it difficult to provide water for people living in that region?
   c) How does appropriate technology help local communities provide water for themselves?

2 Look at map **F**.
   a) Describe the location of places where fewest people have access to clean water.
   b) Why may water supplies be contaminated in urban areas such as Calcutta?

3 **a)** Give three reasons why the north and west of Britain usually has a water supply surplus.

   b) Give three reasons why the south and east of Britain often has a water supply deficit.
   c) Why do most people living in the UK have a guaranteed supply of clean water?

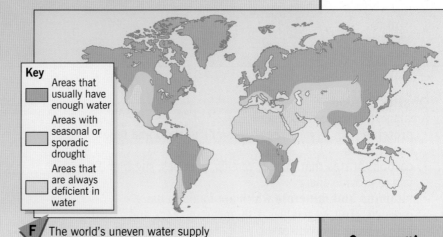

Key
- Areas that usually have enough water
- Areas with seasonal or sporadic drought
- Areas that are always deficient in water

F The world's uneven water supply

Summary

There is a big difference in the availability and cleanliness of the water supply in LEDCs and MEDCs.

Are there differences in food supply?

There are important differences in food supply between the MEDCs and the LEDCs. These are mainly due to increases in world population and world poverty that have led, in turn, to a sharp rise in the number of people who suffer from **malnutrition**.

A satisfactory diet has two important characteristics:
1 the **quantity** of food consumed
2 the **quality** of food consumed.

Malnutrition is caused by deficiencies in either the amount (quantity) or the type (quality) of food that is eaten. Until recently it was believed that malnutrition resulted from the growth in population being faster than the increase in food supply. Now it is considered to result from poverty. This is because large numbers of people, especially in the LEDCs, are unable to buy either enough food or the right type of food.

Quantity of food

Food is needed to provide energy for the body to work. The amount of food required or consumed by a person is measured in calories. The **dietary energy supply (DES)** is the number of calories needed per person. Like GNP (page 266), the DES can be used to compare countries but fails to recognise differences between individuals or places within a country. It has been estimated that people living in temperate climates, which are mainly the MEDCs, require a minimum of 2600 calories per day (map **A**). This is slightly more than people living in countries with a tropical climate, which are usually the LEDCs, who need 2300 calories. This is partly because:
- there is a greater proportion of adults in the MEDCs and they have greater needs than children
- temperate areas are cooler and so more energy is needed for body heating.

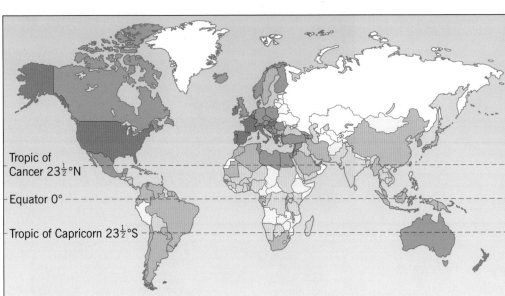

Tropic of Cancer 23½°N

Equator 0°

Tropic of Capricorn 23½°S

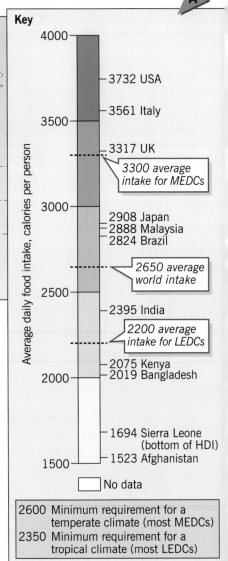

A

Key

Average daily food intake, calories per person

- 3732 USA
- 3561 Italy
- 3317 UK
- *3300 average intake for MEDCs*
- 2908 Japan
- 2888 Malaysia
- 2824 Brazil
- *2650 average world intake*
- 2395 India
- *2200 average intake for LEDCs*
- 2075 Kenya
- 2019 Bangladesh
- 1694 Sierra Leone (bottom of HDI)
- 1523 Afghanistan

☐ No data

| 2600 | Minimum requirement for a temperate climate (most MEDCs) |
| 2350 | Minimum requirement for a tropical climate (most LEDCs) |

Quality of food

A balanced diet should contain:
- **proteins** such as meat, eggs and milk to build and renew body tissues
- **carbohydrates** that include cereals, potatoes, sugar and fats to provide energy
- **vitamins and minerals** which are found in dairy products, fruit, vegetables, fish, meat and eggs and which help prevent many diseases.

In many LEDCs the basis for many people's diet is starchy carbohydrates, e.g. rice, wheat and maize, and a lack of protein, vitamins and minerals.

Diet illnesses

Although malnutrition only leads to starvation and death in extreme conditions, it does reduce people's capacity to work and their resistance to disease. In children it can retard mental and physical development and can cause illness.

In the richer MEDCs, many health problems result from:
- either eating too much (quantity) = obesity and overweight
- or unhealthy diets (quality) = heart disease.

In the poorer LEDCs, health problems can be caused by:
- a lack of calories and therefore energy (quantity) = marasmus (diagram **B**)
- a predominance of cereals and a deficiency of protein (quality) = kwashiokor (diagram **B**)
- a lack of vitamins or minerals (quality) = beri-beri (a wasting of the limbs due to insufficient vitamin B) and rickets (deformities in bones, legs and the spine due to a lack of vitamin D).

B

Marasmus

Symptoms
- The child is very thin and has no fat but only skin covering the bones.
- The child does not grow.
- the child has a runny tummy (diarrhoea)

The causes
- Mainly lack of calories.
- Child does not get enough food.

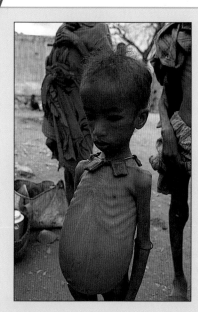

Kwashiorkor

Symptoms
- The body of the child looks swollen.
- The skin looks dark and is peeling off, especially on limbs.
- The child looks unhappy.

The causes
- Lack of protein and too many cereals.
- Shortage of 'building' foods like milk, beans, ground peanuts, fish, eggs or meat.
- Too much reliance on ordinary maize-meal porridge.

Food supply in the LEDCs

Apart from sub-Saharan Africa, there has been an increase in available food supplies in every developing region since the 1970s (table **C**). However:
- Some parts of these regions have unfavourable climatic and soil conditions that limit the growth of crops. Such places are also likely to lack the money to buy, or the transport to move, food from better-off regions.
- Other places have been encouraged by MEDCs and transnational companies to grow cash crops. While this is meant to help a country's trade balance, it actually means that fewer crops are grown for home consumption.

C Percentage of people suffering from malnutrition

Region	1970	2000
Sub-Saharan Africa	35	42
Near East and North Africa	23	12
Central America and Caribbean	24	11
South America	17	9
South Asia	34	22
East Asia	35	16
China	46	14
All developing regions	36	19

Activities

1 **a)** Why is:
   - the quantity of food
   - the quality of food

   important in a good diet?

   b) What is the average food intake, in calories, of:
   - the MEDCs
   - the LEDCs?

   c) Why do you think food supplies in many LEDCs have increased but have declined in sub-Saharan Africa (see also pages 204–205)?

2 **a)** What is *malnutrition*?

   b) What health problems are likely to be caused by a poor diet in:
   - an MEDC
   - an LEDC?

   Give reasons for your answer.

Summary
There are considerable differences in the quantity and quality of food supplies between MEDCs and LEDCs. A poor diet can cause numerous illnesses.

Are there regional differences in the UK?

So far we have seen that there are considerable differences in development *between* countries and especially between the MEDCs and the LEDCs. However, there are also considerable differences in development *within* countries. For example, in an MEDC such as the UK, it is perceived that:

- There are regional differences, with regions towards the south and east being more economically developed than those to the north and west (map **A** and table **B**). These differences, based on GNP, population and social characteristics (pages 266–269) are summarised in map **C**.
- London, the capital city, is more developed than other urban areas.
- Urban areas are more developed than rural areas.
- Within urban areas, the outer suburbs are more developed than inner city areas.

It has long been recognised that there are economic, social and political differences both between and within the various regions. Indeed, successive British governments have, since the 1930s, tried (without too much success) to even out these differences. As a member of the EU, Britain is not alone in having to face this problem. The situation is even worse in Italy where there is the 'rich, industrialised North' and the 'poor, less developed South'.

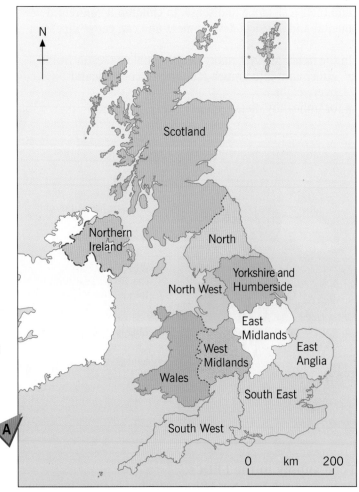

A

B

UK Standard Economic Regions	GNP (EU average = 100)	Average annual income (£)	Unemployment (%)	Infant mortality (per 1000)	Average house price (£)	One car per family (%)	5 or more GCSEs (%)
South East	116	23 640	2.8	5.1	96 000	71	48
East Anglia	101	19 970	2.9	5.2	69 000	76	48
South West	94	20 700	3.3	5.2	74 000	75	51
East Midlands	93	19 520	4.7	6.3	58 300	69	43
West Midlands	91	17 230	5.3	7.4	61 000	69	42
Yorks & Humberside	91	18 350	5.1	7.3	53 100	67	40
North West	90	18 230	5.4	5.8	55 800	59	44
North	89	16 300	8.7	6.4	46 400	64	40
Wales	84	15 200	6.3	5.9	53 000	67	41
Scotland	97	19 370	6.0	6.0	53 200	62	50
Northern Ireland	79	17 470	6.5	6.1	43 100	65	51
UK average	**99**	**19 700**	**6.9**	**5.9**	**65 700**	**68**	**46**

The EU, recognising the problem, has developed a regional policy that is aimed at reducing disparities between regions. The EU hope to achieve this by means of various funds that include the European Regional Fund and the European Social Fund. Those parts of the UK that qualify for financial help under the EU policy are shown on map **D**. This map again highlights the fundamental differences between the south-east of England and the remainder of the UK.

Finally you should remember that a definition of development no longer applies just to wealth and standards of living. It has now been widened to include people's quality of life. However, quality of life is much harder to measure than other characteristics of development. It also differs according to the perceptions and aspirations of individual people. For example:

- Many urban dwellers believe that urban areas provide a higher quality of life, with their jobs, shops, schools, hospitals and various forms of entertainment.
- Most rural dwellers believe that small towns, villages and the countryside provide a less congested, quieter and cleaner environment in which to live.

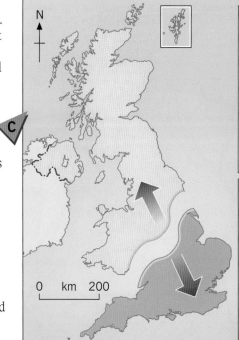

C

North and West
often has:
Higher: • unemployment
• infant mortality
• employment in manufacturing
Lower: • GNP
• weekly earnings
• house prices
• car ownership
• employment in services

South and East
often has:
Lower: • unemployment
• infant mortality
• employment in manufacturing
Higher: • GNP
• weekly earnings
• house prices
• car ownership
• employment in services

0   km   200

EU regional policy **D**

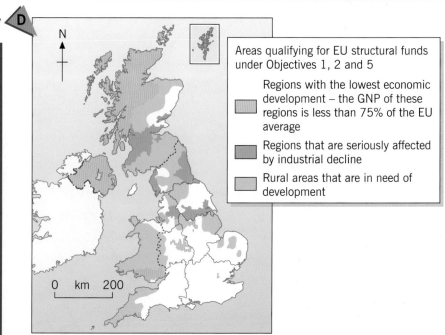

N

Areas qualifying for EU structural funds under Objectives 1, 2 and 5

Regions with the lowest economic development – the GNP of these regions is less than 75% of the EU average

Regions that are seriously affected by industrial decline

Rural areas that are in need of development

0   km   200

Activities

1 Using table **B** and map **C**:
   a) i) Name three regions in the south and east of England.
   ii) Name three regions in the north of England.
   b) i) List six differences between the two groups of regions.
   ii) For each of the six, suggest a reason for the difference.
   c) How is the EU trying to even out differences between the two groups of regions?

2 Why do you think that in the UK:
   a) urban areas may be considered to be more developed than rural areas
   b) outer suburbs of cities may be considered to be more developed than inner city areas?

3 With which of the perceptions of 'quality of life' given in diagram **E** do you agree? Give reasons for your answer.

Summary   Differences in development occur within countries as well as between them.

E

There are better jobs and more things to do in cities.

We prefer to live in a quieter, less polluted place.

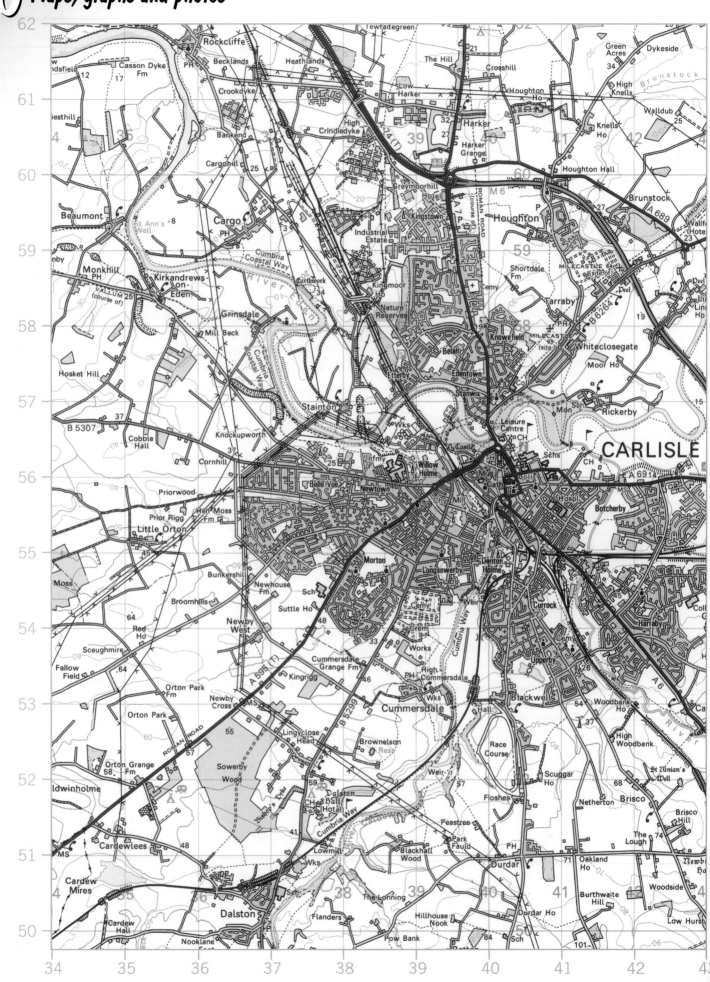

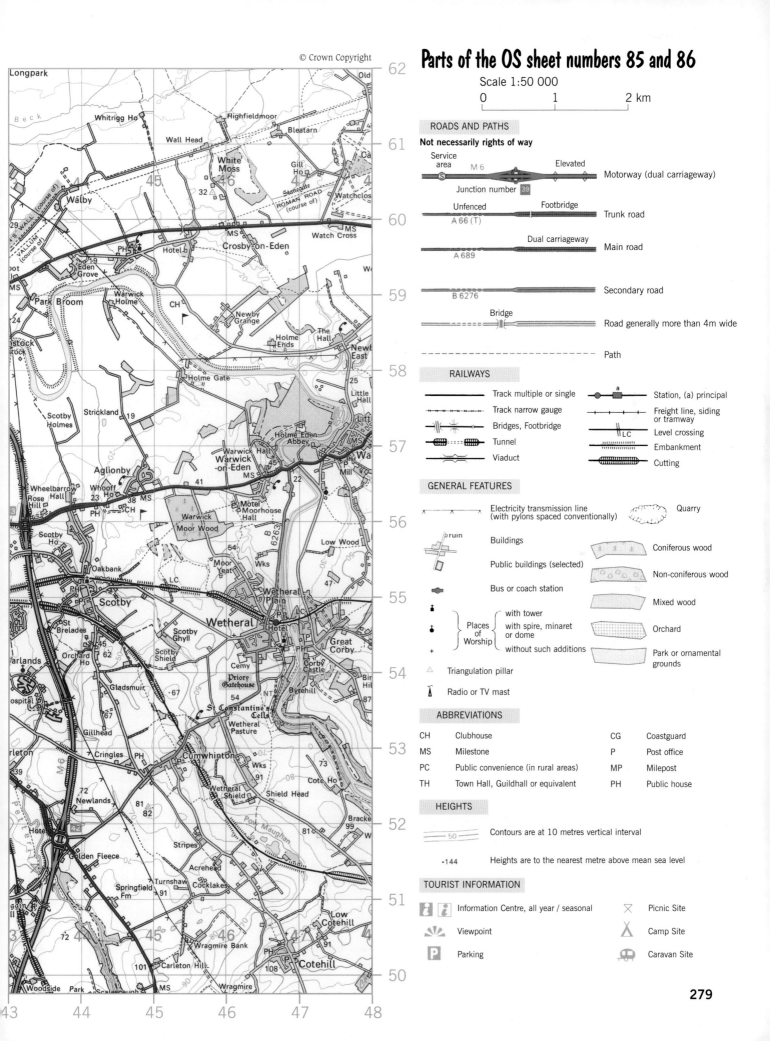

© Crown Copyright

Parts of the OS sheet numbers 85 and 86

Scale 1:50 000

0   1   2 km

ROADS AND PATHS

Not necessarily rights of way

Service area M 6 Elevated Junction number 39	Motorway (dual carriageway)
Unfenced Footbridge A 66 (T)	Trunk road
Dual carriageway A 689	Main road
B 6276	Secondary road
Bridge	Road generally more than 4m wide
– – – – –	Path

RAILWAYS

▬▬▬	Track multiple or single	●▬▬▬ a	Station, (a) principal
┼┼┼┼	Track narrow gauge	┼┼┼┼	Freight line, siding or tramway
▦▦	Bridges, Footbridge	LC	Level crossing
▦▦	Tunnel	▥▥▥▥	Embankment
▦▦	Viaduct	▨▨▨	Cutting

GENERAL FEATURES

✕ ✕ ✕	Electricity transmission line (with pylons spaced conventionally)
ruin	Buildings
	Public buildings (selected)
	Bus or coach station
♟ { Places of Worship { with tower / with spire, minaret or dome / + without such additions
△	Triangulation pillar
♟	Radio or TV mast

	Quarry
	Coniferous wood
	Non-coniferous wood
	Mixed wood
	Orchard
	Park or ornamental grounds

ABBREVIATIONS

CH	Clubhouse		CG	Coastguard
MS	Milestone		P	Post office
PC	Public convenience (in rural areas)		MP	Milepost
TH	Town Hall, Guildhall or equivalent		PH	Public house

HEIGHTS

— 50 —	Contours are at 10 metres vertical interval
•144	Heights are to the nearest metre above mean sea level

TOURIST INFORMATION

🅸 🅸	Information Centre, all year / seasonal	✕	Picnic Site
☀	Viewpoint	⛺	Camp Site
🅿	Parking		Caravan Site

279

Ordnance Survey map skills

Symbols

Ordnance Survey (OS) maps use symbols rather than words to describe features. This helps make the map less crowded and easier to read. Symbols can be shown as drawings, coloured areas, letters or lines. Each map is provided with a key which lists what each symbol means. In a GCSE examination you will be given a key but it is helpful if you can recognise at least some of the symbols without looking them up.

Look at the key to the 1:50 000 map on page 279.

1 Draw and label the following:
   a) five symbols that look like pictures
   b) three symbols that are lines
   c) three symbols that are coloured areas.

2 Copy and label the following symbols:

Grid references

To help locate areas and features on a map, a grid of squares may be drawn. On Ordnance Survey maps, these lines are drawn in blue and each has its own special number or reference. The blue lines form **grid squares**. **Grid references** are the numbers that give the position of the grid square or feature inside it.

- **Four-figure grid references** give the reference for a grid square which equals an area on the ground of one square kilometre.
- **Six-figure grid references** are more accurate and locate a 100 metre square area within a grid square.

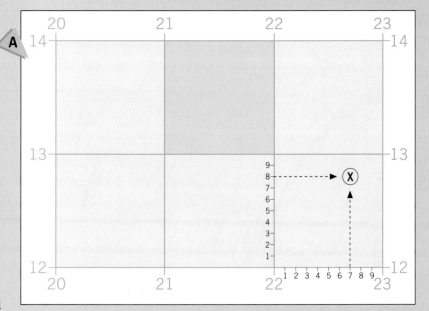

Four-figure reference
1 Give the number of the line forming the left-hand side of the square.
2 Give the number of the line forming the bottom of the square.
3 Put the numbers together.

The four-figure reference for the dark green square is **2113**.

Six-figure reference
1 Give the number of the line forming the left-hand side of the square.
2 Estimate the number of tenths **across** the grid square.
3 Give the number of the line forming the bottom of the square.
4 Estimate the number of tenths **up** the grid square.
5 Put the numbers together.

The six-figure reference for point ⊗ is **227128**.

Look at the OS map on pages 278 and 279.
a) Name the main features in grid squares 3651, 4256 and 4657.
b) In which grid square is the village of Dalston?
c) Draw the symbol and name the feature at 366506, 377545, 400599, 399619 and 381521.

Direction

Direction may be given on a map using the points of a compass. There are four main points. These are north, south, east and west. In a GCSE examination you will need to know eight points of the compass although it would be better to know the 16 shown on drawing **B**. Direction is given with the compass point away from where you are. For example, moving in a west direction means that you go towards the west.

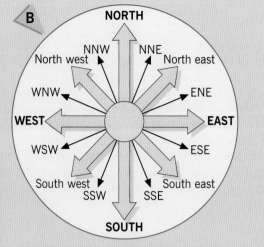

Which is west and which is east? Remember they spell **WE** across from left to right.

> 1 Look at the OS map on pages 278 and 279. Give the directions from Wetheral (4654) to:
> **a)** Warwick-on-Eden (4656) **b)** Houghton (4059)
> **c)** motorway intersection (4351) **d)** Cotehill (4650).

Scale and distance

A map may be used to find out how far one place is from another. Maps have to be drawn smaller than real life to fit on a piece of paper. How much smaller they are is shown by the **scale**. In a GCSE examination, the Ordnance Survey map may have a scale of either 1:50 000 or 1:25 000.

1:50 000 This means that 1 centimetre on the map is equal to 50 000 centimetres on the ground. (2 cm on the map represents 1 km on the ground.)

1:25 000 This means that 1 centimetre on the map is equal to 25 000 centimetres on the ground. (4 cm on the map represents 1 km on the ground.)

An Ordnance Survey map will always have a scale drawn on it. Drawing **C** shows the scale for a 1:50 000 map.

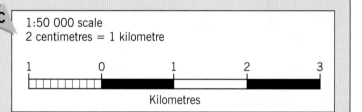

C
1:50 000 scale
2 centimetres = 1 kilometre

Kilometres

Measuring distance

Both straight and winding distances can be measured accurately on a map by using the edge of a piece of paper, as shown in drawing **D**.

Another way is to use a ruler to measure the distance then change the cm to km using the **scale**. For a quick and rough estimate of distance you can use the grid lines, which are 1 km apart on both the 1:50 000 and 1:25 000 maps.

> Look at the OS map on pages 278–279 and measure the distances by road from intersection 43 (432560) to:
> **a)** Aglionby (448564)
> **b)** Wetheral station (467547)
> **c)** Cotehill (469502).

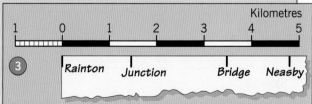

1 Break the distance to be measured into sections. Mark the start and end of the first section.

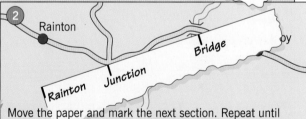

D
2 Move the paper and mark the next section. Repeat until the entire length has been measured.

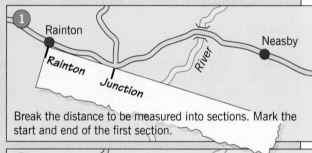

3 Place the paper on the **scale line**, putting the left-hand mark on the 0. Read the distance off the scale next to the right-hand mark. It is 4.8 km.

Cross-sections

A cross-section is a diagram that gives a cut-away or side view of the landscape. It shows the landscape as it would appear if it was sliced open, rather like cutting a large piece of cake in half.

Cross-sections are useful to geographers because they show the shape and height of the land. They can also be used to show the location of other important physical and human features such as vegetation cover, land use and settlement. The diagrams below show how to draw an accurate cross-section.

1 Look at map **B** and draw a cross-section from Carrock Fell (342337) to Thanet Well (398349).
  • Use the same horizontal scale as on the map.
  • Use a vertical scale of 1 cm = 100 m.
  • Mark and name the features along the cross-section.

2 Look at map **D** on page 47 and draw a cross-section from 622532 to 660510. Use the same scale as in Activity **1** above, and mark and name the features.

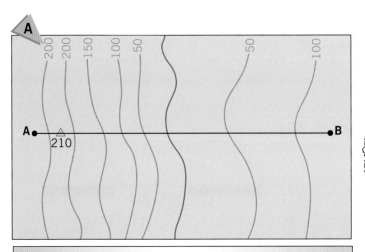

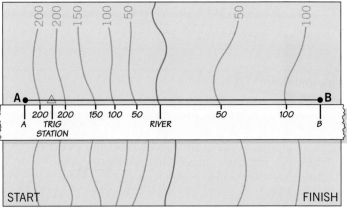

Stage 1
Locate the section line A–B on the map. Look carefully at the contours to find the highest and lowest points. These are needed for when you choose your vertical scale. Notice where the land is rising or falling and where any features such as rivers, hilltops or roads are located.

Stage 2
Place the straight edge of a piece of paper between the two end-points of the section. Carefully mark the point where each contour crosses the paper. Label the contour heights on the paper and mark any other important features. On steeper slopes you may only be able to mark alternate contours.

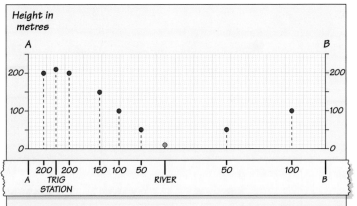

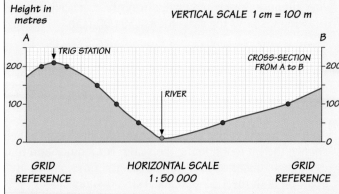

Stage 3
Place the straight edge of the paper against a horizontal line on graph paper. Use a vertical scale of 1 cm = 100 metres for a 1:50 000 map, or 1 cm = 50 metres for a 1:25 000 map. Mark the graph paper with small dots to show the height of the contours and positions of other features.

Stage 4
Join the dots together with a smooth, freehand curve. Do not use a ruler. Notice the smooth curve for hilltops and valley floors. Use arrows to help label any features. Add six-figure grid references for the two end points. Remember to add a title and the horizontal and vertical scales.

B Part of OS map sheet 90 (Penrith and Keswick), scale 1:50 000

© Crown Copyright

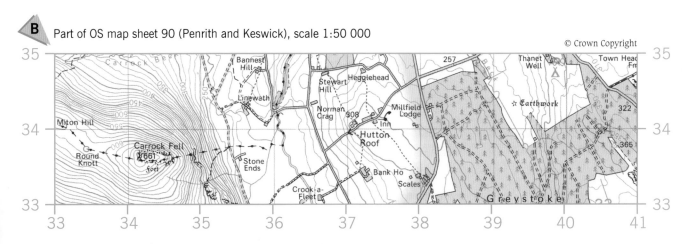

Sketch cross-sections

Cross-sections need not always be drawn as accurately as the one in diagram **A**. **Sketch cross-sections** are drawn freehand and do not use measured heights and distances. Their shape represents the shape of the landscape but relief features are not exact and they only give a general impression of the area. Sketch cross-sections are very useful, however. They are a visual way of presenting information and can help to describe and explain the location of important geographical features.

An example of a sketch cross-section is shown below in diagram **D**. Notice that it is not too crowded and most of the text is positioned away from the section itself. This helps make the diagram clearer and easier to interpret. Further examples may be found on pages 56 and 62.

Find a physical map of the UK in your atlas. Draw a sketch cross-section across the country from Wales to East Anglia. Add colour to the section and name and label the main features as described below (**C**).

C

① Draw a rough outline sketch of the section.

② Add appropriate colours and symbols to show vegetation.

③ Name the main features. Always print clearly.

④ Annotate the section with short, carefully worded labels.

⑤ Give the section a title.

D

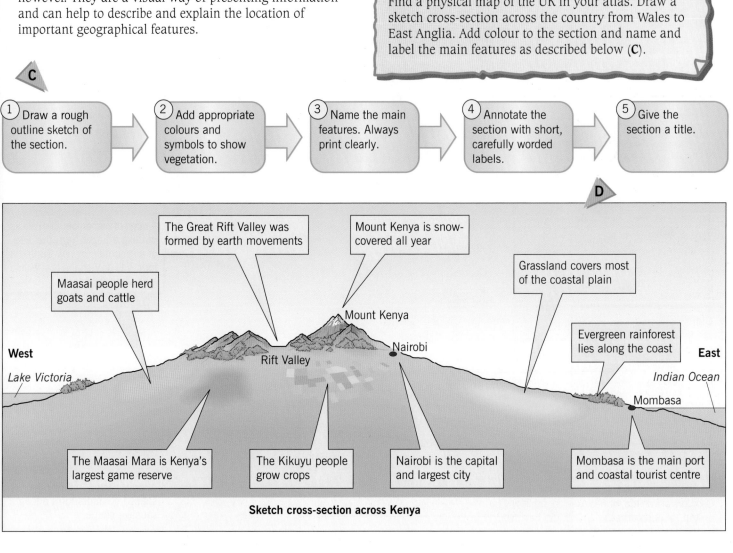

Sketch cross-section across Kenya

The Great Rift Valley was formed by earth movements

Mount Kenya is snow-covered all year

Grassland covers most of the coastal plain

Maasai people herd goats and cattle

Evergreen rainforest lies along the coast

West

Lake Victoria

Mount Kenya

Nairobi

Rift Valley

East

Indian Ocean

Mombasa

The Maasai Mara is Kenya's largest game reserve

The Kikuyu people grow crops

Nairobi is the capital and largest city

Mombasa is the main port and coastal tourist centre

Contours and relief

Lines on a map that join places of the same height are called **contours**. They show both the height of the land and its shape. The shape of the land is called **relief**. The difference in height between one contour and the next is called the **contour interval**. On 1:50 000 Ordnance Survey maps the contour interval is 10 metres and on 1:25 000 it is 5 metres.

On a contour map, equally spaced contours show an even slope, close contours a steep slope, widely spaced contours a gentle slope, and an absence of contours flat land. Contours usually form recognisable patterns. From these patterns we can recognise the shape of the landscape and identify features such as hills, mountains and valleys. Some of these shapes are shown below.

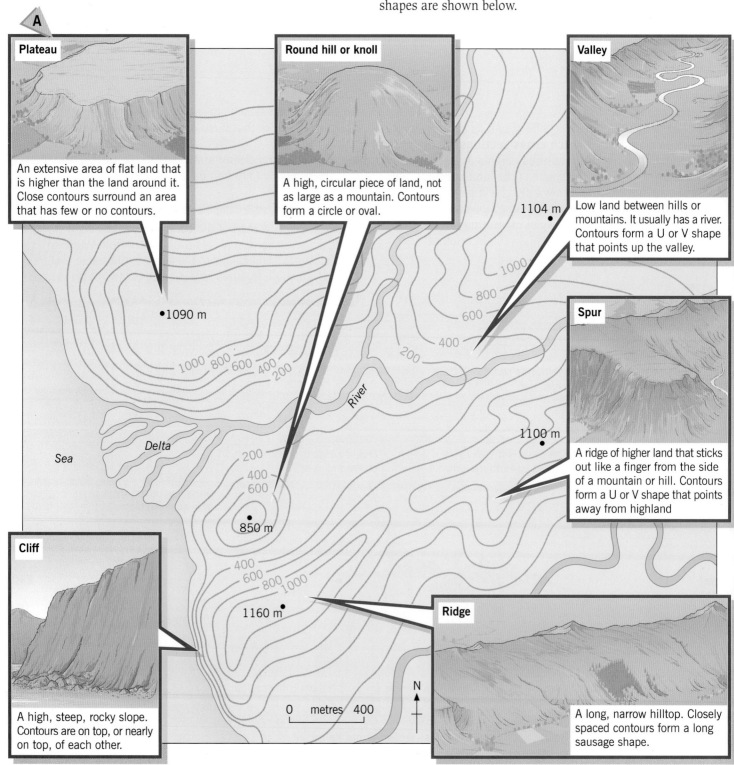

A

Plateau

An extensive area of flat land that is higher than the land around it. Close contours surround an area that has few or no contours.

Round hill or knoll

A high, circular piece of land, not as large as a mountain. Contours form a circle or oval.

Valley

Low land between hills or mountains. It usually has a river. Contours form a U or V shape that points up the valley.

Spur

A ridge of higher land that sticks out like a finger from the side of a mountain or hill. Contours form a U or V shape that points away from highland

Cliff

A high, steep, rocky slope. Contours are on top, or nearly on top, of each other.

Ridge

A long, narrow hilltop. Closely spaced contours form a long sausage shape.

•1090 m

1104 m

1100 m

850 m

1160 m

River

Delta

Sea

1000 800 600 400 200

1000 800 600 400 200

200 400 600

400 600 800 1000

0   metres   400

N

Describing physical features on a map

Knowing how to describe places using maps is important in geography. Maps, however, show a large amount of information and it can be difficult to identify and describe the main features of an area. Using a simple checklist or set of key questions can make the task easier.

In physical geography we need to describe the relief, drainage and vegetation, as shown in drawing **C**. Look at the key questions below and see how you could apply them to map **A** opposite.

B

Relief is the height and shape of the land.

Drainage includes surface water features such as rivers, lakes and marshes.

Vegetation describes the plant life and includes woodland, rough pasture and farmed areas.

C

4 Are there any rivers? If there are, what size and shape are their channels?

5 Are there any other surface water features such as lakes and marshes? If there are, how many, and what are they like?

6 If there is a coast, what is its direction and is it smooth or uneven?

3 Are there any valleys? If there are, how many, and what shape are they?

7 Is the coastline flat or steeply sloping? Are there any beaches or cliffs?

2 What height is the land? What are the highest and lowest points?

8 Does the coast have bays or headlands? Is it broken by river mouths or estuaries?

1 What is the overall relief of the area? Is it mainly flat, hilly or mountainous?

9 Are there any other coastal features to identify?

D Part of OS map sheet 90 (Penrith and Keswick), scale 1:50 000

© Crown Copyright

1 From map **D**, give grid references for the following features: steep slope, gentle slope, narrow ridge, cliff, valley, round hill, spur.

2 Read the following description of the area shown on map **A**, then copy and complete the paragraph.

The area lies on the coast and is rugged and hilly. The land is generally ...... sloping and is a mixture of plateaux, rounded ...... and narrow .... The highest peaks are over ...... metres in height. A river drains the area from ...... to .... As the river nears the sea, the valley widens and a ...... has developed.

3 Describe the physical features shown on map **D**.

Using photographs in geography

Photos are very useful in geography, as they show what places or features are actually like. They can provide information about an area and help us understand the processes that made it like that. Three main types of photos are used in GCSE Geography.

1 **Ground photos:** show the landscape as we see it from the Earth's surface.
2 **Aerial photos** are taken from aircraft and can be vertical, or oblique. **Vertical photos** are from directly overhead, whilst **oblique photos** are taken at an angle.
3 **Satellite photos** are taken from hundreds of kilometres above the Earth's surface and can cover a large area.

Photos in geography are used for a specific reason and need to be carefully studied and analysed. This means giving more than a quick glance at the photo, as you would if you were flicking through a magazine. It means really getting to know it and becoming something of a detective. Drawing **A** gives some advice on how to use photos in geography.

Photo-response questions are common in geography examinations. In these the answer *must* be based on what can be *observed* in the photo. Proper observation is a very specialised skill which needs lots of practice to develop.

A

Examine the evidence:
• Look at the title for information about what the photo shows.
• Look at the entire photo, not just part of it.
• Look for the main feature or features.
• Look for links and relationships.

Ask the questions:
• What does the photo show?
• What features are relevant to your study?
• What are those features like?
• Why are those features like that?

① Search the photo and look for evidence.

② Select evidence to solve your problem.

③ Make deductions based on firm evidence.

B Cliff collapse near Scarborough, Yorkshire

1 Describe the feature shown in photo **B**. Using photo evidence, suggest what has happened.

2 Draw a sketch of photo **B** and add labels to show the main features.

3 **a)** Draw a sketch of photo **A** on page 230 and add labels to show the main features.
   b) Describe the resort and, using photo evidence, suggest reasons for its popularity with tourists.

4 Find a photo of a topic that you are studying.
   a) Draw a labelled sketch of the photo.
   b) Describe the main features in the photo and suggest why they are like that.

Some examples of labelled photos and sketches may be found on pages 114, 145, 172 and 271.

Sketching photos

Whilst photos show what places are actually like, they often give too much information and can be difficult to interpret. A labelled photo sketch can overcome this problem by simplifying the picture and highlighting only the important features. It is not necessary to be a good artist to draw a sketch.

The following are some guidelines for developing this skill. You could either draw the sketch freehand or use tracing paper. A grid can help you position key features.

1

First look carefully at the photo. Are there any obvious landmarks or features such as large buildings or hills?

Next draw a broad outline to provide a framework for your sketch. Use a pencil so that it may be changed easily.

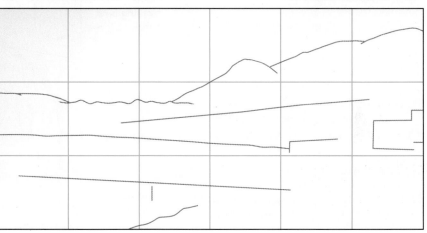

2

Once the skyline and landmarks have been drawn, add more detail to your sketch by drawing in key features such as vegetation and small buildings.

Don't try to draw in everything. Keep the sketch clear and simple.

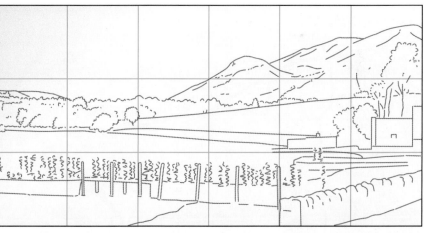

3

Draw labelled arrows to identify the main features that are important or difficult to draw. Further labels can be used for description and explanation.

Finally add a title, and colour your sketch to make it more attractive and easier to understand.

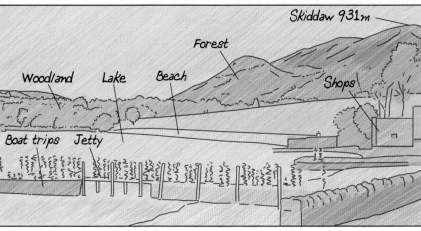

Skiddaw 931m
Forest
Woodland   Lake   Beach   Shops
Boat trips   Jetty

Drawing sketch maps and ...

Maps show a large amount of information and can be complicated and difficult to interpret. A **sketch map** simplifies what is shown on a map and only shows the features that are of particular interest. Sketch maps do not need to be perfect drawings that are accurate and correct in every detail. They must, however, be clear and tidy and their features drawn in roughly the correct positions. Labels are important on sketch maps. They should be short, carefully worded and suitably positioned. Adding longer labels which identify features and explain why they are significant is called **annotating** a sketch map.

Drawing **B** is a sketch map of Berwick-upon-Tweed based on map **C** below. A sketch map of Carlisle may be found on page 151.

A

Guidelines for drawing a sketch map

1 Draw a box the same shape as the map area that you are using.

2 Draw in relief features such as hills, rivers and coastline.

3 Draw in the human features such as roads, towns and villages.

4 Label the main features that the sketch map needs to show.

5 Add a title and north sign.

   Use colour and remember: the most important thing is to keep your sketch map simple.

B Berwick-upon-Tweed: site and situation

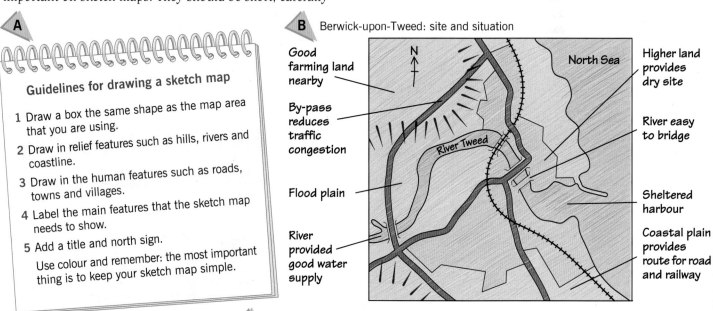

Good farming land nearby

By-pass reduces traffic congestion

Flood plain

River provided good water supply

N

North Sea

River Tweed

Higher land provides dry site

River easy to bridge

Sheltered harbour

Coastal plain provides route for road and railway

C

... using maps and photos together

When used together, maps and photos can give us a very good idea of what an area is like. The map helps us see the overall pattern of the area, whilst the photo provides added detail and gives us a visual picture of the area. **Oblique aerial photos** are particularly helpful when used with maps. They are taken from an aircraft with the camera pointing at an angle to the ground. Photos like this are now available for all parts of the UK.

To link a map with a photo you must first determine the direction the camera was facing and from where the picture was taken. This can best be done by locating an obvious feature, like a bridge, road or river, on both the photo and the map. Then locate another feature and move, or **orientate**, the photo until it is lined up with the same features on the map. The drawings in **D** show how to orientate the map and photo of Berwick.

1 Draw a sketch map of Berwick-upon-Tweed to show communications and tourist facilities.

2 Look at the map and photo of Berwick.
   a) In which direction was the camera facing?
   b) From where was the photo taken? Give the approximate map reference.
   c) Match the letters A, B, C, D, E and F on the photograph with the following features:
   • *River Tweed* • *Royal Border Bridge* • *A698*
   • *golf course* • *railway line* • *ramparts*.

3 Look at photo **C** on page 157 and the map on pages 278–279, which both show Carlisle. Orientate the photo using two obvious features.
   a) In which direction was the camera facing?
   b) From where was the photo taken? Give the approximate grid reference.

D

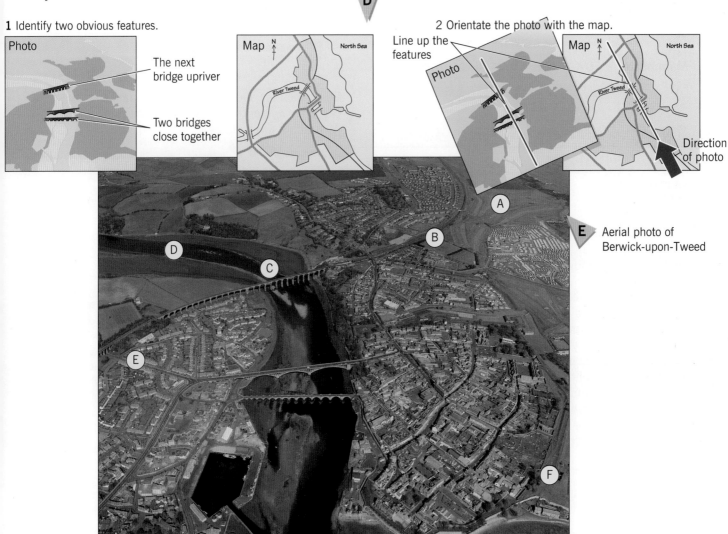

1 Identify two obvious features.

Photo

The next bridge upriver

Two bridges close together

Map N North Sea

River Tweed

2 Orientate the photo with the map.

Line up the features

Photo

Map N North Sea

River Tweed

Direction of photo

E Aerial photo of Berwick-upon-Tweed

What different types of map are there?

A **map** is a drawing of a place seen from above. It is very simplified and drawn at a reduced scale. Maps are useful to geographers because they show where places are and what they are like. They also show the spread or distribution of features and can be used to compare places. Extracting and using information from maps is one of the most important skills in GCSE Geography. It is always tested in examinations.

Atlas maps

An **atlas** is a book containing many different maps. Some maps show very large areas such as the entire world or whole continents. Others show smaller areas such as a country or part of a country. Atlases also contain a number of different types of map including **political maps** which show countries and cities, and **physical maps** which show mountains, lowlands and rivers. Other maps may include information on population, land use, climate and a variety of other topics.

The **contents** page at the front of the atlas shows on which page each map can be found. The **index** at the back of the book shows exactly where a particular place is located. Drawing **A** shows how to use the index.

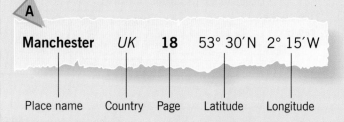

A

Manchester	*UK*	**18**	53° 30′N	2° 15′W
Place name	Country	Page	Latitude	Longitude

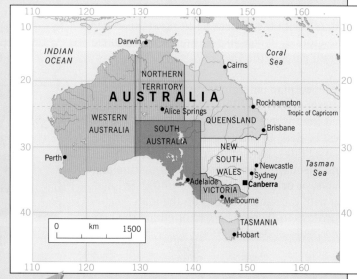

B Political map of Australia

Using the map above and an atlas, name and give the index reference for one town in each of Australia's seven states.

Dot maps

Dot maps are used to show distribution, or the way in which something is spread over an area. In geography, they are most often used to show population distribution.

On map **C**, each dot represents 1 million people. The map gives a good general impression of world population distribution. It clearly shows that some places are densely populated whilst others have very few people. The map is not so good, however, for obtaining accurate figures on population. In China, for example, the dots are just too close together to count. Other single dots, as in Canada for example, represent a population that is spread over a huge area.

C World population distribution

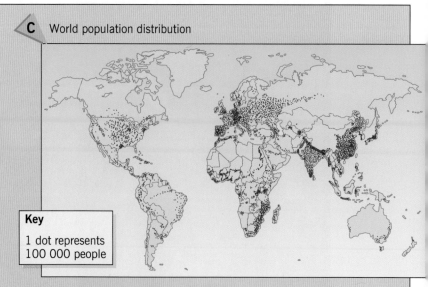

Key

1 dot represents
100 000 people

1 Look at map **C** and name three areas that are densely populated and three areas that have very few people.

2 Why is it difficult to obtain accurate figures from a dot map?

Choropleth maps

Choropleth maps use different colours or shading to show variations between places. Examples include population density, development (page 269) and rainfall. Map **D** shows differences in rainfall totals in Britain. Notice how clearly the pattern of rainfall is shown and how easy it is to compare rainfall amounts in different areas. The limitations of choropleth maps are that they are based on areas and do not provide exact figures for particular locations.

Hints for drawing choropleth maps

- Data should be divided into four, five or six groups.
- Shading should be in just one or two colours.
- Areas with the highest values should be darkest and those with the lowest should be lightest.

1. Give the rainfall amounts for each of the towns named on the map.
2. Describe the distribution of rainfall across Britain.

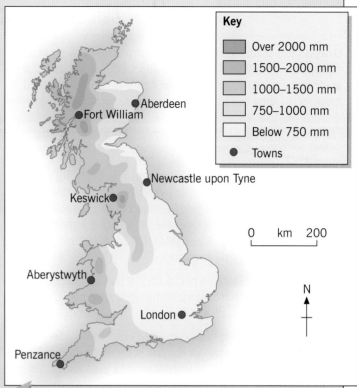

Key

	Over 2000 mm
	1500–2000 mm
	1000–1500 mm
	750–1000 mm
	Below 750 mm
●	Towns

D Annual rainfall for Britain

Isopleth maps

Isopleths, or **isolines** as they are sometimes called, join places of the same value. Contours on an OS map are examples of isopleths – they join places of equal height and are shown on pages 284 and 285. Other examples include isobars which join places of equal pressure (page 80), and isotherms which join areas of equal temperature (page 78).

Isopleth mapping is a good method to use if there is data available for a large number of points. The data can be plotted on a map as numbered points, and isopleths drawn to join up those of equal value. Map **E** is a partially completed map of the San Francisco area to show how the isopleths are drawn. The isopleths show the effects of the 1989 earthquake as a percentage of the earthquake's original force. The map can be used to help prepare disaster plans for future earthquakes (page 66).

Look carefully at map **E**.
a) Give the percentage of the earthquake's original force measured at each of the named towns.
b) How far from the epicentre is San Francisco?
c) Describe the pattern shown by the map.

E 1989 San Francisco earthquake

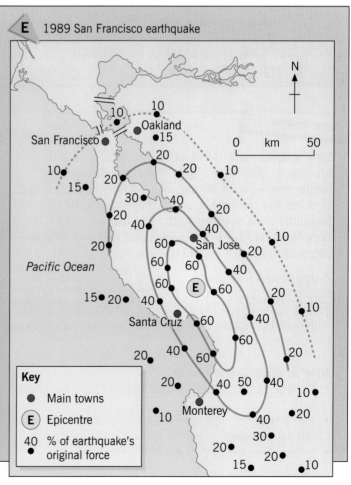

Key

- ● Main towns
- Ⓔ Epicentre
- 40 % of earthquake's
- ● original force

How can we use graphs in geography?

Graphs are diagrams that show information in a clear and simple way. They show patterns and trends and can be used to describe a situation and show how one thing is related to another. Graphs are important in geography and will always be included in an examination. You should know the names for different types of graph and be able to draw and interpret each one.

Bar graphs

Bar graphs present information in the form of a bar or column. The bars can be drawn either horizontally or vertically. Each bar shows the number or value of something. Graph **A** shows the results of a survey which asked how pupils got to school. Notice how easy the graph is to read and interpret.

Bar graphs, or **bar charts** as they are sometimes called, are used to compare different things or quantities. Examples may be found on pages 14, 127 and 226.

A divided bar chart can be used to show how a total is broken up. The length of the bar represents the total value which can be a real number or a percentage. Graph **B** compares employment structures in two countries.

How to draw a bar chart
- Check the size of the values. Note the largest and smallest.
- Draw a frame with two axes.
- Label each axis.
- Decide on an appropriate scale for each axis.
- Draw bars of equal width.

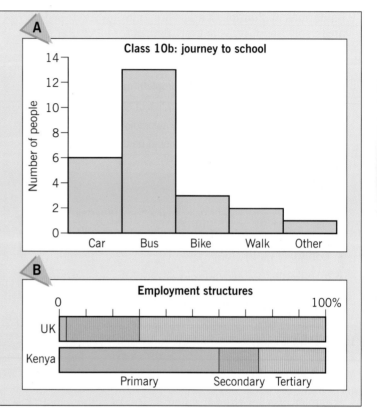

A — Class 10b: journey to school

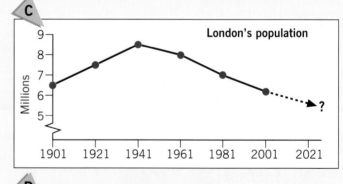

B — Employment structures

Line graphs

Line graphs show information as a series of points that are joined up to form a line. They show changes and trends over a period of time and can help forecast future changes. For example, graph **C** shows how London's population has changed since 1901 and what may happen in the future. Other examples may be found on pages 23, 130, 131 and 231.

A **compound line graph** has more than one line drawn on the same graph outline and can show how a total is broken up. Graph **D** shows how the percentage of people employed in primary, secondary and tertiary industries has changed over time.

How to draw a line graph
- Draw a frame, label each axis and add an appropriate scale.
- Plot a series of clear dots or crosses at the points where the values meet.
- Join up the dots or crosses with a smooth line.

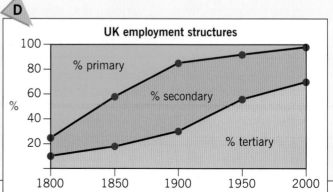

C — London's population

D — UK employment structures

Pie graphs

A **pie graph** is drawn as a circle which is then divided into several pieces or sectors. The whole circle is always equal to 100 per cent.

Pie graphs, or **pie charts** as they are also called, show proportions and help us to see how something is divided up. They are used whenever a total can be divided into separate parts. For example, employment structures (page 188), exports and imports (page 262) and energy sources as shown in graph **E**.

How to draw a pie graph
- Convert the values into percentages.
- Multiply each percentage figure by 3.6 to give degrees.
- Using a protractor, plot the largest segment first.
- Plot the rest of the segments from largest to smallest. Always leave 'Others' to last.
- Colour or shade the segments and add a key or labels.

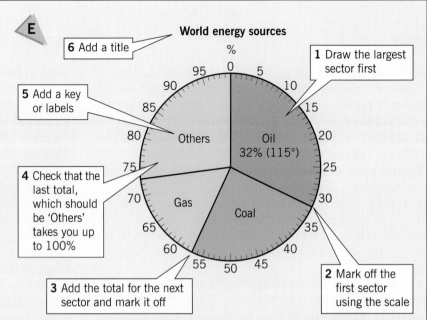

E

6 Add a title

World energy sources

5 Add a key or labels

1 Draw the largest sector first

4 Check that the last total, which should be 'Others' takes you up to 100%

Oil 32% (115°)

Others

Gas

Coal

3 Add the total for the next sector and mark it off

2 Mark off the first sector using the scale

Another method is shown in graph **E**. Here the percentage scale is already drawn on the graph outline. This makes the task much easier.

Scatter graphs

A **scatter graph** has data plotted as a number of dots or crosses. They are used to see if information about two different things is linked or related, for example between temperature and height of land, or average income and literacy.

To show if there is a relationship between the two variables a **best fit line** should be drawn. This is a line that comes as close to as many points as possible.

The three main types of relationship or correlation are shown in graphs **F**, **G** and **H**.

How to draw a scatter graph
- Draw a frame, label each axis and add an appropriate scale.
- Plot clear dots and crosses at the points where the values meet.
- Do not join up the dots or crosses.

F

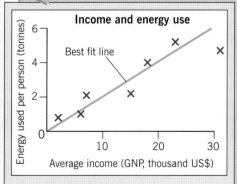

Income and energy use

Best fit line

Energy used per person (tonnes)

Average income (GNP, thousand US$)

G

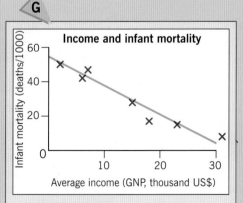

Income and infant mortality

Infant mortality (deaths/1000)

Average income (GNP, thousand US$)

H

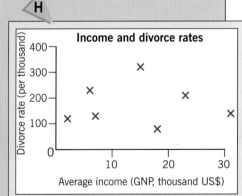

Income and divorce rates

Divorce rate (per thousand)

Average income (GNP, thousand US$)

Positive correlation
- Slopes up from left to right.
- As one value increases, so does the other.
- As average income increases, so does use of energy.

Negative correlation
- Slopes down from left to right.
- As one value increases, the other decreases.
- As average income increases, infant mortality decreases.

No correlation
- Points are scattered and no best fit line can be drawn.
- No relationship can be seen.
- There is no link between average income and divorce rates.

Glossary

A

Abrasion Erosion caused by the rubbing and scouring action of rock fragments carried by rivers, glaciers, waves and the wind.

Accessibility How easy a place is to get to.

Acid rain Rainwater containing chemicals that result from the burning of fossil fuels.

Ageing population The increasingly higher proportion of the population living to a greater age.

Aid Help usually given by the richer countries of the world, or by international charities, to the poorer countries. It may be short-term aid such as food given for an emergency, or long-term aid such as training in health care.

Alluvium Fine soil left behind after a river floods; also called 'silt'.

Altitude The height of a place above sea-level.

Anticyclone An area of high pressure usually associated with fine, settled weather.

Appropriate technology Development schemes that meet the needs of the local people and the environment in which they live.

Arch A natural opening through a rock caused by erosion.

Arête A narrow, knife-edged ridge caused by glacial erosion.

Aspect The direction towards which a slope or building faces.

Attrition Erosion caused when rocks and boulders, transported by rivers and waves, bump into each other and break up into smaller pieces.

B

Bar A barrier of sand stretching across a sheltered bay.

Backwash Water running back down a beach.

Biological weathering The breakdown of rock through the actions of plants and animals.

Biome A large ecosystem containing the same types of vegetation and animal life. Examples include the tropical rainforest and savanna grasslands.

Birth rate The number of live births per 1000 people per year.

Brownfield site An area of land that has been built on and is ready for redevelopment.

Burgess model An urban land use model of concentric zones with the oldest part of the city at the centre and the newest part on the outer edge.

Business park A group of new offices or modern factories built in pleasant surroundings, usually on the edge of a city.

Bustee An Indian term for a shanty town.

C

Central Business District (CBD) The commercial and business centre of a town or city where land values are at their highest.

Chemical weathering The breakdown of rocks by chemical action.

Climate The average weather conditions of a place over many years.

Clint A small ridge or flat-topped block of found on an exposed limestone surface.

Collision margin A boundary between two plates that are moving towards each other. As both consist of continental crust, fold mountains form.

Commercial farming The growing of crops and rearing of animals for sale and profit.

Common Agricultural Policy (CAP) An approach set up by the European Union (EU) to control farm production and to make its members more self-sufficient in food.

Commuter A person who travels some distance from their home to their place of work.

Condensation The process by which water vapour changes to a liquid (rain) or a solid (snow) when cooled.

Conflict Disagreement over the use of resources.

Confluence The point where two rivers meet.

Conservation The care and protection of resources and the environment.

Conservative margin A boundary between two plates that are sliding past each other and where crust is neither being formed nor destroyed; earthquakes are common here.

Constructive margin A boundary between two plates that are moving apart and where new crust is being formed.

Continental crust A part of the thin outer layer of the Earth that is relatively light, so it cannot sink and cannot be destroyed.

Contour ploughing Ploughing along the contours of a slope (not up and down).

Convectional rainfall Rain that is produced when the sun heats the ground causing warm air to rise.

Corrasion Erosion caused by the rubbing and scouring effect of material carried by rivers, glaciers, waves and the wind.

Corrie (cirque or cwm) A deep, rounded hollow with a steep backwall and sides caused by glacial erosion.

Corrosion Erosion caused by acids in rivers and the sea dissolving rocks by chemical action.

Counter-urbanisation The movement of people and employment away from large cities to smaller settlements within the countryside.

Countryside Stewardship Scheme An approach to environmental conservation in which farmers take on responsibility for the countryside and are paid for their efforts.

Crust The thin outer layer of the Earth.

D

Death rate The number of deaths per 1000 people per year.

Deforestation The complete clearance of forested land.

Delta An area of silt deposited by a river where it enters the sea or a lake.

Demographic transition model A model that tries to show how changes in birth and death rates over a period of time may be related to different stages of development.

Densely populated An area that is crowded with people.

Deposition The laying down of material carried by rivers, sea, ice or wind.

Depression An area of low pressure usually associated with cloud, rain and strong winds.

Desertification The gradual change of land into desert.

Destructive margin A boundary between two plates that are moving towards each other. One is oceanic crust and the other is continental crust.

Developed countries Countries that are usually quite rich, have many services and a high standard of living. Also called 'more economically developed countries' (MEDCs).

Developing countries Countries that are often quite poor, have few services and a low standard of living. Also called 'less economically developed countries' (LEDCs).

Development The use of resources and technology to increase wealth and improve standards of living – a measure of how rich or how poor a country is.

Dew point The temperature at which moist air becomes saturated.

Diet The amount and quality of food needed to keep a person healthy and active.

Discharge The amount of water in a river at a given time, usually measured in cumecs (cubic metres per second).

Drainage basin The area of land drained by a main river and its tributaries. Also called a 'river basin'.

Drought A prolonged period of weather that is drier than usual.

Dry valley A valley that has no running water on its surface.

E

Earthquake A sudden movement within the Earth's crust, usually close to a plate boundary.

Economic conditions Factors that affect people's job opportunities and wealth.

Ecosystem A system where plants and animals interact with each other and their natural surroundings.

Ecotourism A sustainable form of tourism aimed at protecting the environment and local cultures. Also called 'green tourism'.

Embankment A raised river bank built to prevent or reduce flooding.

Emigrant A person who leaves a country with the intention of living elsewhere.

Employment structure The proportion of people working in primary, secondary and tertiary occupations.

Environment The surroundings in which people, plants and animals live.

Environmental conditions Factors that affect the quality of our surroundings.

Epicentre The point on the Earth's surface immediately above the focus of an earthquake.

Erosion The wearing away of the land by material carried by rivers, glaciers, waves and the wind.

Erratic A large boulder transported by ice and deposited in an area of totally different rock.

Escarpment A ridge that has a steep slope on one side and a gentle slope on the other.

Ethnic group A group of people with common characteristics related to race, nationality, language, religion or culture.

Evaporation The process by which liquid, such as water, changes to water vapour when it is warmed.

Evapotranspiration The loss of moisture from water surfaces and the soil (evaporation) and vegetation (transpiration).

Exfoliation A form of physical weathering where, due to extreme changes in temperature, the surface of a rock peels away in layers.

Exports Goods and services produced by a country and sold to other countries.

F

Flood The flow of water over an area that is usually dry.

Flood plain The wide, flat area at the bottom of a valley which is often flooded.

Focus The point in the Earth's crust where an earthquake has its origin.

Fold mountains Mountains formed when two tectonic plates move together causing rock layers to bend and fold as they are forced upwards.

Food chain or food web The transfer of energy through an ecosystem from primary producers to consumers and decomposers.

Footloose industry An industry that is not tied to raw materials and so has a wide choice of location.

Formal sector employment Work that provides a regular income. It may be in an office, a shop or an organised factory.

Fossil fuels Energy resources such as coal, oil and natural gas which come from the fossilised remains of plants and animals.

Fragile environment Parts of the natural world that can be easily damaged.

Free trade When governments neither restrict nor encourage the movement of goods between countries.

Freeze–thaw weathering (frost shattering) A process of physical weathering in which rock is broken up due to water in cracks repeatedly freezing and thawing.

Front The boundary between two masses of air, one of which is colder and drier than the other.

Frontal rainfall Rain that occurs where warm air rises over cold air in a depression.

Function The main purpose of a town or settlement. Functions include markets, industry, port and resort facilities.

G

Glacial trough A valley with steep sides, a flat floor and a U-shaped cross-section carved by a glacier.

Glacier A slow-moving mass of ice flowing down a valley.

Global citizenship An appreciation that we all live in the same world and should help each other and try to improve the quality of life for everyone.

Global warming The increase in the world's average temperature, believed to result from the release of carbon dioxide and other gases into the atmosphere by the burning of fossil fuels.

Green belt An area of land around a city where the development of housing and industry is severely restricted and the countryside is protected for farming and recreation.

Green Revolution The introduction of high-yielding varieties of rice and wheat into less economically developed countries.

Greenfield site An area of land that has not previously been built upon.

Greenhouse effect The way that gases in the atmosphere trap heat from the Sun.

Gross national product (GNP) per capita The total value of goods and services produced by a country in a year divided by its total population. It is used as a measure of economic development.

Groundwater Water stored underground in permeable rocks.

Groyne A barrier built out into the sea to slow the movement of material along a beach.

Gryke A deep and narrow groove caused by chemical weathering on an exposed limestone surface.

H

Hanging valley A tributary valley left high above the main valley when its glacier was unable to erode downwards as quickly as the larger glacier in the main valley.

Heavy industry The manufacture of goods that require large amounts of bulky or heavy raw materials.

Hierarchy A ranking of settlements or shopping centres according to some measure of their importance, e.g. number of services, size, population, etc.

High order goods Products that are usually expensive and only bought occasionally.

High-tech industry An industry using advanced techniques to make high-value goods. Examples include computing, biotechnology and telecommunications.

Honeypot A place of attractive scenery or special interest which attracts tourists in large numbers.

Human Development Index (HDI) A measure of development adopted by the United Nations to compare countries. It uses health, education and wealth to measure both social and economic progress.

Hurricane A severe tropical storm with low pressure, heavy rainfall, and winds of extreme strength which can cause widespread damage. Also called a 'tropical cyclone'.

Hydraulic action Erosion caused by the sheer force of water breaking off small pieces of rock.

Hydrograph A graph showing changes in the discharge of a river over a period of time.

Hydrological cycle (water cycle) The continuous recycling of water between the sea, air and land.

I

Igneous rock A volcanic rock formed by the cooling of molten magma or lava.

Immigrant A person who arrives in a country with the intention of living there.

Impermeable A rock or soil that does not let water pass through it.

Imports Goods and services bought by a country from other countries.

Industry Any type of economic activity, or employment, that produces goods or provides services.

Infant mortality The average number of deaths of children under 1 year of age per 1000 live births.

Informal sector employment Self-employed work that is irregular and has little or no security. Examples include street-trading and shoe-shining.

Inner city The part of an urban area next to the city centre characterised by older housing and industry.

Interdependence When countries work together and rely on each other for help.

Interlocking spur Ridges of high ground that project into V-shaped valleys. They occur on alternate sides of a valley and interlink rather like the teeth of a zip fastener.

Isobar A line joining points of equal pressure.

Isohyet A line joining points with the same rainfall.

Isotherm A line joining points with the same temperature.

K

Karst An area of Carboniferous limestone scenery with extensive underground drainage. The unique landforms are a result of chemical weathering processes.

Lag time The period of time between peak rainfall and peak river discharge.

Lateral moraine A narrow band of rock debris along the side of a glacier.

Latitude The distance of a place north or south from the Equator.

Lava Molten rock flowing out of the ground, usually from a volcano.

Levée (dyke) An artificial embankment built to prevent flooding by a river or the sea.

Life expectancy The average number of years a person born in a particular country might be expected to live.

Light industry The production of high-value goods such as car stereos and fashion clothing.

Limestone A sedimentary rock formed from the remains of shells and marine skeletons on the sea bed. It consists mainly of calcium carbonate.

Limestone pavement Flat areas of bare limestone. The surface is usually made up of small blocks or ridges (clints) and narrow grooves (grykes) caused by chemical weathering.

Literacy rate The proportion of people who can read and write.

Location The position of a place or feature.

Longshore drift The movement of material along a coast by breaking waves.

Low order goods Products that are usually low cost and bought often.

Magma Molten rock below the Earth's surface.

Malnutrition Ill-health caused by a poor diet that is lacking in important nutrients such as carbohydrates or proteins.

Market A place where raw materials and goods are sold; or a group of people who buy raw materials and goods.

Meander The winding course of a river.

Medial moraine A narrow band of rock debris down the middle of a glacier.

Metamorphic rock A rock that has been altered by extremes of heat and/or pressure.

Migration The movement of people from one place to another to live or work.

Million cities Places with a population of more than 1 million.

Moraine Rock debris that is usually angular and is transported and later deposited by a glacier.

Multinational company A large company which, by having factories and offices in several countries, is global because it operates across national boundaries. Also called a 'transnational corporation'.

National Park An area of attractive countryside where scenery, vegetation and wildlife are protected so that they may be enjoyed by visitors and people living and working there, both now and in the future.

Natural increase The growth in population resulting from birth rates being greater than death rates.

Natural resources Raw materials like water, coal and soil, which are obtained from the environment.

Negative factors Things that discourage people from living in an area.

Newly industrialised countries (NICs) Countries, mainly in the Pacific Rim of Asia, that have undergone rapid and successful industrialisation since the early 1980s.

Non-renewable resource A resource such as coal or oil which, once used, cannot be used again or replaced.

Nutrient recycling The process by which minerals necessary for plant growth are constantly re-used. They are taken up from the soil by plants then returned when the plants shed their leaves or die.

Ocean currents The flow of water in certain directions within the sea.

Oceanic crust A part of the thin outer layer of the Earth that is heavy, can sink and is continually being destroyed or replaced.

Old age dependency ratio A measure of the number of retired people or pensioners per 1000 people of working age.

Onion weathering A form of physical weathering where, due to extreme changes in temperature, the surface of a rock peels away in layers.

Overcultivation The exhaustion of the soil by growing crops – especially the same crop – on the same piece of land year after year.

Overgrazing When there are too many animals for the amount of food available, which may lead to the destruction and loss of the protective vegetation cover.

Overpopulation When the resources of an area are unable to support the number of people living there.

Oxbow lake A crescent-shaped lake which has been cut off from the main river channel and abandoned.

Pacific Rim Places that surround the Pacific Ocean.

Permeable A rock or soil that allows water to pass through it.

Photosynthesis The process by which green plants turn sunlight into plant growth.

Physical or mechanical weathering The breakdown of rocks by changes of temperature.

Plate boundary or margin The place where two plates meet. It is here where most of the world's earthquakes occur and volcanoes may be found.

Plate tectonics The theory that the surface of the Earth is divided into several plates consisting of continental and oceanic crust.

Plates Huge slabs of the Earth's crust.

Plucking A process of glacial erosion by which ice freezes onto weathered rock and, as the glacier moves, large pieces of rock are pulled away with it.

Pollution Noise, dirt and other harmful substances produced by people and machines which spoil an area.

Population density The number of people living in a given area, usually one square kilometre (1 km²).

Population distribution How people are spread out over an area.

Population explosion A sudden rapid rise in the number of people in an area.

Population growth The increase in the number of people in an area.

Population growth rate A measure of how quickly the number of people in an area increases.

Population pyramid A type of horizontal bar graph used to show the population structure of an area.

Population structure The proportion of males and females within selected age groups.

Porous A rock or soil containing tiny pores either through which water can pass, or in which it can be stored.

Positive factors Things that encourage people to live in an area.

Precipitation The deposition of moisture usually from clouds. It includes rain, hail, snow, sleet, dew, frost and fog.

Prevailing wind The direction from which the wind usually comes.

Primary industries Industries that extract raw materials directly from the land or sea. Examples include farming, fishing, forestry and mining.

Pull factors Things that attract people to live in an area.

Push factors Things that make people want to leave an area.

Pyramidal peak (horn) A triangular-shaped mountain formed by glacial action.

Quality of life A measure of how happy and content people are with their lives.

Quaternary industries Industries that provide information and advice or are involved in research. Examples of quaternary occupations include financial advisers and research scientists.

Raw materials Natural resources that are used to make things.

Recycling Turning waste materials into something useful that can be used again.

Redevelopment Attempts to improve an area.

Refugees People who have been forced to move away from their own country and are therefore homeless.

Regeneration Renewing or improving something that has been lost or destroyed.

Relief The shape and height of the land.

Relief rainfall Rain caused by air being forced to rise over hills or mountains.

Renewable resource A resource like waves or wind power that can be used over and over again.

Resource Any material or product that people find useful.

Resurgence The reappearance of an underground stream onto the surface; commonly found in limestone areas.

Retailing The sale of goods individually or in small quantities, usually to shoppers.

Run-off Rainfall carried away from an area by streams and rivers.

Rural–urban fringe The area where the city and countryside meet. There is often competition for land use here.

Rural–urban migration The movement of people from the countryside to towns and cities where they wish to live.

Saltation A process of transportation by rivers in which small particles bounce along the bed in a 'leap-frog' movement.

Science park An estate of modern offices and high-tech industries having links with a university.

Secondary industries Industries that make, or manufacture, things. They process raw materials or assemble components to make a finished product. Examples include steelmaking and car assembly.

Sedimentary rocks Rocks formed from material laid down millions of years ago. They are usually found in layers and consist of either small particles of other rocks or the remains of plants and animals.

Seismometer An instrument used to measure the strength of an earthquake.

Self-help scheme A method of improving shanty town areas by encouraging and helping people to improve their own housing.

Service industries Occupations such as health, education, transport and retailing that provide a service for people. They may also be called 'tertiary industries'.

Shanty town A collection of shacks and poor-quality housing which often lack electricity, a water supply or any means of sewage disposal. They are common in developing countries and may also be called 'squatter', 'spontaneous' or 'informal' settlements.

Silt Fine soil left behind after a river floods. Also called 'alluvium'.

Site The actual place where a settlement (farm or factory) is located.

Situation The location of a settlement, farm or factory, in relation to the places surrounding it.

Social conditions Factors that directly affect people's quality of life, such as housing, shopping and recreational facilities.

Soil erosion The wearing away and loss of topsoil, mainly by the action of wind, rain and running water.

Solution A type of chemical weathering in which water dissolves minerals in rocks.

Sparsely populated An area that has few people living in it.

Spit An extended beach that grows by deposition across a bay or river mouth.

Stack An isolated pillar of rock detached from the mainland by wave erosion.

Stalactite A column of calcium carbonate hanging from the roof of a limestone cave where water drips down from the roof.

Stalagmite A cone-shaped feature of calcium carbonate that builds upwards from the floor of a limestone cave where water drips down from the roof.

Standard of living How well-off a person or country is.

Storm surge A rapid rise in sea-level caused by storms – especially tropical cyclones – forcing water into a narrowing sea area.

Subsidies Money given to farmers and industry by the government.

Subsistence farming Where all farm produce is needed by the farmer's family or village, and there is no surplus for sale.

Suburbanised villages Small settlements which have grown in size and become urban areas in countryside surroundings. Also called 'dormitory towns' or 'commuter settlements', as many residents who live and sleep there travel to nearby towns for work.

Suburbs A zone of housing around the edge of a city.

Suspension A process of transportation by rivers in which material is picked up and carried along within the water itself.

Sustainable development A way of improving people's standard of living and quality of life without wasting resources or harming the environment.

Swallow hole An opening in the surface of a limestone area down which a river may disappear.

Swash The movement of material up a beach after a wave breaks.

Synoptic chart A map showing the state of the weather at a given time.

Tectonic processes Movements within the Earth's crust.

Terminal moraine A ridge of rock debris dumped at the end of a glacier.

Terracing Steps cut onto the hillside to create level land for cultivation.

Tertiary industries Occupations such as health, education, transport and retailing which provides a service for people. They may also be called 'service industries'.

Traction A process of transportation by rivers in which material is rolled along the bed.

Trade The movement and sale of goods and services between one country and another.

Trade balance The difference between the cost of imports and the value of exports.

Trade deficit When a country spends more on its imports than it earns from its exports.

Trade surplus When a country earns more money from its exports than it spends on its imports.

Trading bloc A groups of countries that have joined together to improve trade.

Transnational corporation A company which, by having factories and offices in several countries, is global because it operates across national boundaries. Also called a 'multinational company'.

Transpiration The process by which water from plants changes into water vapour.

Transportation The movement of material by rivers, glaciers, waves and wind.

Tributary A small river that flows into a larger river.

Tropical cyclone A severe tropical storm with low pressure, heavy rainfall, and winds of extreme strength which can cause widespread damage. Also called a 'hurricane'.

Truncated spur A ridge that originally extended into a valley and which has had its end removed by a glacier.

Urban growth The increase in the size of towns and cities.

Urban land use model A simple map to show how land is used in a city.

Urban sprawl The unplanned, uncontrolled growth of urban areas into the surrounding countryside.

Urbanisation The increase in the proportion of people living in towns and cities.

Volcano A cone-shaped mountain or hill formed from lava and ash.

V-shaped valley A narrow, steep-sided valley formed as a result of rapid erosion by a stream or river.

Water cycle The continuous recycling of water between the sea, air and land.

Waterfall A sudden fall of water over a steep drop.

Watershed The boundary separating two river basins.

Weather The day-to-day conditions of the atmosphere, including temperature, sunshine, rainfall and wind.

Weathering The breakdown of rocks by physical and chemical processes.

Index

The names of places are printed in **bold** text.